# CONTROLLED RELEASE OF BIOACTIVE MATERIALS

*Academic Press Rapid Manuscript Reproduction*

Based on the Symposium at the 6th International Meeting of the Controlled Release Society in New Orleans, Louisiana, August 1979

# CONTROLLED RELEASE OF BIOACTIVE MATERIALS

EDITED BY Richard Baker

*Bend Research*
*Bend, Oregon*

**ACADEMIC PRESS 1980**

*A Subsidiary of Harcourt Brace Jovanovich, Publishers*

NEW YORK LONDON TORONTO SYDNEY SAN FRANCISCO

ACADEMIC PRESS, INC.
111 Fifth Avenue, New York, New York 10003

*United Kingdom Edition published by*
ACADEMIC PRESS, INC. (LONDON) LTD.
24/28 Oval Road, London NW1 7DX

**Library of Congress Cataloging in Publication Data**

Main entry under title:

Controlled release of bioactive materials.

Based on the proceedings of the Sixth International Symposium on Controlled Release Materials.
1. Delayed-action preparations—Congresses.
2. Pesticides, Controlled release—Congresses.
I. Baker, Richard W. II. Controlled Release Society. III. International Symposium on Controlled Release Materials, 6th, New Orleans, 1979.
RS201.D4C65 615'.191 80-19872
ISBN 0-12-074450-3

PRINTED IN THE UNITED STATES OF AMERICA

80 81 82 83 9 8 7 6 5 4 3 2 1

# CONTENTS

# CONTRIBUTORS

*Numbers in parentheses indicate the pages on which authors' contributions begin.*

M. D. Ahmed (449, 461), Schistosomiasis Control Project, Ministry of Health, Mogadishu, Somalia

Anthony J. Allen (399), Environmental Management Laboratory, University of Akron, Akron, Ohio 44325

Christopher G. Anderson (45), Department of Chemical Engineering, Twente University of Technology, Enschede, The Netherlands

James M. Anderson (45), Department of Macromolecular Sciences, Case Western Reserve University, Cleveland, Ohio 44106

A. H. Awad (449, 461), Schistosomiasis Control Project, Ministry of Health, Mogadishu, Somalia

R. W. Baker (1), Bend Research, Inc., Bend, Oregon 97701

Rajan S. Bawa (83), Department of Nutrition and Food Science, Massachusetts Institute of Technology, Cambridge, Massachusetts; and Department of Surgery, Children's Hospital, Boston, Massachusetts 02115

T. W. Brooks (227), Conrel, An Albany International Company, Needham Heights, Massachusetts 02194

Nate F. Cardarelli (357), Environmental Management Laboratory, University of Akron, Akron, Ohio 44325

John R. Cardinal (123), Department of Pharmaceutics, College of Pharmacy, University of Utah, Salt Lake City, Utah 84112

S. K. Chandrasekaran (99), ALZA Research, Palo Alto, Calfornia 94304

C. Cowley (433), School of Biological Sciences, University of Sussex, Falmer, Brighton

M. A. Dagal (449), Schistosomiasis Control Project, Ministry of Health, Mogadishu, Somalia

Gary E. Daterman (213), Pacific Northwest Forest and Range Experiment Station, Forestry Sciences Laboratory, Corvallis, Oregon 97330

C. C. Doane (227), Conrel, An Albany International Company, Needham Heights, Massachusetts 02194

Dee S. Dundee (387), Departments of Chemistry and Biological Sciences, University of New Orleans, Lakefront, New Orleans, Louisiana 70122

Charles Ebert (107), Department of Pharmaceutics, University of Utah, Salt Lake City, Utah 84112

B. Eckenhoff (61), ALZA Research, Palo Alto, California 94304

William Evans (357), Environmental Management Laboratory, University of Akron, Akron, Ohio 44325

Sen-Maw Fang (45), Department of Pharmaceutics, College of Pharmacy, University of Utah, Salt Lake City, Utah 84112

Jan Feijen (45), Department of Chemical Engineering, Twente University of Technology, Enschede, The Netherlands

M. M. Fooladi (317), Department of Polymer Science, University of Southern Mississippi, Hattiesburg, Mississippi 39401

P. A. Giang (237), Livestock Insects Laboratory Agricultural Research, Science and Education Administration, USDA, Beltsville, Maryland 20705; and US Livestock Insects Laboratory Agricultural Research, Science and Education Administration, USDA, Kerrville, Texas 78028

Mary L. Good (387), Departments of Chemistry and Biological Sciences, University of New Orleans, Lakefront, New Orleans, Louisiana 70122

W. R. Good (155), CIBA-GEIGY Corporation, Pharmaceuticals Division, Suffern, New York 10901

Donald E. Gregonis (45), Department of Pharmaceutics, College of Pharmacy, University of Utah, Salt Lake City, Utah 84112

D. V. Satyanarayana Gupta* (189), Biological Transport Laboratory, Department of Chemical Engineering, Washington University, St. Louis, Missouri 63130

J. K. Haworth (227), Conrel, An Albany International Company, Needham Heights, Massachusetts 02194

D. K. Hayes (237), Livestock Insects Laboratory Agricultural Research, Science and Education Administration, USDA, Beltsville, Maryland 20705; and US Livestock Insects Laboratory Agricultural Research, Science and Education Administration, USDA, Kerrville, Texas 78028

J. Heller (1), SRI International, Menlo Park, California 94025

Dean S. T. Hsieh (83,177), Department of Nutrition and Food Science, Massachusetts Institute of Technology, Cambridge, Massachusetts 21613; and the Department of Surgery, Children's Hospital Medical Center, Boston, Massachusetts 02115

*Present address: Polymer Department, Gulf South Research Institute, New Orleans, Louisiana

H. Jaffe (237), Livestock Insects Laboratory Agricultural Research, Science and Education Administration, USDA, Beltsville, Maryland 20705; and US Livestock Insects Laboratory Agricultural Research, Science and Education Administration, USDA, Kerrville, Texas 78028

Sung-Ho Kim (123), Department of Pharmaceutics, College of Pharmacy, University of Utah, Salt Lake City, Utah 84112

Sung Wan Kim (45, 107), Department of Pharmaceutics, College of Pharmacy, University of Utah, Salt Lake City, Utah 84112

F. W. Knapp (267), Department of Entomology, University of Kentucky, Lexington, Kentucky 40536

M. Koura (449, 461), Schistosomiasis Control Project, Ministry of Health, Mogadishu, Somalia

Robert S. Langer (83, 177), Department of Nutrition and Food Science, Massachusetts Institute of Technology, Cambridge, Massachusetts 02141; and Department of Surgery, Children's Hospital, Boston, Massachusetts 02115

P. I. Lee (135), Central Research, CIBA-GEIGY Corporation, Ardsley, New York 10502

Danny H. Lewis (251), Southern Research Institute, Birmingham, Alabama

Clifford S. Lofgren (251), Insects Affecting Man and Animals Research Laboratory, Agricultural Research, Science and Education Administration, USDA, Gainesville, Florida

Thomas A. Marks (19), Research Triangle Institute, Research Triangle Park, North Carolina 27709

C. L. McCormick (317), Department of Polymer Science, University of Southern Mississippi, Hattiesburg, Mississippi 39401

James McRea (107), Department of Pharmaceutics, University of Utah, Salt Lake City, Utah 84112

William E. Meyers (251), Southern Research Institute, Birmingham, Alabama 35202

J. A. Miller (237), Livestock Insects Laboratory Agricultural Research, Science and Education Administration, USDA, Beltsville, Maryland 20705; and US Livestock Insects Laboratory Agricultural Research, Science and Education Administration, USDA, Kerrville, Texas 78028

Sumita Mitra (45), Department of Macromolecular Sciences, Case Western Reserve University, Cleveland, Ohio 44106

K. F. Mueller (155), CIBA-GEIGY Corporation, Central Research, Ardsley, New York 10502

Chad Nontapan (267), Department of Entomology, University of Kentucky, Lexington, Kentucky 40506

J. Ofusu-Barko (433), Zoology Department, University of Ghana, Legon, Accra, Ghana

D. G. Osborn (227), Conrel, An Albany International Company, Needham Heights, Massachusetts 02194

Robert V. Petersen (45), Department of Pharmaceutics, College of Pharmacy, University of Utah, Salt Lake City, Utah 84112

Colin G. Pitt (19), Research Triangle Institute, Research Triangle Park, North Carolina 27709

Bernadette M. Quitter (399), Environmental Management Laboratory, University of Akron, Akron, Ohio 44325

Colleen M. Radick (399), Environmental Management Laboratory, University of Akron, Akron, Ohio 44325

William D. Rhine (83, 177), Department of Nutrition and Food Science, Massachusetts Institute of Technology, Cambridge, Massachusetts 02141; and the Department of Surgery, Children's Hospital Medical Center, Boston, Massachusetts 02115

Charles Sartwell (213), Pacific Northwest Forest and Range Experiment Station, Forestry Sciences Laboratory, Corvallis, Oregon 97330

Anton Schindler (19), Research Triangle Institute, Research Triangle Park, North Carolina 27709

K. G. Seymour (331), Agricultural Products Department, The Dow Chemical Company, Midland, Michigan 48640

J. E. Shaw (99), ALZA Research, Palo Alto, California 94304

K. N. Somasekharan (415), Department of Materials Science and Engineering, Washington State University, Pullman, Washington 99163

Suk-Zu Song, (123), Department of Pharmaceutics, College of Pharmacy, University of Utah, Salt Lake City, Utah 84112

Lonne L. Sower (213), Pacific Northwest Forest and Range Experiment Station, Forestry Sciences Laboratory, Corvallis, Oregon 97330

R. E. Sparks (189), Biological Transport Laboratory, Department of Chemical Engineering, Washington University, St. Louis, Missouri 63130

R. V. Subramanian (415), Department of Materials Science and Engineering, Washington State University, Pullman, Washington 99163

Vikas Sukhatme (177), Department of Nutrition and Food Science, Massachusetts Institute of Technology, Cambridge, Massachusetts 02141; and the Department of Surgery, Children's Hospital Medical Center, Boston, Massachusetts 02115

Gary Swindler (387), Departments of Chemistry and Biological Sciences, University of New Orleans, Lakefront, New Orleans, Louisiana 70122

F. Theeuwes (61), ALZA Research, Palo Alto, California 94304

J. D. Thomas (433), School of Biological Sciences, University of Sussex, Falmer, Brighton

E. S. Upatham (449, 461), Department of Biology, Faculty of Science, Mahidol University, Bangkok, Thailand

Robert K. Vander Meer (251), Insects Affecting Man and Animals Research Laboratory Agricultural Research, Science and Education Administration, USDA, Gainesville, Florida 32601

Richard M. Wilkins (343), Department of Agricultural Biology, The University of Newcastle Upon Tyne, England

# PREFACE

This book had its origins in the Sixth International Symposium on Controlled Release Materials, sponsored by the Controlled Release Society and held in August 1979 at New Orleans, Louisiana. Research on controlled release materials has accelerated in the past few years. The variety and number of controlled release applications described in the book testifies to the growing importance of this new technology. We hope this volume will receive the attention of this wider audience.

I would like to thank Prof. Mary Good for her assistance during the organization of this symposium, Dr. H. K. Lonsdale for his advice and assistance, and Miss Kimberly Plummer for her excellent job in typing the manuscript.

# THEORY AND PRACTICE OF CONTROLLED DRUG DELIVERY FROM BIOERODIBLE POLYMERS

*J. Heller*

SRI International
Menlo Park, California

*R. W. Baker*

Bend Research, Inc.
Bend, Oregon

*Drug release from bioerodible polymers is discussed in terms of three dissolution mechanisms. These are (1) water-soluble polymers insolubilized by degradable cross-links, (2) water insoluble polymers solubilized by ionization of pendant groups, and (3) water insoluble polymers solubilized by backbone cleavage.*

## I. INTRODUCTION

This paper will discuss various methods that can be used to release active agents from bioerodible polymeric matrices to a surrounding aqueous environment. It is limited to devices in which the active agent is dissolved or dispersed in a polymer, and does not cover the important method in which the active agent is chemically bound to the polymer and is released to the surrounding medium by hydrolysis of a bond between the active agent and the polymer chain (1,2). Also, it does not cover the extensive work of Kincl and coworkers on the release of steroids from cholesterol implants (3).

For the purposes of this discussion it is convenient to systematize polymer erosion according to the three mechanisms shown in Figure 1, where ⓧ denotes a hydrolytically




ISBN 0-12-074450-3

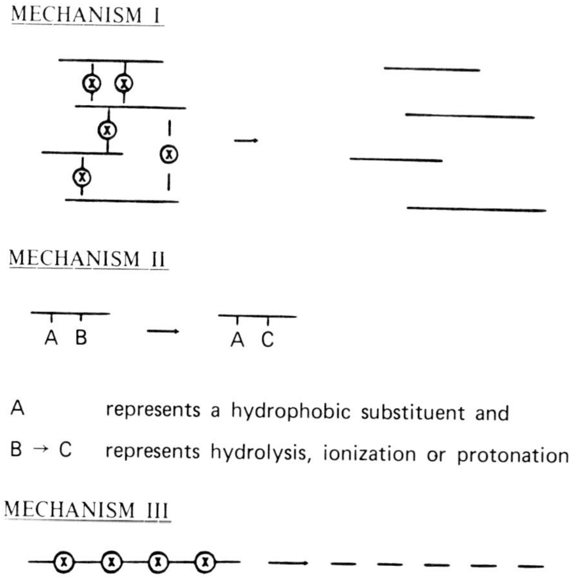


*FIGURE 1. Schematic representation of degradation mechanism.*

unstable bond. In general terms, mechanism I encompasses water-soluble polymers that have been insolubilized by hydrolytically unstable cross-links; mechanism II includes polymers that are initially water-soluble and are solubilized by hydrolysis, ionization, or protonation of a pendant group; and mechanism III includes hydrophobic polymers that are converted to small water-soluble molecules by backbone cleavage. Clearly, these represent extreme cases, and erosion by a combination of mechanisms is possible.

## II. DISCUSSION

### *A. Mechanism I - Solubilization by Cross-link Cleavage*

In these systems, water-soluble polymers are insolubilized by means of hydrolytically unstable cross-links. Consequently, the resulting matrix is highly hydrophilic and completely permeated by water. Since the active agent is located in an aqueous environment, its water solubility becomes an important consideration, and compounds with appreciable water solubility will be rapidly leached out, independent of the matrix erosion rate.

A useful application of these systems is for the slow release of water-soluble macromolecules that can be entangled in these systems and consequently are unable to escape until a sufficient number of cross-links have cleaved and the matrix cross-link density has been reduced.

It is possible to construct two general types of erodible hydrogels, shown schematically in Figure 2. In the first type, the hydrolytic instability resides in the cross-links so that the ultimate degradation products are high molecular weight, water-soluble polymers. In the second type, the hydrolytic instability resides in the polymer backbone and those systems can degrade to low molecular weight, water-soluble fragments.

Hydrogels with hydrolytically unstable cross-links have been prepared by copolymerizing acrylamide or N-vinyl pyrrolidone with N,N'-methylenebisacrylamide (4,5). The structure of these hydrogels and their hydrolytic degradation is shown in Figure 3.

We have briefly investigated the release of bovine serum albumin (BSA) from such hydrogels and the results for varying amounts of cross-linking comonomer in poly(N-vinyl pyrrolidone) gels are shown in Figure 4. At low cross-linker levels,

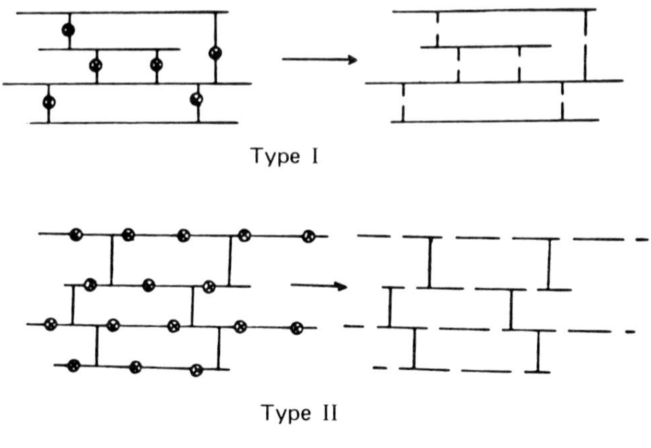


*FIGURE 2. Schematic representation of hydrogel degradation.*

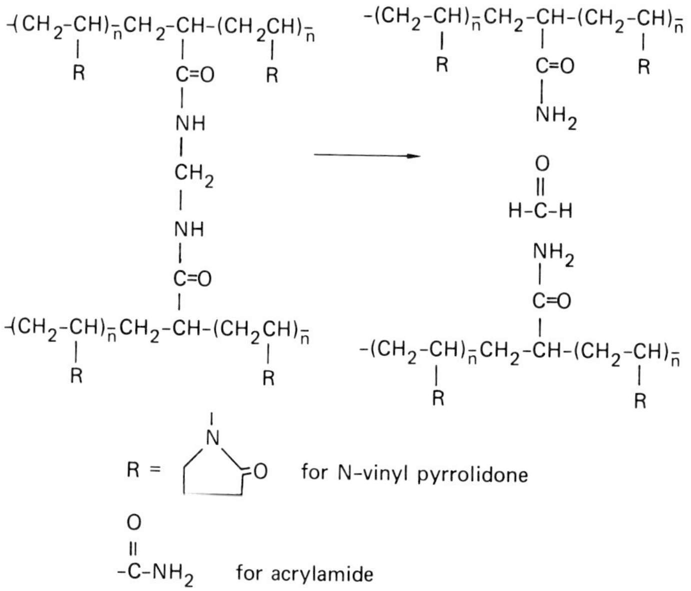


*FIGURE 3. Chemistry of acrylamide and N-vinylpyrrolidone hydrogel degradation.*

the hydrogel swells extensively and BSA is released very rapidly. At higher cross-linker concentrations, the degree of swelling was reduced and the release of BSA slowed. However, no observable polymer degradation took place in the several weeks during which the release rate was being measured.

Similar results were observed with polyacrylamide gels shown in Figure 5. Because of more favorable reactivity ratios, less cross-linker is required to produce highly cross-linked networks.

A relationship between cross-linker concentration and equilibrium water uptake for a polyacrylamide hydrogel is shown in Figure 6. At low cross-linker concentrations, the hydrogels sorb several times their weight in water and BSA can thus readily diffuse out of this loose structure. There is no correlation between polymer erosion and BSA release. Unfortunately, the methylenebisacrylamide linkage is not sufficiently labile to completely immobilize BSA and allow

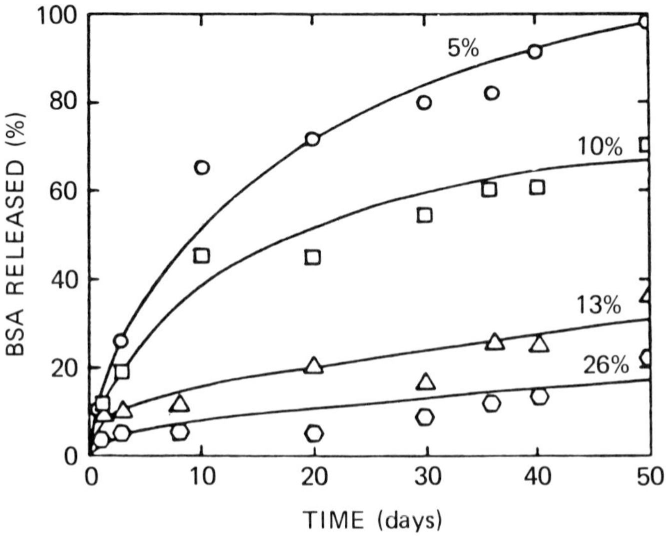


*FIGURE 4. Release of BSA from N-vinylpyrrolidone hydrogels as a function of cross-link density.*

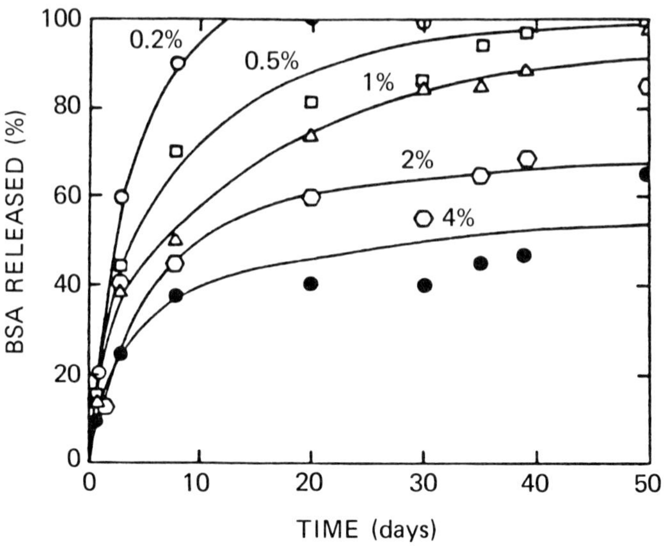


*FIGURE 5. Release of BSA from acrylamide hydrogels as a function of cross-link density.*

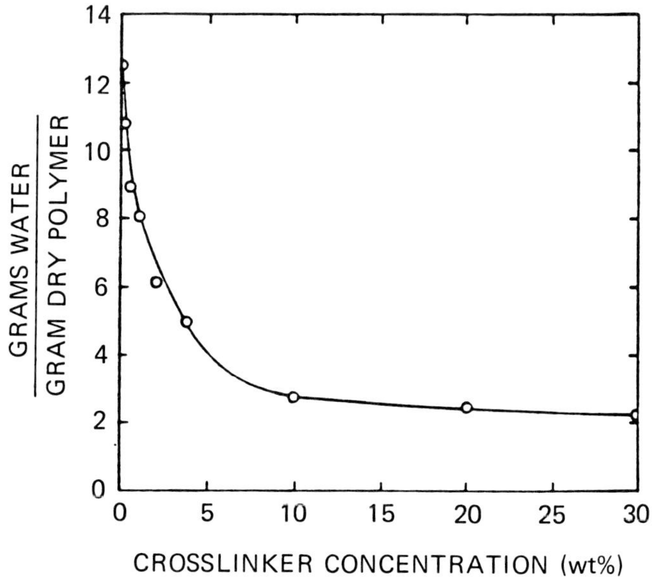


*FIGURE 6. Water uptake for acrylamide hydrogels as a function of cross-link density.*

the macromolecule to escape only when cross-linkers cleave. Therefore, when BSA is entangled in a tightly cross-linked network no release is observed for many weeks.

Clearly, we need to develop a hydrogel which has linkages considerably more labile than methylenebisacrylamide so that we can produce tight networks from which BSA can only be released by cleavage of bonds and consequent loosening of cross-link structure.

We have recently developed such materials. However, because patents are now being prepared we cannot divulge exact structures until the patents have been filed. These hydrogels belong to type II hydrogels in that the hydrolytic instability resides in the polymer backbone rather than cross-links. Release of BSA from these materials as a function of degree of hydrolytic instability is shown in Figure 7. We expect to describe these systems in detail as soon as the patents have been filed.

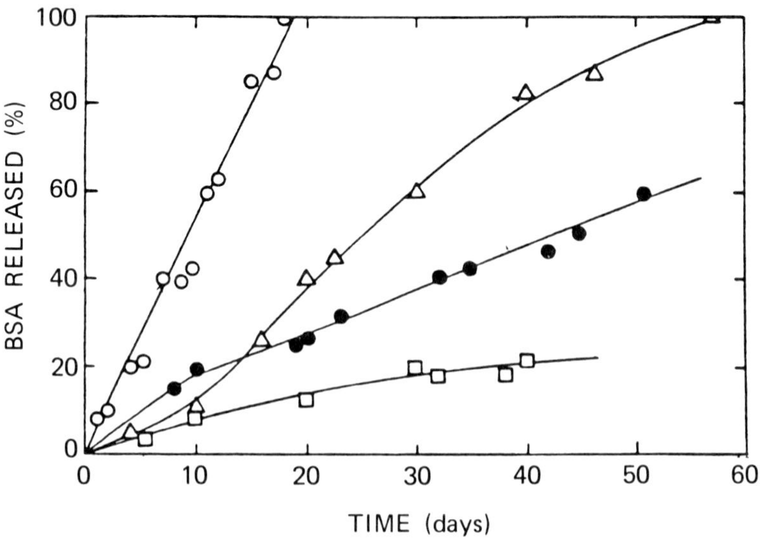


*FIGURE 7. Release of BSA from type II hydrogels as a function of backbone hydrolytic instability.*

*B. Mechanism II - Solubilization by Hydrolysis, Ionization or Protonation of Pendant Groups*

Systems in this category include all polymers that are initially water-insoluble but become water-soluble as a consequence of hydrolysis, ionization, or protonation of pendant groups. Because no backbone cleavage takes place, the solubilization does not result in any significant changes in molecular weight.

Even though the major emphasis in the developments of such polymers has been related to enteric coatings, we have found that partially esterified copolymers of methyl vinyl ether and maleic anhydride can, in a constant pH environment, release hydrophobic drugs incorporated therein by excellent zero order kinetics (6).

Figure 8 shows the polymer dissolution rate and the rate of hydrocortisone release for n-butyl half-ester polymer films containing the dispersed drug. Each pair of points represents a separate device in which the amount of drug released by the device into the wash solution was determined by UV measurements and the amount of polymer dissolved was calculated from the total weight loss of the device. The excellent linearity of both polymer erosion and drug release

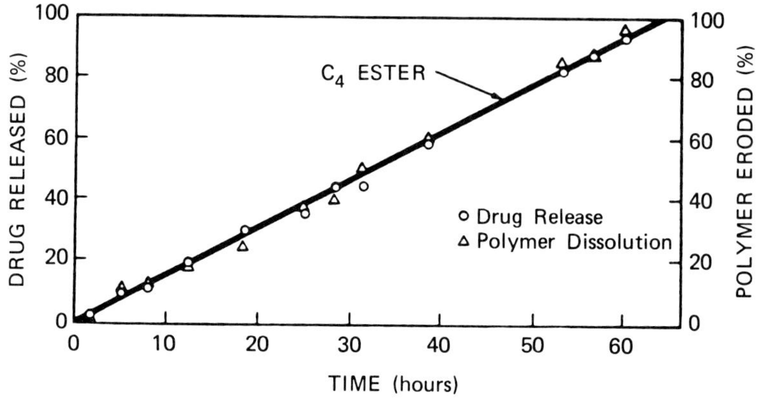


*FIGURE 8. Rate of polymer dissolution and rate of release of hydrocortisone for the n-butyl half-ester of methyl vinyl ether-maleic anhydride copolymer containing 10 wt% drug dispersion. Reproduced with permission from J. Appl. Polymer Sci., 22, 1991 (1978).*

over the lifetime of the device provides strong evidence for a surface-erosion mechanism and for negligible diffusional release of the drug. The latter result was independently verified by placing a drug-containing film in water at a pH low enough that no dissolution of the matrix took place and periodically analyzing the aqueous solution for hydrocortisone. None was found over several days.

The polymer becomes soluble by ionization of the carboxyl group as shown below:

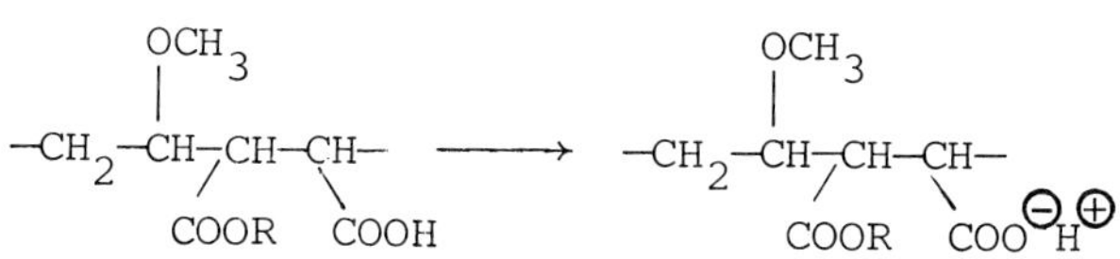


It was found (7) that these materials characteristically exhibit a pH range above which they are soluble and below which they are insoluble. This pH range is quite sharp, about 0.25 pH units, and changes with the number of carbon atoms in the ester side group of the copolymers.

This behavior can be understood readily by considering the number of ionized carboxyls necessary to drag the polymer chain into solution. With relatively small ester groups, only a low degree of ionization is sufficient to solubilize the polymer; hence the dissolution pH is low. As the size of the alkyl group increases, so does the hydrophobicity, and progressively more ionization is necessary to solubilize the polymer, resulting in an increasingly high dissolution pH. The same argument holds for polymers having the same ester grouping but different degrees of esterification. The higher the degree of esterification, the more hydrophobic the polymer and consequently the higher the dissolution pH.

Because the dissolution pH depends on the size of the alkyl group at any given pH, the polymer dissolution rate and hence the drug release rate can be modified by altering the size of the alkyl group. This effect is shown in Figure 9.

Alternatively, for any given alkyl group, the erosion rate and drug release can be readily altered by changing the outside pH. This effect is shown in Figure 10.

The usefulness of this polymer system for delivery of a therapeutic agent *in vivo* has been briefly investigated by cutting hydrocortisone-containing films into small circles and placing them in the lower forniceal cul de sac of New Zealand rabbits (8). Kinetic plots of drug release were obtained by placing weighed devices in eyes and removing the devices at intervals. The amount of drug released was determined by measuring residual drug in the devices and subtracting that from the known original amount. The results are shown in Figure 11.

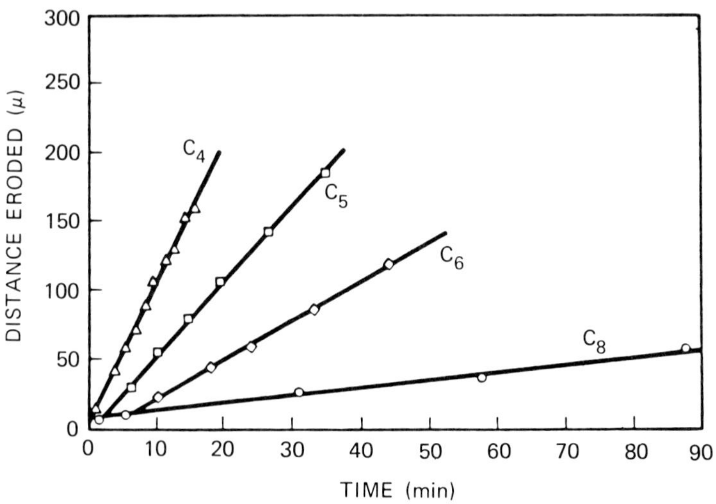


*FIGURE 9. Effect of size of ester group in half-esters of methyl vinyl ether-maleic anhydride copolymers on rate of erosion at pH 7.4. Reproduced with permission from J. Appl. Polymer Sci., 22, 1991 (1978).*

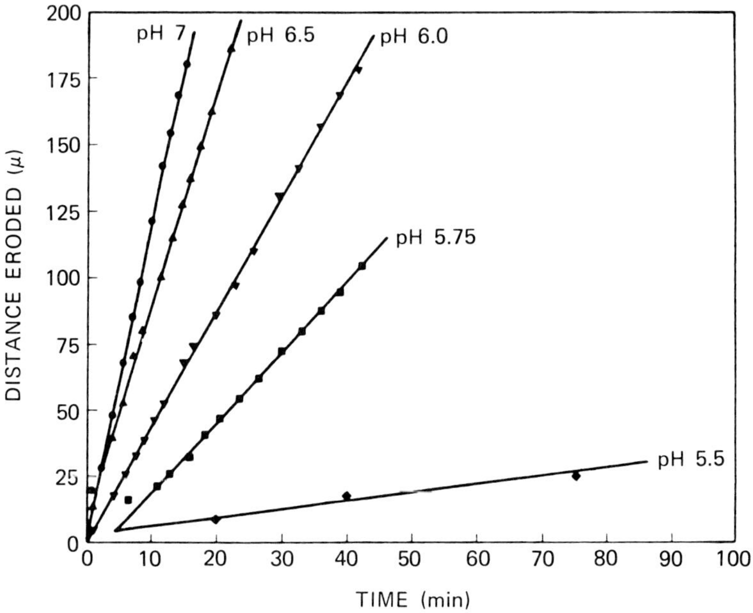


*FIGURE 10. Effect of erosion medium pH on erosion rate of half-esters of methyl vinyl ether-maleic anhydride copolymers. Reproduced with permission from J. Appl Polymer Sci., 22, 1991 (1978).*

Clearly, the devices are highly functional and release hydrocortisone by excellent zero-order kinetics. Furthermore, considering that each point represents a different device and a different rabbit eye, there is very little scatter, indicating a high degree of reproducibility.

We have recently used the partial ester copolymer as a model for a bioerodible drug delivery system that releases a therapeutic agent only in response to the presence of a specific external molecule (9). In this model, hydrocortisone was incorporated into an n-hexyl half ester of a methyl vinyl ether-maleic anhydride copolymer and the polymer-drug mixture fabricated into discs. These discs were then coated with a hydrogel containing immobilized urease. In a medium of constant pH and in the absence of external urea, the rate of release of hydrocortisone was that normally expected for the polymer at a given pH. In the presence of external urea, ammonium bicarbonate and ammonium hydroxide were generated within the hydrogel, which accelerated polymer erosion and hence drug release. Drug delivery rate increase was proportional to the amount of external urea and was reversible;

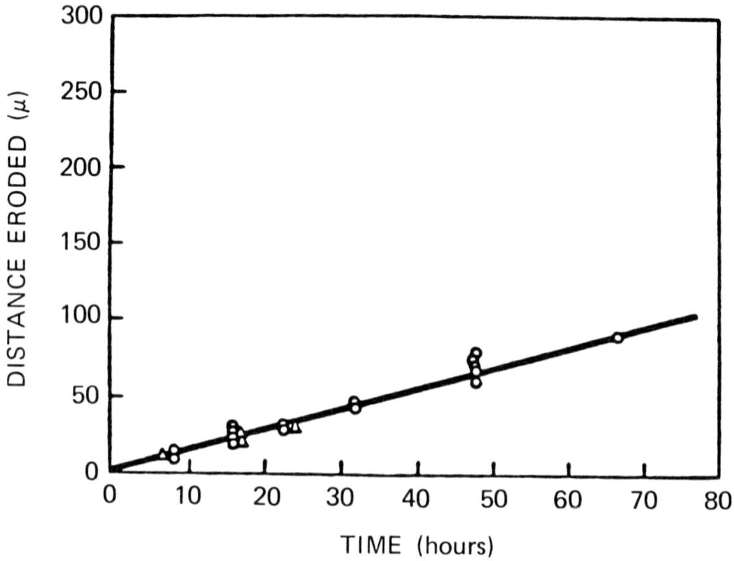


*FIGURE 11. In vivo release of hydrocortisone from ocular inserts in rabbits.*

that is, when external urea was removed, the rate of drug release gradually returned to its original value. The effect of urea on polymer erosion and drug release is shown in Figures 12 and 13.

### *C. Mechanism III - Solubilization by Backbone Cleavage*

This category includes all water insoluble polymers that undergo hydrolytic backbone cleavage and are solubilized by conversion to small, water-soluble molecules.

The hydrolytic erosion of a solid polymer matrix can occur by two extreme mechanisms. In one, referred to as heterogeneous, the process is confined to the surface of the device. In the other, called homogeneous, the hydrolysis occurs at a uniform rate throughout the matrix.

In a purely homogeneous process, the degradation occurs at a uniform rate through the matrix and the matrix will remain essentially intact until all parts reach some critical degree of reaction, at which point the matrix dissolves.

Drug release from such a matrix is complicated because it is a combination of diffusion and erosion. Diffusional drug release in the absence of erosion can be described by the

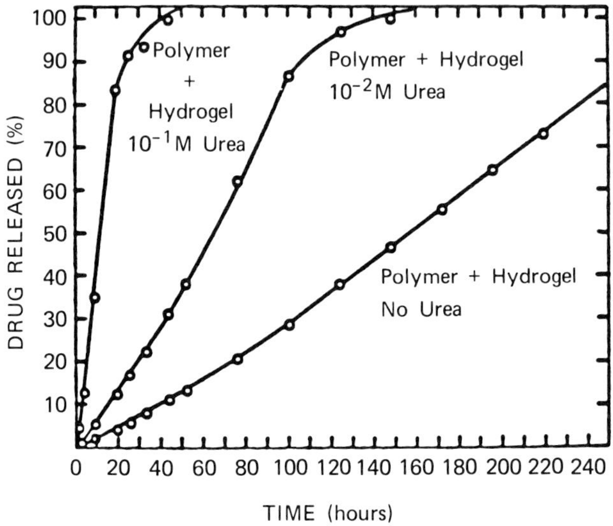


*FIGURE 12. Rate of hydrocortisone release at 35°C from a n-hexyl half-ester of methyl vinyl ether-maleic anhydride copolymer of pH 6.25 in the absence and presence of external urea. Reproduced with permission from J. Pharm. Sci., 68, 919 (1979).*

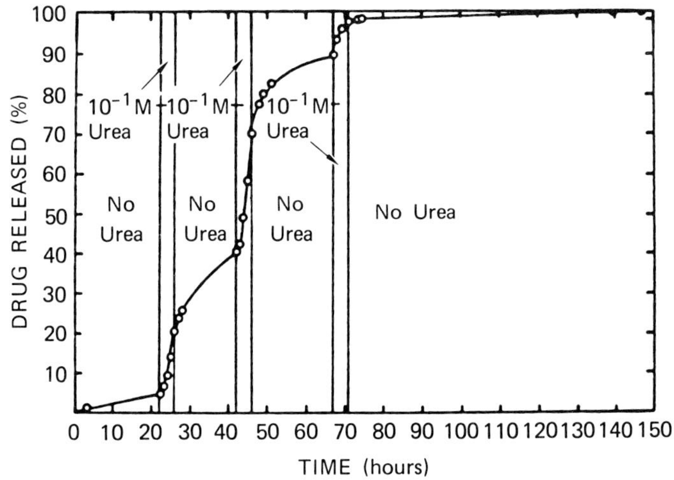


*FIGURE 13. Rate of hydrocortisone release at 35°C from a n-hexyl half-ester of methyl vinyl ether-maleic anhydride copolymer at pH 6.25 as a function of sequential addition and removal of $10^{-1}$ urea. Reproduced with permission from J. Pharm. Sci., 68, 919 (1979).*

Higuchi model (10) which proposes that drug is initially removed from the surface regions of the polymer and consequently a progressively thicker drug-depleted layer forms adjacent to the surface of the device.

Using this model the following expression for drug release has been derived:

$$\frac{dM_t}{dt} = \frac{A}{2}\left[\frac{2PC_o}{t}\right]^{\frac{1}{2}} \tag{1}$$

where $M_t$ is drug released at time t, P is the permeability of the polymer to the drug, A is the surface of both sides of the film and $C_o$ is the initial concentration of the drug in the polymer. The key feature of this equation is the drug release declines continuously proportional to $t^{-\frac{1}{2}}$. In an eroding polymer the situation is more complex because the permeability P increases with time because the matrix is gradually loosened by hydrolytic cleavages.

The change of permeability with time is diffucult to estimate accurately, but the following simplified assumption can be made

$$\frac{P}{P_o} = \frac{\text{Initial number of bonds}}{\text{Number of bonds remaining}} = \frac{N}{N-Z} \tag{2}$$

where $P_o$ is the original permeability and P is the permeability after Z cleavages.

If we further assume simple first order kinetics where the rate of bond cleavage dZ/dt is proportional to the number of cleavable bonds present, then

$$\frac{dZ}{dt} = K(N-Z) \tag{3}$$

where K is the first order rate constant. Then by simple integration

$$Kt = \ln\left(\frac{N}{N-Z}\right) \tag{4}$$

Equations (2) and (4) can now be combined to yield

$$P = P_o \exp(Kt) \tag{5}$$

Substituting this equation into the Higuchi expression (1) yields

$$\frac{dM_t}{dt} = \frac{A}{2}\left[\frac{2P_o \exp(Kt)\, C_o}{t}\right]^{\frac{1}{2}} \qquad (6)$$

A plot of the integrated form of this equation along with diffusional release in the absence of erosion is shown in Figure 14. With no erosion, drug release slowly declines following the so-called $t^{-\frac{1}{2}}$ kinetics. However, because of polymer degradation which causes permeability to rise above the initial value, the normal decline is slowed and then completely reversed.

Figure 15 shows the release of hydrocortisone from a polyester prepared from diglycolic acid and trans cyclohexane

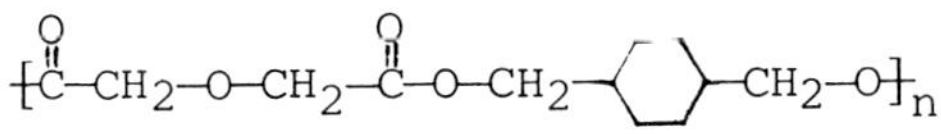

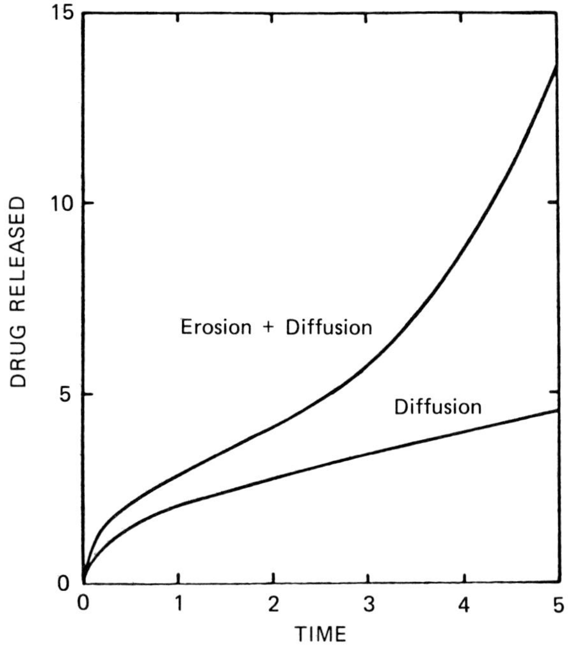


*FIGURE 14. Theoretical curve of drug release from a polymer slab by diffusion and diffusion + erosion.*

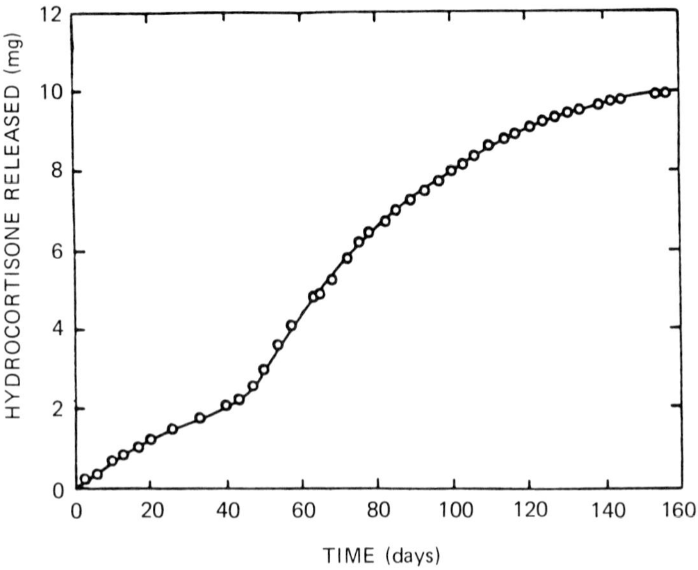


*FIGURE 15. Release of hydrocortisone from a diglycolic acid-trans cyclohexanedimethanol polyester.*

dimethanol (11). As can be seen, this curve is similar to the theoretical curve, indicating that drug release occurs by both diffusion and erosion.

A much preferred case are polymers that erode by a purely heterogeneous process, because drug release from devices that maintain constant geometry will be zero order.

Clearly, the principal requirement for achieving surface erosion is to confine the degradation process to the surface of the bulk polymer. While surface erosion can probably be achieved by constructing a highly hydrophobic polymer, so that water penetration and consequent bulk hydrolysis is much less probable than surface reaction, enough bulk degradation may take place to seriously degrade molecular weight and mechanical properties will deteriorate. Consequently, a much preferred approach is one in which a pH-sensitive reaction is selected and the interior of the matrix is buffered so that bulk hydrolysis is prevented and reaction can take place only at the surface of the device, where the buffer is neutralized.

Ideally, the active agent should be totally immobilized within the matrix and be released only by matrix dissolution. In real devices there will be some diffusional loss. The diffusion of active agents from surface-eroding slabs has been treated mathematically (12); it was shown that after a certain time, the length of which depends on the rate of diffusion of the active agent from the slab, diffusion and erosion became coupled and zero-order drug release was predicted.

The only polymers thus far described in detail that undergo true surface erosion are the partial esters of methyl vinyl ether/maleic anhydride copolymers (6). However, these represent a special case because solubilization does not involve backbone cleavage and consequently the only polymer chains that can escape into the aqueous environment are at the surface of the device.

None of the described backbone-degradable polymers undergo surface erosion; rather, they undergo bulk erosion. This is not suprising; they were originally designed as bioerodible suture materials and were not intended as drug carriers capable of releasing a drug by zero-order kinetics. However, work in progress at SRI International and Bend Research, Inc. (11) and elsewhere (13,14) is devoted specifically to the development of bioerodible polymers that undergo surface erosion and are capable of releasing drugs incorporated therein by zero-order kinetics.

## REFERENCES

1. F. W. Harris, A. E. Aulabaugh, R. D. Case, M. K. Dykes, and W. A. Feld, Controlled Release Polymeric Formulations, D. R. Paul and F. W. Harris, (Eds.), ACS Symp. Series No. 33, p. 222 (1976).
2. A. N. Neogi and G. G. Allan, Controlled Release of Biologically Active Agents, A. C. Tanquary and R. E. Lacey, (Eds.), Plenum Press, New York, p. 195 (1976).
3. A. A. Joseph, J. L. Hill, J. Patel, S. Patel, and F. A. Kincl, J. Pharm. Sci., 66, 490 (1977).
4. B. K. Davis, Experentia, 28, 348 (1972).
5. V. P. Torchilin, E. G. Tischenko, V. N. Smirnov, and E. I. Chazov, J. Biomed. Mater. Res., 11, 223 (1977).
6. J. Heller, R. W. Baker, R. M. Gale, and J. O. Rodin, J. Appl. Poly. Sci., 22, 1991 (1978).
7. L. C. Lappas and W. McKeehan, J. Pharm. Sci., 54, 176 (1965).
8. J. Heller and R. W. Baker, paper presented at the University of Southern California Symposium on Applications of Polymers in Dentistry and Medicine, November 1973.
9. J. Heller and P. V. Trescony, J. Pharm. Sci., 68, 919 (1979).

10. T. Higuchi, J. Pharm. Sci., 50, 874 (1961).
11. J. Heller and R. W. Baker, work in progress under contract 1-HD-7-2826.
12. R. W. Baker, unpublished results.
13. N. S. Choi and J. Heller, U.S. Patents 4,093,709, June 6, 1978, 4,131,648, December 26, 1978 and 4,138,344 February 6, 1979.
14. R. C. Capozza, L. Sendelbeck, and W. J. Balkenhol, Polymeric Delivery Systems, R. J. Kostelnick, (Ed.), Gordon and Breach Scientific Publ., 1978, p. 59.

# BIODEGRADABLE DRUG DELIVERY SYSTEMS BASED ON ALIPHATIC POLYESTERS: APPLICATION TO CONTRACEPTIVES AND NARCOTIC ANTAGONISTS

*Colin G. Pitt*
*Thomas A. Marks*
*Anton Schindler*

Research Triangle Institute
Research Triangle Park, North Carolina

*The classes of polymer which form the basis of different types of drug delivery systems are discussed, and the relationship between the chemical structure of the polymer and its permeability, morphology, biodegradability, and mechanical properties are considered using polyesters as specific examples. Studies of the permeability and biodegradability of poly(ε-caprolactone), poly(DL-lactic acid), and various copolymers are described and used to illustrate how these properties may be varied by the choice of polymer structure. An induction period prior to bioerosion of these polymers, coupled with high permeability, permits their use as reservoir devices (capsules) which exhibit constant, diffusion-controlled drug release rates and which erode after the drug is exhausted. The applications of this approach to the long term delivery (1 year) of levonorgestrel, a contraceptive agent, and the short term delivery (1-2 months) of naltrexone, a narcotic antagonist, are described.*

## I. INTRODUCTION

There are three fundamentally different approaches to the design of polymeric drug delivery systems—control by osmotic pressure, by diffusion, and by erosion (1). Only the last,




ISBN 0-12-074450-3

by definition, necessarily involves polymer biodegradation. Regardless of the approach selected, it is necessary to carefully control the properties of the polymer, the most important of which are its permeability, biodegradability (or stability), biocompatability, and mechanical properties. Since these properties are often interdependent, their optimization can present a challenging exercise in polymer design.

Polymers used for drug delivery may be separated into two distinct classes, those that are in the glassy state (glass transition temperature $T_g > 37^oC$) and those that are in the rubbery state ($T_g < 37^oC$) at body temperature. If the polymer is in the glassy state it will generally be tough and inflexible with a low permeability, and drug release by leaching and polymer erosion will be favored relative to diffusion. The only possibility of achieving a constant rate of drug release from this class of polymer is by constant bioerosion of a device with a constant surface area, i.e., a slab, or by diffusion controlled release from a microcapsular reservoir device where the very thin walls and increased surface-to-volume ratio compensate for the reduced permeability. Poly(lactic acid), $T_g$ $57^oC$, formulated as microcapsules, is an example of the latter system (2-4).

Polymers in the rubbery state will have a high permeability but will often lack sufficient mechanical strength and form stability unless reinforced by the presence of cross-links, crystalline domains, or strong interactions between chains, e.g. hydrogen bonding. Silicone rubber is an example of a non-degradable rubbery polymer ($T_g$ $-120^oC$) which must be strengthened by both filler and light cross-linking. If the polymer is to be biodegradable, the means of reinforcement must also be biodegradable. This can present some difficulties, for if diffusion controlled drug release is operative, degradation of filler or cross-links may change the polymer morphology and hence its permeability. Changes in device dimensions and crystallinity produced by biodegradation will also affect the rate of drug diffusion. These problems can be minimized if bioerosion occurs after the useful life of the device, i.e., after drug depletion. This is the approach which we have preferred in our laboratory.

## II. POLYMER DESIGN

Excluding polymers which are bioerodible by virtue of their water solubility, e.g., poly(vinyl alcohol), the classes of polymer which have been shown to be degraded at an appreciable rate *in vivo* are:

1. Activated Carbon-Carbon Polymers

e.g. $[CH_2-C(CN)_2]_n \rightarrow CH_2O + CH_2(CN)_2$

$[CH_2-C(CN)COOR]_n \rightarrow CH_2O + CN\cdot CH_2COOR$

2. Polyamides, Polyurethanes

e.g. $[CH_2-NH-CO-CH_2]_n \rightarrow NH_2CH_2COOH$

3. Polyesters, Polycarbonates

e.g. $[CH_2-O-CO-CH_2]_n \rightarrow HOCH_2COOH$

4. Polyacetals, Polyketals, Poly orthoesters

e.g. $[CH(OR)-O-CH_2]_n \rightarrow ROH + HOCH_2CHO$

5. Inorganic Polymers

e.g. $[P(OR)_2=N]_n + H_2O \rightarrow ROH + NH_3$

6. Natural Polymers Subject to Enzymatic Degradation

e.g. Polysaccharides, Collagen

The selection of any one of these classes for study is somewhat subjective, for within each class one can find a spectrum of properties. The permeability, biodegradability and mechanical properties may all be changed substantially by logical but rather small changes in the chemical structure of the substituent or chain. The polyesters (Figure 1) provide a good illustration. Thus, on the basis of electronic effects and the properties of monomeric esters, one can anticipate that the rate of non-enzymatic hydrolysis of polymers of α-hydroxy acids, e.g., poly(glycolic acid), will be greater than polymers of β, γ, δ, and ε-hydroxy analogs; aromatic esters will be least readily hydrolyzed. Alkyl or aryl substituents adjacent to the ester group can retard hydrolysis by steric hindrance.

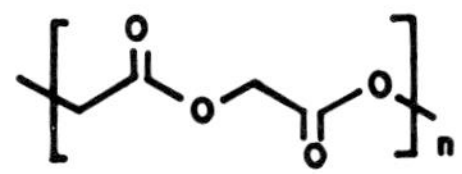

POLY(GLYCOLIC ACID) OR POLY(DIGLYCOLIDE)
CRYSTALLINE, TOUGH, INELASTIC
$T_m$ 230° $T_g$ 36°

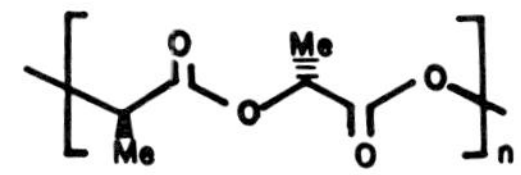

POLY(D-LACTIC ACID) OR POLY(DD-DILACTIDE)
CRYSTALLINE, TOUGH, INELASTIC
$T_m$ 180° $T_g$ 67°

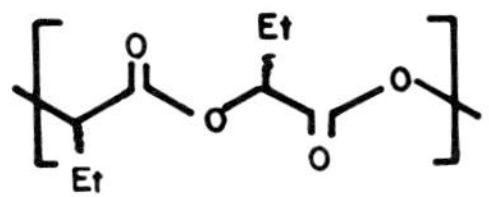

POLY(DL-ETHYLGLYCOLIC ACID)
AMORPHOUS, RUBBERY
$T_g$ 16°

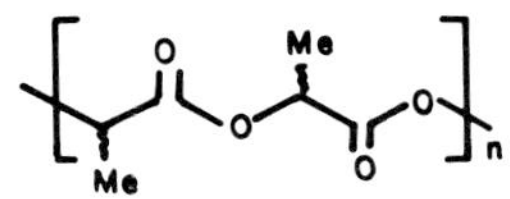

POLY(DL-LACTIC ACID) OR POLY(DL-DILACTIDE)
AMORPHOUS, TOUGH, INELASTIC
$T_g$ 57°

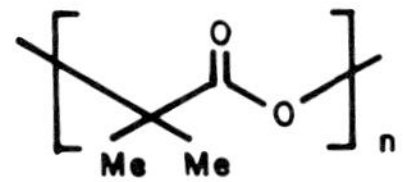

POLY(DIMETHYLGLYCOLIC ACID)
CRYSTALLINE, TOUGH, INELASTIC
$T_m$ 240°

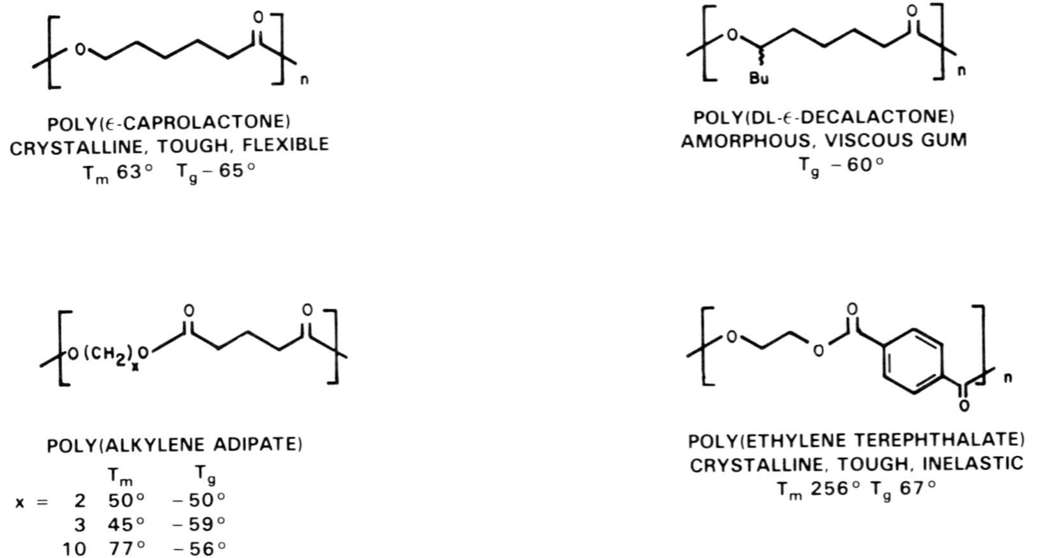


*FIGURE 1. The chemical structures and relevant physical properties of some typical polyesters.*

The polymer morphology and permeability are superimposed upon these chemical effects, determining the accessibility of the bulk of the polymer to aqueous hydrolysis and the loss of cleaved chain fragments, as well as the rate of drug diffusion and mechanical properties. Here the $T_g$ and the melting point ($T_m$) provide a good indication of structure-morphology relationships (Figure 1). Poly(glycolic acid), the simplest aliphatic polyester, is a practically insoluble, high melting, crystalline polymer with a $T_g$ of $36^{o}C$. These properties reflect the polarity and structural order of this polymer. Stereospecific introduction of a single methyl group, as in poly(D or L-lactic acid), has little effect on the $T_g$ and $T_m$. Non-stereospecific introduction of a methyl group, as in racemic poly(DL-lactic acid) does not change $T_g$ greatly but completely eliminates the crystallinity because of the random arrangement of the asymmetric centers. Similarly, nonspecific introduction of two different alkyl groups will eliminate both the crystallinity and reduce the $T_g$, and a polymer such as poly(DL-methylethylglycolic acid) is rubbery at $37^{o}C$. However, if both the alkyl groups are identical, as in poly(dimethylglycolic acid), the symmetry and crystallinity are restored.

It is possible to reduce the $T_g$ substantially without changing the crystallinity by increasing the length of the methylene repeat unit, e.g. poly(alkylene adipates), poly (ε-caprolactone), and their homologs. These polymers are permeable (low $T_g$) and mechanically strong (crystallinity). The crystallinity can be reduced or eliminated by introduction of an asymmetric center, as in poly(DL-ε-decalactone). This modification increases the permeability but reduces the mechanical strength to the extent that the latter polymer is a viscous gum.

Introduction of aromatic groups into the polymer backbone increases the $T_g$ and the crystallinity, and a polymer such as poly(ethylene terephthalate) is a tough, high melting, inflexible, impermeable material.

Plasticization, blending, and copolymerization represent other ways in which the polymer properties may be tailored, and specific examples are given below.

## III. POLYMER PERMEABILITY

There is good correlation between the $T_g$ and the permeability of those polyesters which have been tested. Poly (ε-caprolactone) and its copolymers have low $T_g$'s and are as permeable as silicone rubber; poly(DL-lactic acid) and its copolymers with poly(glycolic acid) are in the glassy state and are relatively impermeable to drugs such as steroid hormones. Using progesterone as an example, the maximum steady state flux ($J_{lim}$), solubility, partition and diffusion coefficients of this steroid in poly(ε-caprolactone) and its copolymers with DL-lactic acid are compared with silicone rubber in Table I. The permeability of poly(ε-caprolactone), as reflected by the value of $J_{lim}$, is almost equal to that of silicone rubber, largely because of the greater drug solubility in poly(ε-caprolactone).[1] While poly(DL-lactic acid) is $10^4$ times less permeable than poly(ε-caprolactone), copolymers of DL-lactic acid and ε-caprolactone have the same permeability as poly(ε-caprolactone). This is understandable, for copolymerization affords a polymer with a $T_g$ between the $T_g$'s of the two homopolymers, as illustrated in Figure 2, and with less crystallinity than poly(ε-caprolactone).

Progesterone is released very rapidly from thin films (100-300 μm) of poly(ε-caprolactone) and its copolymers with lactic acid and glycolic acid, release being essentially complete within 24 hours (6). The fraction of drug load released is proportional to $(\text{time})^{1/2}$, although the kinetics are moderated by an aqueous boundary layer which produces some curvature in a plot of the fraction released vs $t^{1/2}$ (Figure 3).

Release of progesterone from poly(DL-lactic acid) films is very much slower, even from thinner films (10 μm), and leaching may be partly responsible for the rate which was observed (Figure 4). Similar results were obtained with copolymers of glycolic acid and DL-lactic acid except that in some instances there was a marked increase in rate after several weeks (Figure 5). This coincided with the mechanical deterioration of the film and is an example of erosion-enhanced release (6).

The rate of diffusion could also be enhanced by plasticization of poly(DL-lactic acid) with tributyl citrate (5). This served to reduce the glass transition below 37° and transformed the polymer to the rubbery state. However, the effect was short lived (Figure 6) as the plasticizer was rapidly leached into the external aqueous medium. The permeability could also be enhanced (200 fold) by casting

[1]*The maximum steady state flux, $J_{lim}$, is the product of the diffusion coefficient and the solubility of the drug in the polymer and is equal to the maximum drug diffusion rate through a membrane unit area and thickness.*

TABLE I. The Maximum Steady State Flux ($J_{lim}$), Partition Coefficient (K), Solubility ($C_s$) and Diffusion Coefficient (D) of Progesterone in Poly(ε-caprolactone, Poly(DL-Lactic Acid), Poly(ε-caprolactone-co-DL-lactic acid), and Silicone Rubber.

| Polymer | $J_{lim} \times 10^{11}$ (g/cm/sec) | Partition coefficient | Solubility (mg/g) | $D \times 10^{9}$ ($cm^2$/sec) |
|---|---|---|---|---|
| Poly(ε-caprolactone) | 6.1 | 1200 | 17 | 3.6 |
| 1:1 copolymer[a] | 5.8 | 1554 | 22 | 2.7 |
| Poly(DL-lactic acid) | $0.33 \times 10^{-3}$ | 46 | 0.65 | $5.1 \times 10^{-3}$ |
| Silicone rubber[b] | 26 | 50 | 0.57 | 450 |

[a] 1:1 (wt%) copolymer of ε-caprolactone and DL-lactic acid.
[b] Data for polymer containing filler (Reference 10).

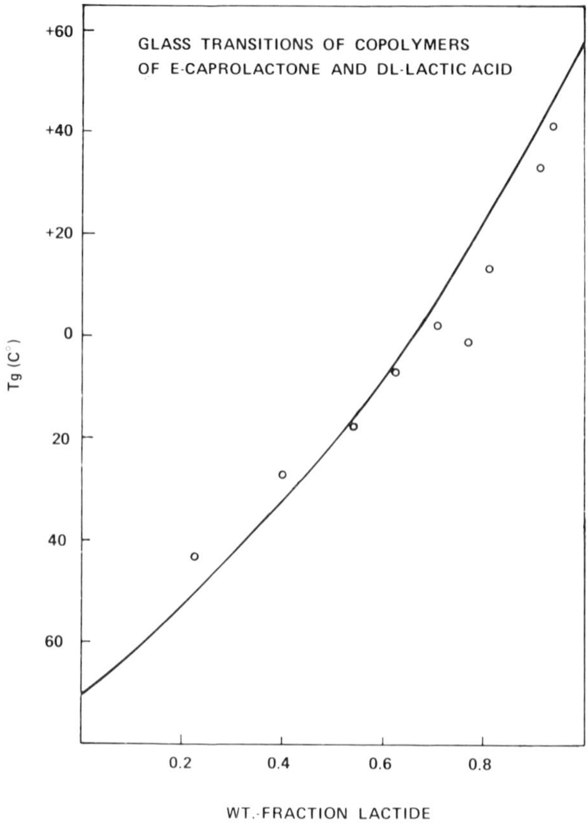


*FIGURE 2. Variation of glass transition temperature ($T_g$) of copolymers of caprolactone and DL-lactide as a function of DL-lactide content. Solid line is calculated relationship, based on Fox equation (from Reference 8).*

poly(DL-lactic acid) from a solution containing glycerine, then soaking the film in water (5). Here the permeability increase was the result of a highly porous structure easily detected by scanning electron microscopy (Figure 7). This structure may mimic that of a film partly depleted of drug.

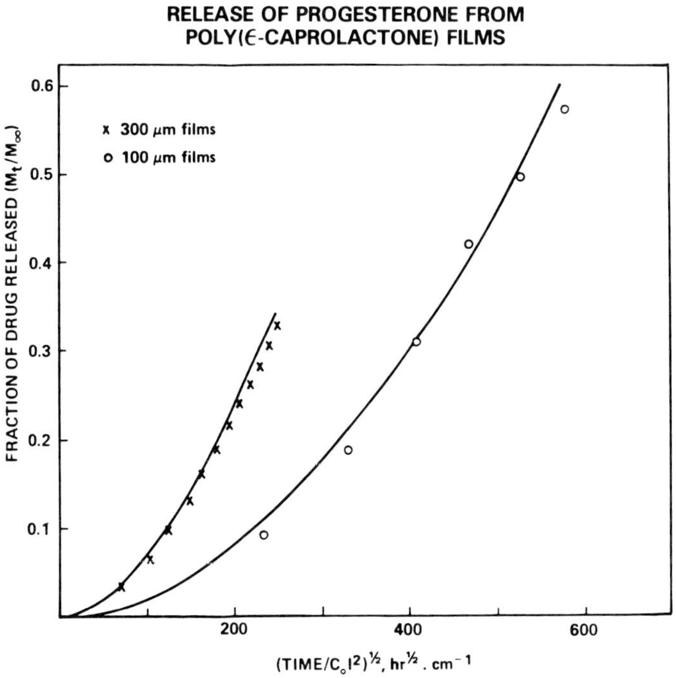


*FIGURE 3. Fraction of progesterone released from poly (ε-caprolactone) films (10% drug load) as a function of time$^{\frac{1}{2}}$, and film thickness. Solid lines are calculated from theoretical relationship (Reference 11) which incorporates effect of aqueous boundary layer (from Reference 6).*

## IV. POLYESTER BIODEGRADATION

Most reports of polymer biodegradation are qualitative, confined to evaluation of a single property, and in many cases the specifications of the polymer are not reported. This is unfortunate because residence times in the body may vary by an order of magnitude depending on the molecular weight, morphology, physical form, etc. As a consequence, the factors which contribute to polymer degradation are still poorly understood.

In these laboratories, poly(ε-caprolactone) has been studied most extensively and changes in viscosity, weight, molecular weight distribution, and tensile properties have been determined *in vivo* (rabbit, rat) and under accelerated

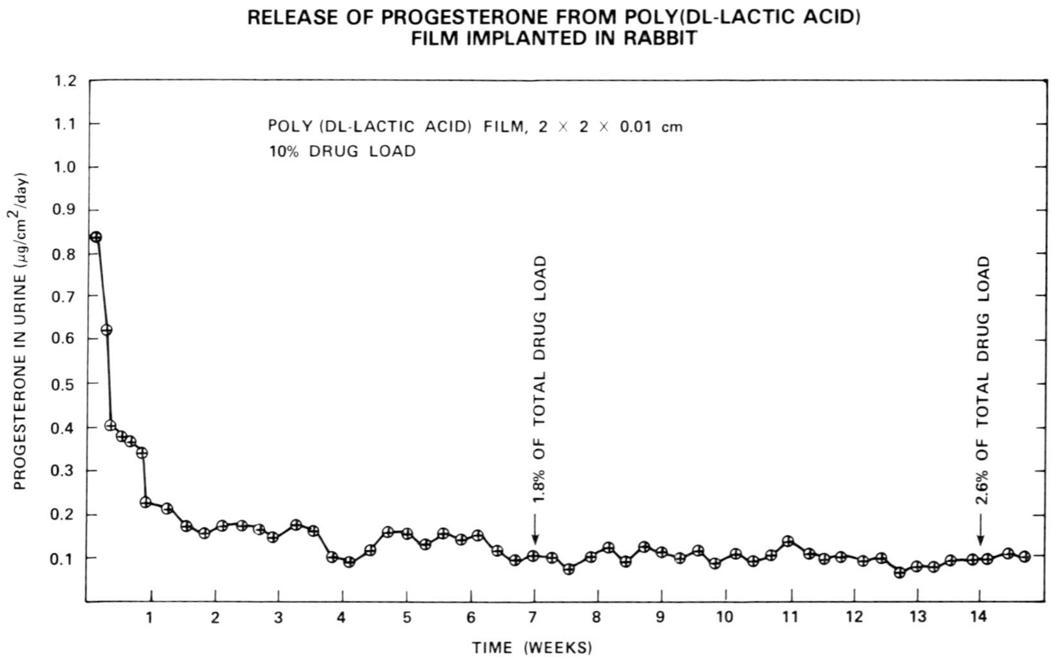


*FIGURE 4. Daily rate of release of progesterone from poly(DL-lactic acid) film implanted in rabbit, measured as radioactivity excreted in urine (from Reference 7).*

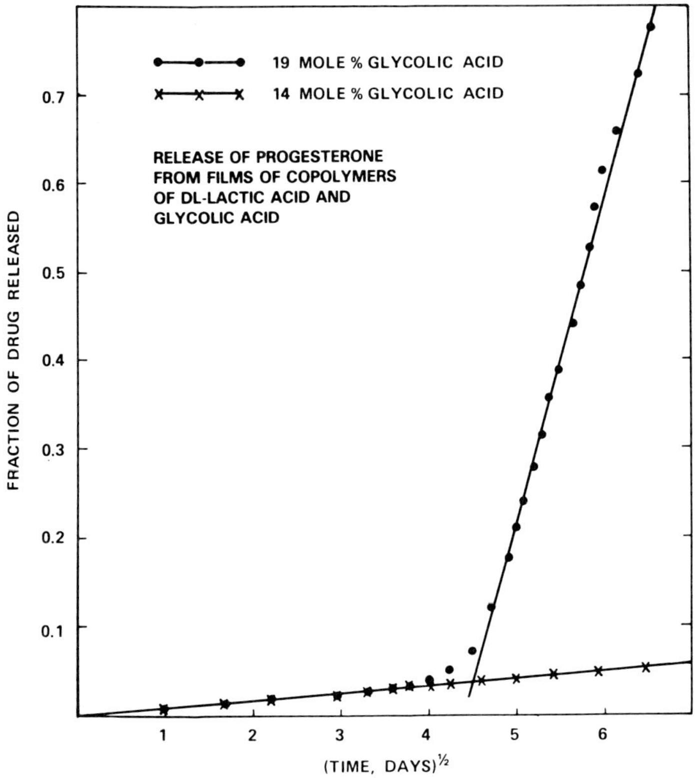


*FIGURE 5. In vitro rates of release of progesterone from two poly(DL-lactic acid-co-glycolic acid) films, thickness 100 μm, drug load 10%, plotted as the fraction of drug released versus the square root of time. Break in line is indicative of fragmentation of one of the films.*

*in vitro* conditions (7,8). In common with other polyesters studied, biodegradation of poly(ε-caprolactone) is characterized by an induction period prior to weight loss. There is a continual decrease in the viscosity and molecular weight (GPC) resulting from hydrolytic cleavage of the ester linkages, and semi-long plots of MW or the intrinsic viscosity versus time are linear until the MW has decreased to about 30,000. Several different polymers are compared in Figure 8, which shows that poly(DL-lactic acid) is degraded more rapidly than poly(ε-caprolactone). This can be attributed to chemical effects, for the same relative rates are observed with the monomeric esters. The greater rate of degradation

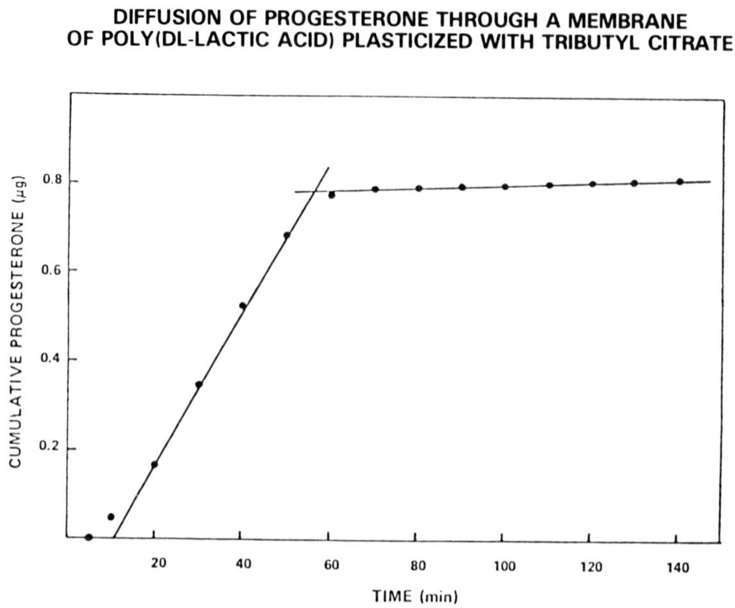


*FIGURE 6. Rate of diffusion of progesterone through a membrane of poly(DL-lactic acid) plasticized with tributyl citrate. Break in plot is indicative of loss of tributyl citrate (from Reference 5).*

of the copolymers must be attributed to morphological changes. Others have observed that copolymers of lactic acid and glycolic acid are degraded more rapidly than the homopolymers (9), presumably for the same reason.

The semi-logarithmic relationship between the molecular weight and time *in vivo* is consistent with bulk hydrolytic cleavage of the ester linkages autocatalyzed by the generated carboxylic acid end groups. The same rate is observed *in vitro* ($H_2O$, $40^oC$) and for films and cylinders, despite a large surface-to-volume difference. These observations rule out any enzymatic contribution to the hydrolysis rate.

For poly(ε-caprolactone), the decrease in molecular weight is accompanied by a slow increase in crystallinity, elastic modulus, and a decrease in the % elongation at break and ultimate tensile strength. Weight loss commences when

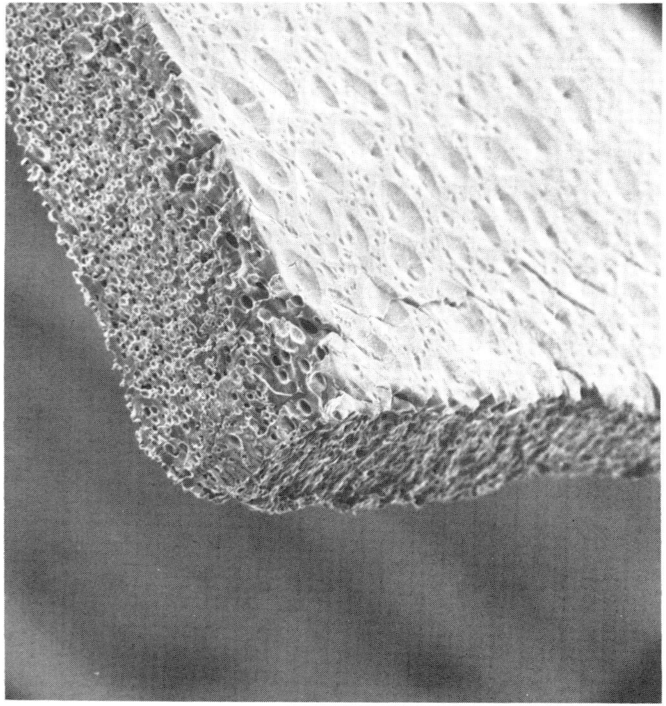

Magnification: 330

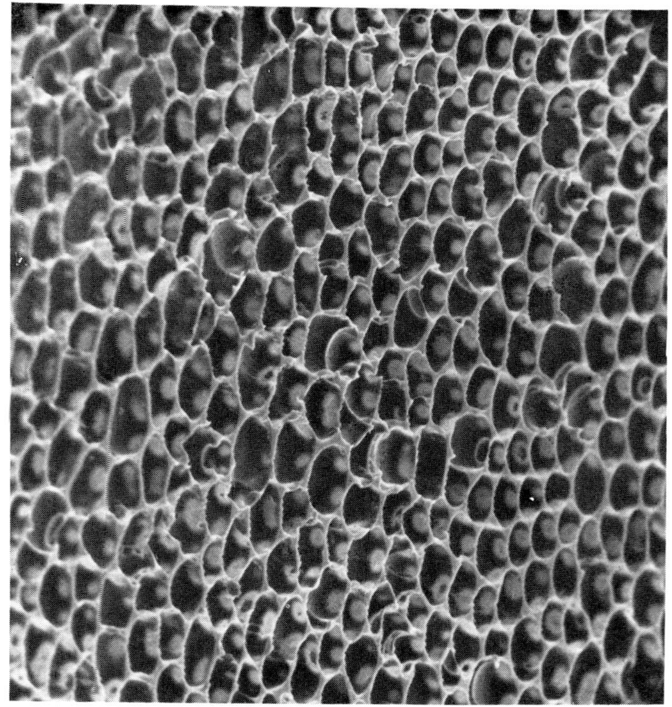

Magnification: 1020

*FIGURE 7. Scanning electron micrographs of films prepared from poly(DL-lactic acid) and 10% glycerine (from Reference 5).*

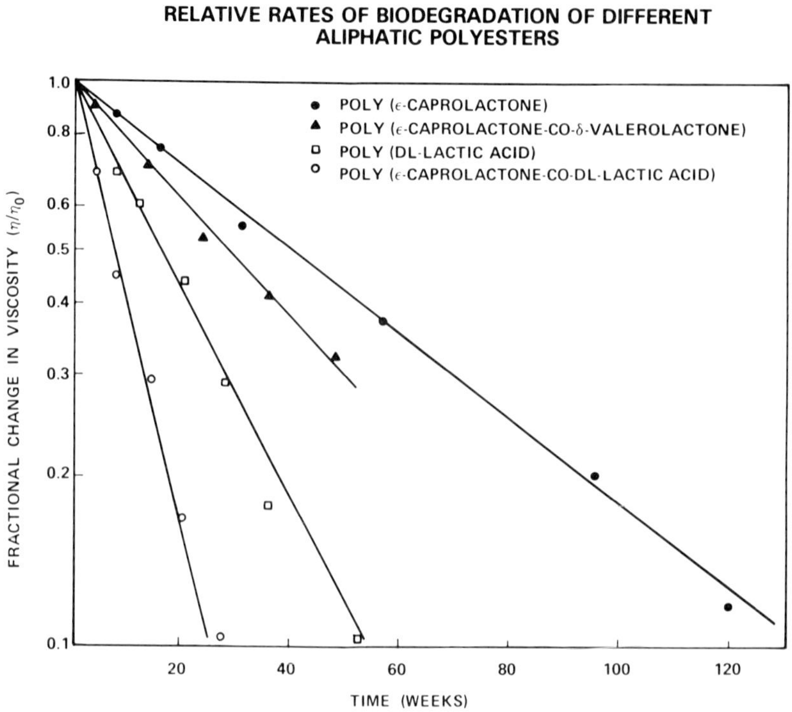


*FIGURE 8. Fractional change in the intrinsic viscosity of different polyesters as a function of time implanted in rabbit, illustrating dependence of degradation rate on chemical structure and morphology.*

the MW is less than about 30,000. Changes in both the weight and the viscosity of various copolymers of ε-caprolactone and DL-lactic acid are shown in Figure 9.

## V. APPLICATION TO CONTRACEPTION AND NARCOTIC ANTAGONIST THERAPY

The substantial induction periods prior to weight loss, coupled with the high permeability of poly(ε-caprolactone) and its copolymers, has permitted the development of biodegradable reservoir devices which function by sustained, diffusion-controlled drug delivery and which erode after the drug reservoir is depleted. These devices retain the

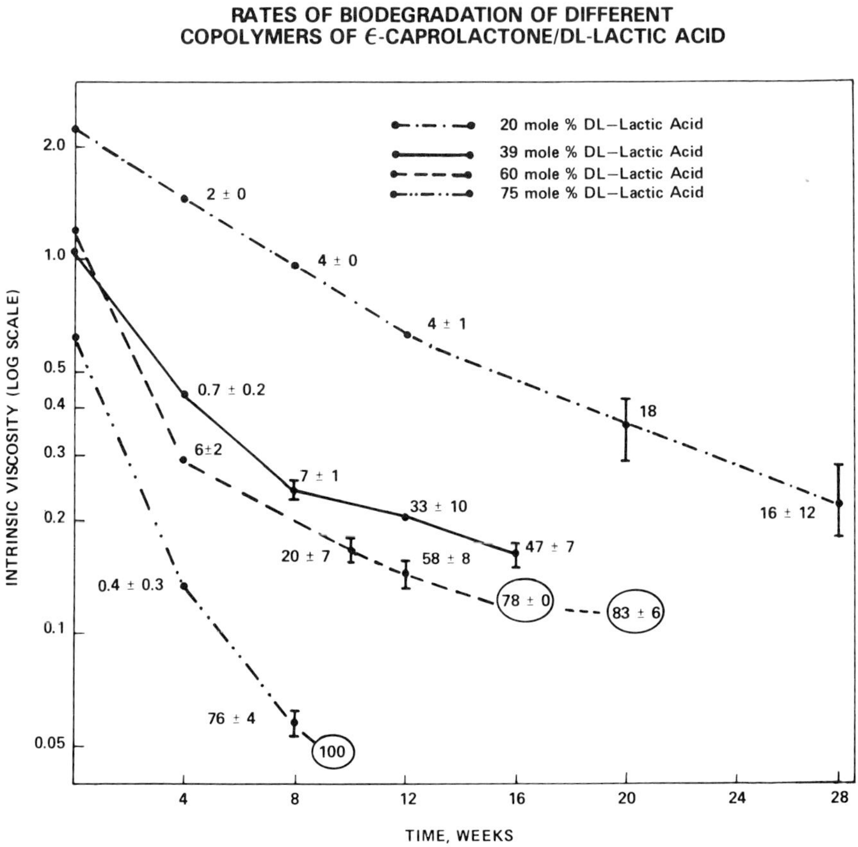


*FIGURE 9. Semi-logarithmic plots of intrinsic viscosities of poly(ε-caprolactone-co-Dl-lactic acid) capsules versus time in vivo (rabbit). Numbers at individual data points show averaged magnitude of weight loss.*

advantages of subdermal poly(dimethylsiloxane) systems, i.e., constant and sustained drug delivery, but the need for surgical removal after drug exhaustion is eliminated.

The primary objective of our studies has been the development of a device capable of subdermal delivery of contraceptive levels, i.e., ∿ 50 μg/day, of levonorgestrel for a period of at least one year. Poly(ε-caprolactone) is the most suitable polymer for this purpose, for it is permeable to levonorgestrel and an induction period of 1 year prior to the onset of bioerosion is easily achieved. Examples of long term levonorgestrel release rates *in vitro* and *in vivo* (rat) are shown in Figures 10 and 11. Sterilization of the device

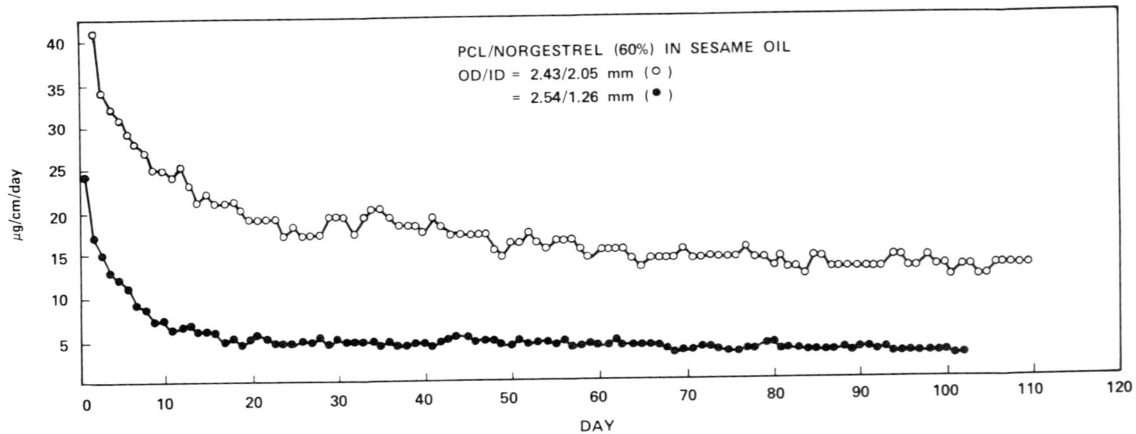


*FIGURE 10. Daily rate of release of norgestrel from poly(ε-caprolactone) capsules, in vitro, per centimeter length of capsule, using two different capsule wall thicknesses.*

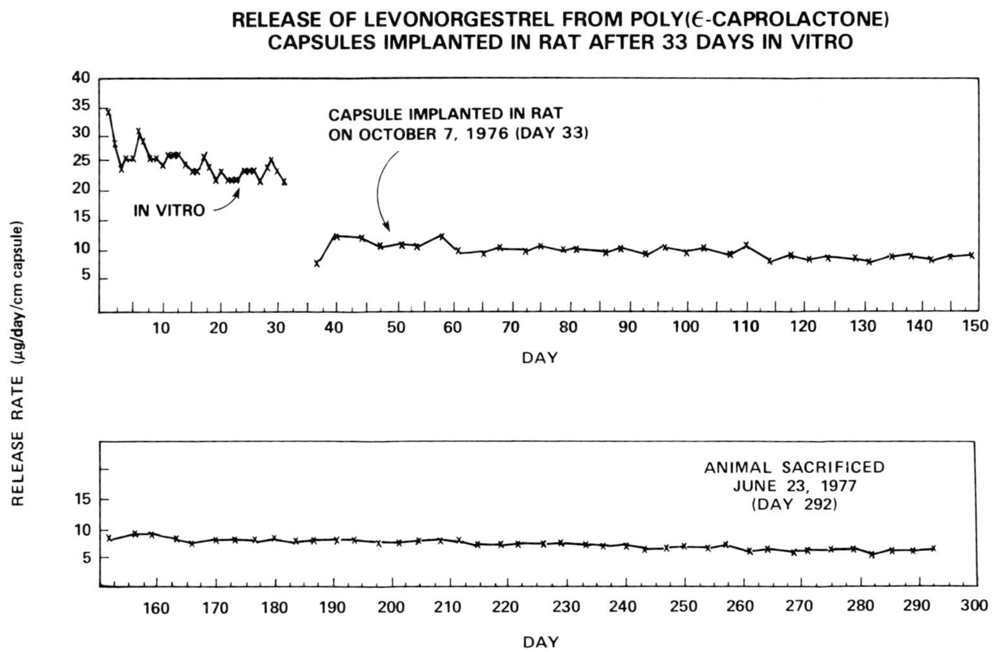


*FIGURE 11. Daily rate of release of norgestrel from a poly(ε-caprolactone) capsule implanted in rat after 32 day in vitro. Release rate determined by measurement of radioactivity in urine and feces. Difference in the in vitro and in vivo rates is the result of metabolic loss of radiolabel by routes other than urine and feces.*

is achieved by γ-irradiation and toxicology studies have shown no adverse local or systemic effects. Initial pharmacokinetic studies in women are now planned.

The requirements of a narcotic antagonist maintenance program are very different to those of birth control programs. Naltrexone, the most potent narcotic antagonist, is only effective at a dose of at least three mg/day and present treatments are restricted to 1-2 months. The properties of poly (ε-caprolactone-co-Dl-lactic acid) are suitable for a sustained delivery device with these requirements. Thus, by proper choice of molecular weight and chemical composition, it is possible to obtain a polymer which is biostable for a period of anywhere between 4 and 12 weeks (drug delivery phase), and which undergoes rapid bioabsorption subsequently (Figure 9). This copolymer is also in the rubbery state and very permeable.

A copolymer with an intrinsic viscosity of 1.5 dl/g and 20 mole% lactic acid was selected for evaluation. The maximum steady state flux ($J_{lim}$) and the diffusion coefficient (D) of naltrexone in this copolymer were determined from the kinetics of transport through a thin film in a standard diffusion cell.

$$J_{lim} = 2.7\ (\pm\ 0.4) \times 10^{-10}\ \text{g/cm/sec}$$

$$D = 5.9\ (\pm\ 2.4) \times 10^{-9}\ \text{cm}^2\text{/sec}$$

Based on the value of $J_{lim}$ and the kinetic expression for the diffusion controlled release of a cylindrical reservoir (Equation 1),

$$\text{Release rate} = 2\pi DK\Delta C/\ln(od/id) = 2\pi J_{lim}/\ln(od/id) \quad \ldots(1)$$

one can estimate that two 2.5 cm capsules, 2.7 mm outer dimeter and 2.1 mm inner diameter, filled with a 50% aqueous slurry of naltrexone, will release 3 mg of drug per day for 30 days. This is within the range believed desirable for an antagonist maintenance program.

Rates of release were first measured *in vitro* and found to be relatively constant over a 36 day period. Using capsules with a 2.05 mm inner diameter and 2.35 mm outer diameter, the release rate was between 300 and 400 μg/day per cm capsule length for 24 days, declining as the drug was exhausted over the next 20 days (Figure 12a). A lower but more sustained release rate was achieved when capsules with a slightly thicker wall were used (Figure 12b). These rates are lower than the values estimated from $J_{lim}$, but this is not suprising in view of the errors of measurement of the

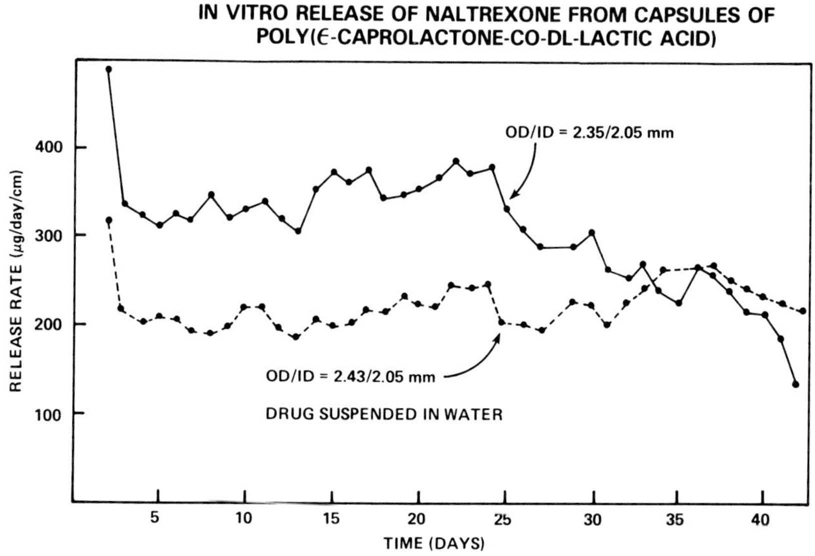


*FIGURE 12. In vitro rate of release of naltrexone-$^3H$ from a poly(ε-caprolactone-co-Dl-lactic acid) capsule containing naltrexone as 50% suspension in water.*

thin walls of the polymer capsule, the sensitivity of the rate to the OD/ID ratio in the range used, and the fact that the method of tube preparation produces some irregularity in the dimensions and possibly some differences in morphology.

Since poly(ε-caprolactone-co-Dl-lactic acid) is permeable to water, the rate of drug released from a capsule containing only naltrexone was measured *in vitro*. There was a slow increase in rate, to ∿300 μg/day/cm length (Figure 13). Presumably the increase was related to water diffusion into the capsule, the water being necessary to maximize dissolution of the drug in the capsule wall; the maximum release rate is dependent on maintaining the drug concentration gradient ($\Delta C$) across the polymer wall. The rate slowly declined after day 30, until the rate was 100 μg/day/cm at day 95. When such capsules, 3 cm length were subdermally implanted in three monkeys, one per animal, morphine self-administration was immediately suppressed and remained so for 33 days.

The naltrexone used in this study was tritium labeled, and blood radioactivity levels were measured to assess the constancy of the drug release rate. The rapid increase in tritium levels after about 20 days in two cases (Figure 14)

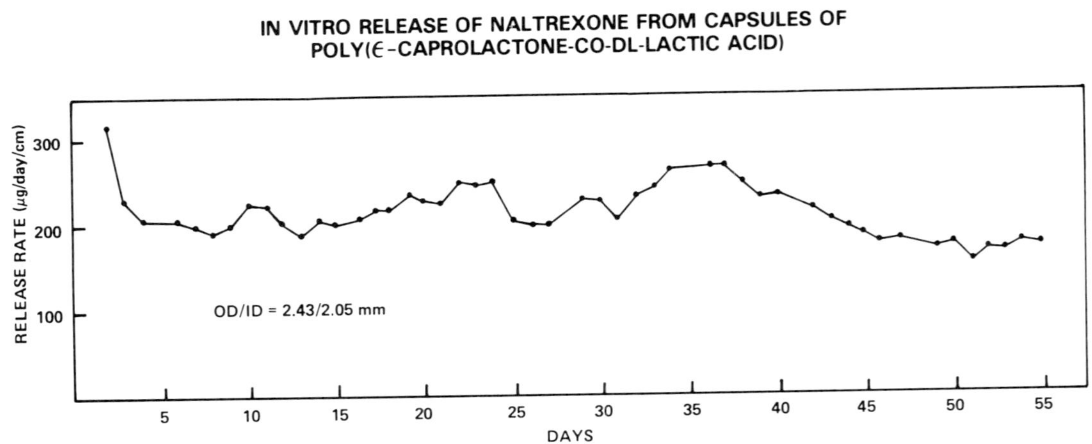

IN VITRO RELEASE OF NALTREXONE FROM CAPSULES OF
POLY(ε-CAPROLACTONE-CO-DL-LACTIC ACID)
RELEASE RATE (μg/day/cm)
300
200
100
OD/ID = 2.43/2.05 mm
5
10
15
20
25
30
35
40
45
50
55
DAYS


*FIGURE 12b.*

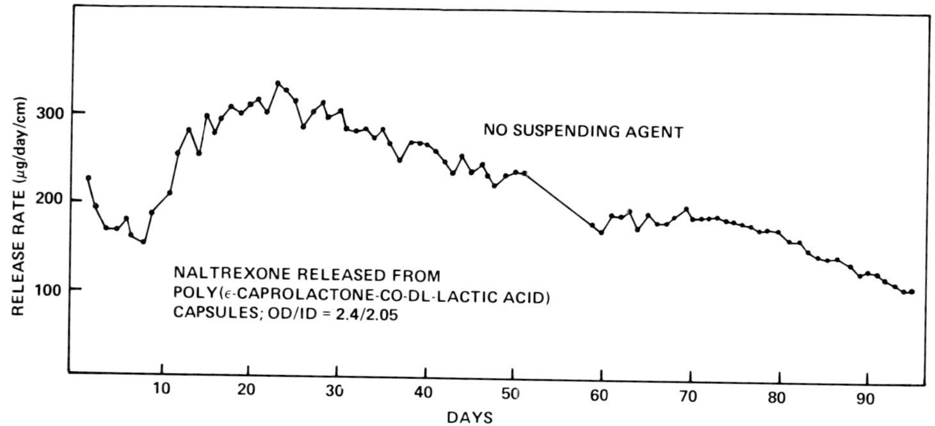


*FIGURE 13. In vitro rate of release of naltrexone-$^{3}H$ from a poly(ε-caprolactone-co-DL-lactic acid) capsule containing dry naltrexone. Capsule dimensions: 2.4 mm OD, 2.05 mm ID, 1 cm length.*

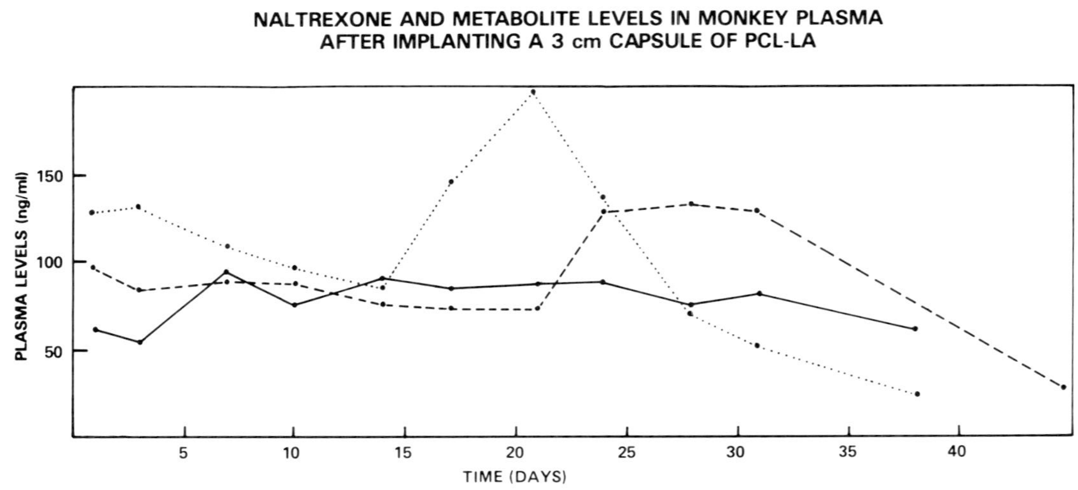


*FIGURE 14. Plasma concentrations of naltrexone and its metabolites resulting from implantation of poly(ε-caprolactone-co-Dl-lactic acid) capsules containing dry naltrexone in three monkeys.*

strongly suggests that capsule degradation commenced at that time. At the end of the study (day 46) the capsules were recovered as partially degraded fragments. Evidently, the time prior to polymer bio-erosion was somewhat less than that estimated from Figure 9, and the morphine blockade could have been maintained for a greater time period by use of a copolymer with a slightly different composition.

These studies with naltrexone only amounted to exploratory experiments, to test the feasibility of using a polyester in the form of a biodegradable reservoir device. It would appear that the approach is feasible and that it is only necessary to optimize the various components to achieve a practical device.

## ACKNOWLEDGMENTS

This work was supported in part by the Contraceptive Development Branch, Center for Population Research, National Institute of Child Health and Human Development, NIH, under Contract No. N01-HD-3-2741.

This paper has also been published in a research monograph by the National Institute on Drug Abuse.

## REFERENCES

1. R. W. Baker and H. K. Lonsdale, Chem. Tech., 5, 668-674 (1975).
2. D. L. Gardner, A. J. Patanus, and D. J. Fink, in Proc. Drug Delivery Systems, H. L. Gabelnick, (Ed.), DHEW Publication No. (NIH) 77-1238, August 1976, pp. 265-278.
3. E. S. Nuwayser, D. L. Williams, P. M. Meier, T. C. Wu, S. Z. Merchant, and J. K. Kerrigan, in Proc. Drug Delivery Systems, H. L. Gabelnick, (Ed.), DHEW Publication No. (NIH) 77-1238, August 1976, pp. 193-251.
4. C. Thies, in Controlled Release Polymeric Formulations, D. R. Paul and F. W. Harris, (Eds.), ASC Symposium Series 33, American Chemical Society, Washington, D.C., 1976, pp. 190-194.
5. C. G. Pitt, A. R. Jeffcoat, R. A. Zweidinger, and A. Schindler, J. Biomed. Mater. Res., 13, 497-507 (1979).
6. C. G. Pitt, M. M. Gratzl, A. R. Jeffcoat, R. Zweidinger, and A. Schindler, J. Pharm Sci., 68, 1534-1538 (1979).

7. C. G. Pitt, D. Christensen, R. Jeffcoat, G. L. Kimmel, A. Schindler, M. E. Wall, and R. A. Zweidinger, in Proc. Drug Delivery Systems, H. L. Gabelnick, (Ed.), DHEW Publication No. (NIH) 77-1238, August 1976, pp. 141-192.
8. A. Schindler, R. Jeffcoat, G. L. Kimmel, C. G. Pitt, M. E. Wall, and R. Zweidinger, in Contemporary Topics in Polymer Science, Vol. 2, E. M. Pearce and J. R. Schaefgen, (Eds.), Plenum Publishing, New York, 1977, pp. 251-289.
9. R. A. Miller, J. M. Brody, and D. E. Cutright, J. Biomed. Mater. Res., 11, 711-719 (1977).
10. T. J. Roseman, J. Pharm. Sci., 61, 46-50 (1972).
11. T. J. Roseman and S. H. Yalkowsky, J. Pharm. Sci., 63, 1639-1641 (1974).

# CONTROLLED RELEASE OF PROGESTINS FROM poly($\alpha$-AMINO ACID) CARRIERS

*Robert V. Petersen*
*Christopher G. Anderson*
*Sen-Maw Fang*
*Donald E. Gregonis*
*Sung Wan Kim*

Department of Pharmaceutics
College of Pharmacy
University of Utah
Salt Lake City, Utah

*Jan Feijen*

Department of Chemical Engineering
Twente University of Technology
Enschede, The Netherlands

*James M. Anderson*
*Sumita Mitra*

Department of Macromolecular Sciences
Case Western Reserve University
Cleveland, Ohio

## I. INTRODUCTION

The use of polymers in sustained release devices and dosage forms has drawn considerable interest in recent years (1-6). Until now, few examples have been published which show that steroids have been covalently bonded to polymeric systems (7-9) and only very recently did data become available concerning the coupling of steroids to biodegradable




ISBN 0-12-074450-3

polymers (10). It is our purpose here to describe the derivatization of a 17β-ethynyl-17β-hydroxy steroid, namely, the progestin, norethindrone (NE), which is applicable for attachment to biodegradable poly(α-amino acids). The release of the hormone from such systems has been investigated as a possible means of obtaining long-term, injectable contraceptive agents. Such polymeric systems, which release hormone over a prolonged period of time, do not require surgical removal following injection or implantation. Anderson *et al.* (11,12) and Mitra *et al.* (13), have shown that synthetic poly(α-amino acids) which are biodegradable can be developed and the rate of *in vivo* biodegradation can be controlled by varying the hydrophilicity of the side chain groups.

The objective of this research was to synthesize and characterize compounds in which norethindrone was covalently bonded to a poly(α-amino acid) backbone according to the following schematic diagram:

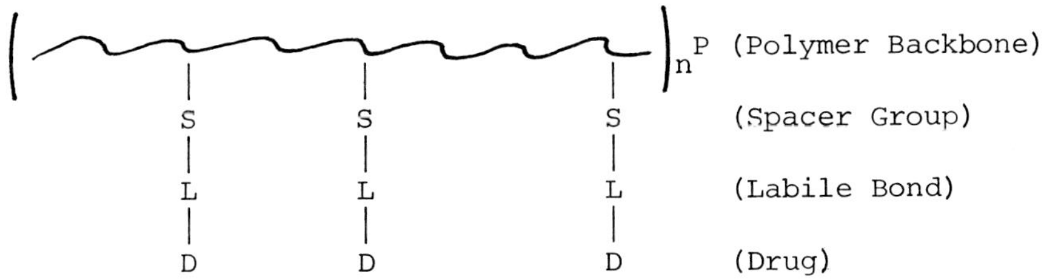


In this system, there are a number of variables which offer potential control in intensity and duration of action. Where the backbone is poly(α-amino acid) the rate of biodegradation is inherently slow so that the relative rates of hydrolysis between the backbone amide bonds and labile bond between the spacer group and the drug favors drug release prior to backbone hydrolysis. However, by variation in the chain length, hydrophilic/hydrophobic character, and chemistry of the spacer group, the rate of release of drug from the polymer can be controlled. Various labile bond types offer another mechanism whereby drug release rate can be controlled so that hydrolytic or enzymatic cleavage will be at a faster rate than biodegradation of the polymer backbone. Some possible bond types include ester, carbonate, amide, acetal, ketal, imino, and others. There are numerous examples of drugs which are appropriate candidates for use in these systems. There are some restrictions of course. Thus it is necessary to utilize compounds which are relatively potent so that the total bulk of a dose of the polymer/drug conjugate is reasonable. In addition, the drug should be one which is used for long-term therapy. Finally, the drug must possess a functional group which can be utilized appropriately for attachment to the polymer.

It is an objective of this research to evaluate these compounds in both *in vitro* and *in vivo* conditions and to determine the biocompatibility characteristics of these compounds by immunogenicity and histopathology tests.

In any system of this type, two processes control drug release, hydrolysis and diffusion. Since most of the useful polymers exhibit a substantial degree of hydrophilic character and some swelling can occur in an aqueous environment, it appears that diffusion can take place fairly rapidly once the drug has been hydrolyzed from the polymer. Thus, hydrolysis appears to be the primary rate controlling factor, in which case it is possible in principle to approach zero-order release kinetics.

## II. SYNTHESES

The polymers selected for this study were the poly(hydroxyalkyl)-L-glutamines (III), described by Lotan *et al.* (14) and Miyake *et al.* (15). These polymers are prepared by base catalyzed polymerization of γ-benzyl-L-glutamate-N-carboxyanhydride (I) giving the polymeric benzyl-L-glutamate (II). The benzyl group is displaced with the desired hydroxyalkyl amines (16) as shown below:

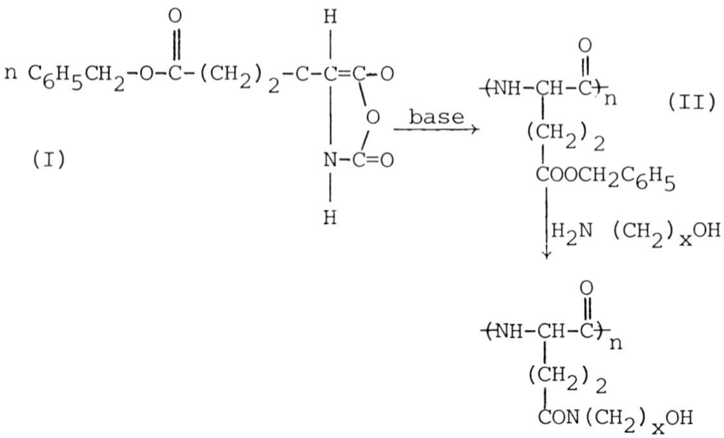


(IIIa, x = 2)
(IIIb, x = 3)

In order to react norethindrone (IV) with the free hydroxyl groups on Compound III, it was first converted to norethindrone-17β-chloroformate V by reaction with an excess of phosgene in methylene chloride. The reaction was catalyzed with pyridine and proceeded according to the following scheme:

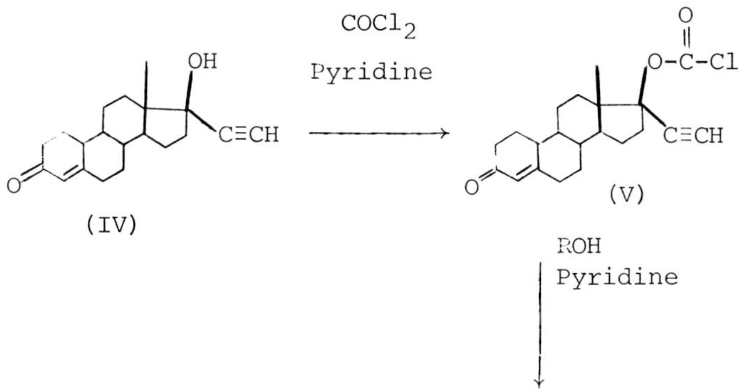


VI, R = $-CH_3$
VII, R = $-(CH_2)_2CH_3$
VIII, R = $-(CH_2)_5CH_3$
IX, R = Poly-N-(3-hydroxypropyl)-L-glutamine

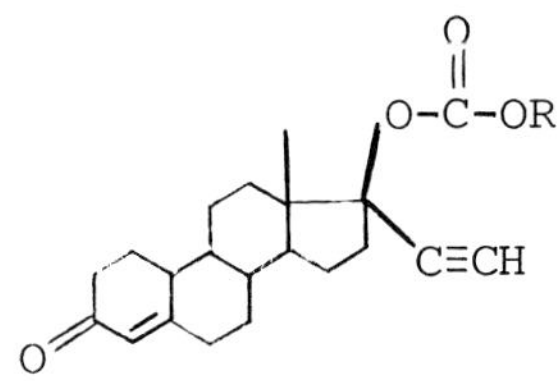


Norethindrone-17β-methyl carbonate (VI), norethindrone-17β-propyl carbonate (VII), and norethindrone-17β-hexyl carbonate (VIII) were prepared by reacting the chloroformate (V) with an excess of methanol, n-propanol, or hexanol, respectively, and a stoichiometric quantity of pyridine.

Coupling reactions of the chloroformate (V) to polymer IIIb to yield Compound IX were monitored using tritiated norethindrone (10). Poly-$N^5$-(3-hydroxypropyl)-dissolved in a mixture of anhydrous DMF and pyridine then reacted with the tritiated chloroformate (V). The polymer was isolated from unreacted products by repeated precipitation in ethyl acetate. The degree of substitution of norethindrone onto the available hydroxyl groups of polymer IIIb was up to 60 percent as determined by liquid scintillation counting techniques. Similar reactions have been carried out using poly-$N^5$-(3-hydroxyethyl)-L-glutamine (IIIa) to obtain a similar conjugate wherein the spacer group is composed of two methylene units instead of the three which are present in compounds prepared from IIIb.

In addition to the 17β-hydroxyl group, norethindrone also possesses a 3-keto function which is capable of derivatization. Easily cleavable functional derivatives of this group were considered for the introduction of the labile bond whose hydrolysis would permit release of the drug. The imine linkage was considered to be particularly suitable for this purpose since its rate of cleavage would be expected to be much faster than the hydrolysis of the amide bonds in the polymeric backbone of the poly(α-amino acids). Thus, it could be expected that a sustained release of the progestin prior to biodegradation of the polymer backbone could be achieved. Reactions were carried out by standard synthetic procedures as shown in the following scheme:

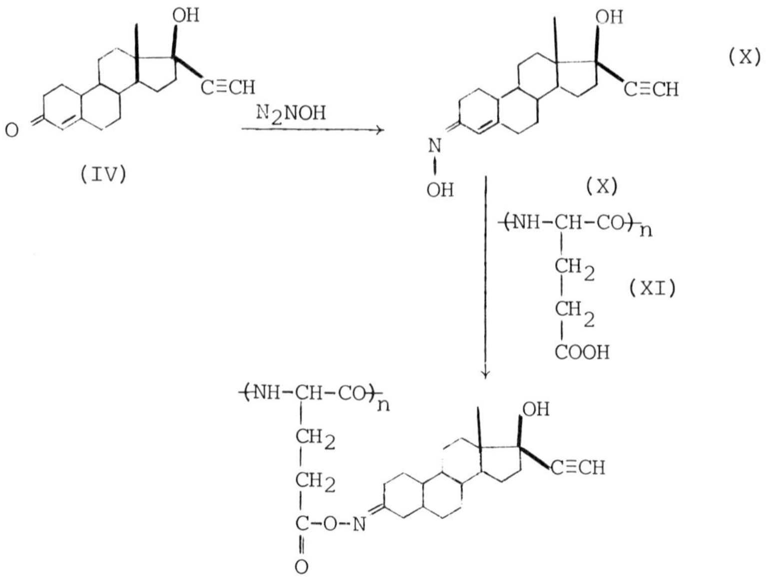


(XII)

In this reaction, norethindrone (IV) was first converted to the oximino derivative, norethindrone oxime (X), by reaction with hydroxylamine. The latter compound was found to exist in two configurations, *syn* and *anti*, as evidenced by the NMR spectra and the TLC behavior. The norethindrone *syn* and *anti* oximes (X) were then esterfied with the side chain carboxyl group of poly(glutamic acid) (XI) using the carbodiimide method of coupling to give the norethindrone/poly(glutamic acid) conjugate (XII). The presence of the oxime ester was proven by IR analyses following rigorous purification by repetitive reprecipitations or dialyses.

Compounds VI, VII, VIII, IX, and XII have been prepared using nonspecific tritiated norethindrone and studied by various *in vitro* and *in vivo* techniques. Norethindrone/polymer conjugates have also been synthesized recently from $^{14}$C-norethindrone in which the label is in the ethynyl side chain for *in vitro* and *in vivo* release studies. These studies are still in progress.

## III. *IN VITRO* RELEASE STUDIES

Several studies have been conducted on norethindrone/polymer conjugates and on model compounds in order to assess the stability of the 17-carbonate bond and the 3-oximino bond of norethindrone. At this time, not all compounds have been studied for *in vitro* release characteristics. The hydrolysis of representative compounds are described below.

### A. *In Vitro Lability Studies on Low Molecular Weight 17-Carbonate Esters of Norethindrone*

The *in vitro* hydrolysis of $^{3}$H-norethindrone-17-methyl carbonate (VI) was measured in pooled plasma of female Sprague-Dawley rats and in 0.02 M Tris-HCl buffer solution (pH 7.4). Test tubes containing 8 ml of plasma or buffer were placed in a constant temperature bath and maintained at 37$^{\circ}$C throughout the study. To each tube was added 50 µl of methanol solution to Compound VI. The reaction mixture was sampled at various intervals and the amount of VI and norethindrone quantified by TLC. Similar tests were performed on Compounds VII and VIII. The results showed that those compounds are hydrolyzed very quickly in plasma but not in the buffer solution. Figure 1 shows the disappearance of VIII and simultaneous appearance of norethindrone. The other two compounds showed a similar pattern, but were not hydrolyzed quite as rapidly as VIII. Table I compares the rates of hydrolysis of Compounds VI, VII, and VIII as determined by this study. From Table I, it can be seen that all of the compounds were hydrolyzed rapidly in rat plasma but very slowly in buffer solution. It may also be noted that within the series tested there was an increase in the rate of hydrolysis in rat plasma with increased molecular weight.

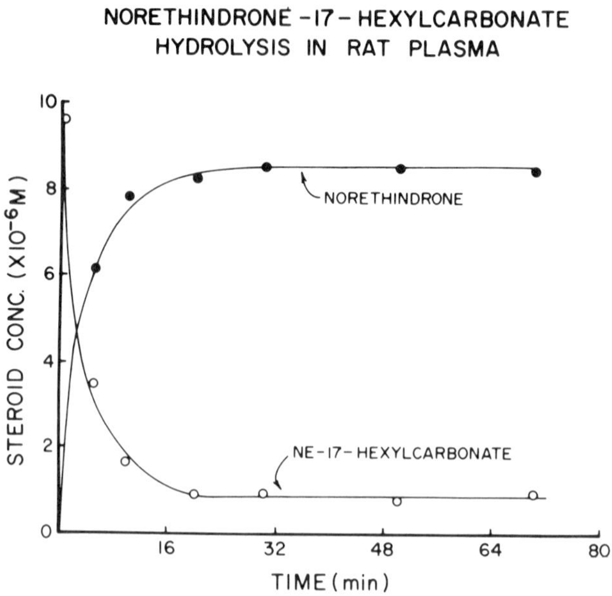


*FIGURE 1. In vitro hydrolysis of norethindrone-17-hexylcarbonate (VIII) in rat plasma.*

The above data provided fairly strong evidence to support the hypothesis that in case small fragments should remain attached to norethindrone following release from the polymer, these fragments will be hydrolyzed rapidly by enzymes present in plasma. Further studies will be conducted to confirm this hypothesis.

### B. *In Vitro Release Studies on $^{3}$H-Norethindrone-17-Carbonate Esters of Poly-N$^{5}$-(3-hydroxypropyl)-L-Glutamine (IX)*

Two studies on *in vitro* hydrolysis of Compound IX were conducted on preparations utilizing high molecular weight IIIb (approx. 53,000) and low molecular IIIb (approx. 15,000). The study utilized a specially designed sample cell which continually circulated the test medium in contact with samples of IX which had been cast into sheets. Tris buffer at pH 7.4 was pumped into the sample cell (approximate volume 45 ml) which contained a sheet sample of IX (2.7 cm$^2$ surface area per side) and the medium was thereafter collected continuously in a collection reservoir. The cell contents were continuously stirred. The pump system supplied approximately 3 ml/hr of fresh medium into the cell. Samples were collected

TABLE I. Hydrolysis of Norethindrone-17-Alkylcarbonate (NEAC)

| | | Half-life ($t_{1/2}$) (min) | | |
|---|---|---|---|---|
| Alkyl group | Conc. (mols) | Rat plasma ln(NEAC)[a] | Rat plasma ln($NE^{\infty}$-NE)[b] | Buffer (0.02 M tris, pH 7.4)[c] |
| Methyl (VI) | $2 \times 10^{-6}$ | 20.0 | 23.7 | >70 |
| Propyl (VII) | $1.12 \times 10^{-5}$ | 4.2 | 4.7 | >70 |
| Hexyl (VIII) | $9.6 \times 10^{-6}$ | 2.7 | 3.9 | >70 |

[a] Determined by the disappearance rate of norethindrone-17-alkylcarbonate (Compounds VI, VII, and VIII).
[b] Determined by the appearance of norethindrone (NE).
[c] No hydrolysis was observed after 70 minutes.

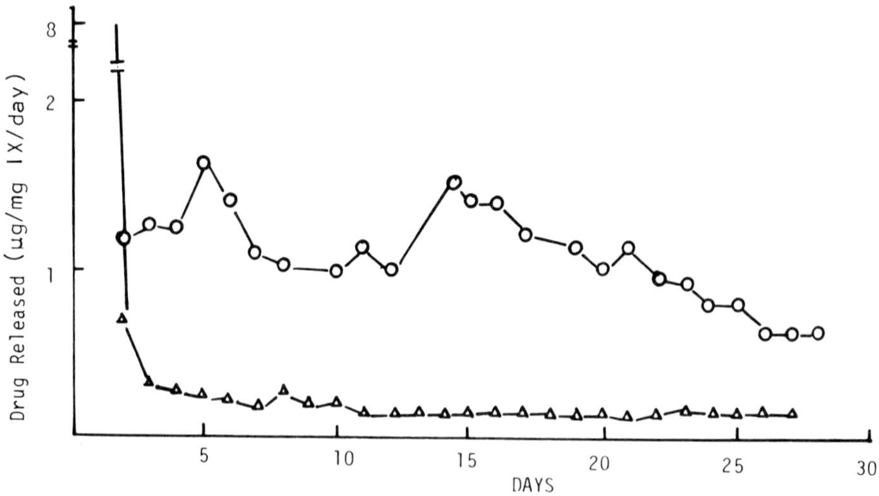


*FIGURE 2. Comparison of in vitro release rates for drug-polymer (IX) prepared from low molecular weight IIIb (○) and high molecular weight IIIb (△) (37°C in tris buffer).*

and counted daily. The results of the study are shown in Figure 2. From this figure it can be seen that there was an initial high release of activity, following which the release pattern became almost level and approached zero-order kinetics by day 25. It will further be noted that the low molecular weight compound released drug considerably more rapidly than did the high molecular weight compound. This may be due to the fact that the low molecular weight product exhibited a substantial degree of swelling, whereas the high molecular weight product did not.

## C. *In Vitro Release Studies on ³H-Norethindrone-3-Oxime Ester of Poly(Glutamic Acid) (XII)*

Hydrolysis studies on the norethindrone oxime (X), the acetyl oxime derivative of norethindrone (XIII) and the polymer bound norethindrone oxime (XII) were carried out in solutions composed of absolute ethanol to which an equivalent volume of phosphate buffer had been added. Aliquots of the hydrolysis solutions were subjected to thin-layer chromatography for separation of the components, the spots were scraped and counted for radiolabel.

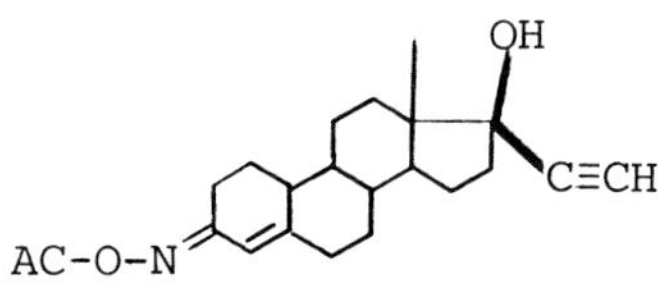


(XIII)

Figure 3 shows the release of the tritium labeled steroid from films of the covalently bound polymer preparation and it is apparent that a steady state release behavior is present after an initial burst. A linear dependence of the cumulative release versus time shows that a zero-order rate was achieved with this system. After about 60 percent of the steroid was released, the polymer was found to swell considerably in the hydrolysis medium and this was attended with an increase in radioactivity in the hydrolysate until the polymer ultimately completely dissolved in the reaction solution. Table II is a comparison of the release behavior of films of XII of different dimensions under different pH conditions. For films of equal loading of steroid and of equal weight, the amount of steroid released per day at steady state was greater for larger surface areas (smaller thicknesses). The release of steroid from XII in pH 2 buffer-50 percent ethanol mixtures also exhibited steady state behavior.

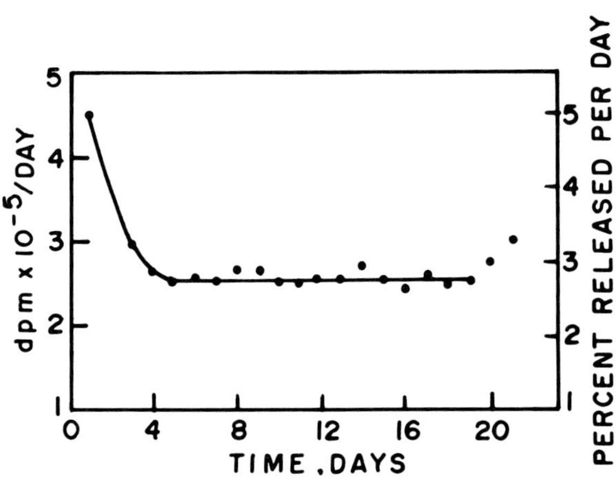


*FIGURE 3. Hydrolysis of XII at 37°C in 50 percent pH 6 buffer-50 percent ethanol.*

TABLE II. *Hydrolysis of $^{3}H$-XII (96 Percent Loading)*

| | | | | Steroid released at steady state | | | |
|---|---|---|---|---|---|---|---|
| pH of elution media | Wt of polymer/film (mg) | Area of film ($cm^2$) | Area of film wt ($cm^2/mg$) | % of total | μg released | % NE in steroid released | % Oxime in steroid released |
| 6 | 10.9 | 0.12 | 0.038 | 2.8 | 220 | --- | --- |
| 6 | 10.9 | 0.56 | 0.052 | 3.7 | 290 | 8 | 92 |
| 2 | 8.3 | 0.30 | 0.036 | 3.7 | 225 | 38 | 62 |

A comparison of the hydrolysis of X with that of XII shows that the cumulative amounts of norethindrone released are similar. However, the release of norethindrone from the polymer-bound material, i.e., XII, would be expected to be slower than that from the free oxime due to steric hindrance of the polymer backbone. This indicates that 0-acetylation of the oxime causes an enhancement in the rate of hydrolysis of the oximino bond. To prove this, the 0-acetyl oxime of norethindrone (XIII) was synthesized as a model compound for XII and its hydrolysis behavior studied. The cumulative amount of norethindrone obtained from XIII was on an average about eight times larger than that from the oxime itself, which indicates that the 0-acyl group does increase the rate of cleavage of the oximino bond. The slow release of norethindrone from XII compared to XIII is indicative of some retarding effect due to the polymer backbone.

## IV. *IN VIVO* RELEASE STUDIES

Although several polymer/drug conjugates have been synthesized, only Compound IX has been studied sufficiently to report at this time. Studies on the effects of varying the extent of drug loading, particle size, and length of spacer group are currently under investigation in this laboratory. Also, the potential value of other derivatives such as the 3-oximino derivatives, has yet to be determined.

In a typical *in vivo* experiment, five or six rats are injected subcutaneously with an appropriate amount of tritium labeled polymer/drug conjugate in a suspension dosage form. Release of radioactivity in urine and feces is subsequently monitored over extended periods of time. In more recent experiments, $^{14}C$-labeled drug has been utilized for such experiments.

### A. *In Vivo Pharmacokinetic Study on Compound IX*

In this experiment, five rats were utilized. Tritiated Compound IX (10 mg polymer/4.9 mg NE/2.92 x $10^7$ dpm/rat) was accurately weighed and placed in normal saline overnight. The next morning, the supernatant was decanted leaving only the wetted polymer.

Mature female Sprauge-Dawley rats weighing 220-260 gm were anesthetized with ether. The wet polymer was resuspended in 1 ml of Ringer's Solution and injected subcutaneously into the dorsal side of the neck. The rats were then placed into a metabolism cage and the feces and urine were collected on a daily basis for at least 10 days following implantation. Thereafter, the biological samples were collected on Mondays, Wednesdays, and Fridays.

The total volume of urine was measured and one ml aliquots of urine were sampled. To each urine sample 0.6 ml of water and 12 ml of New England Nuclear BIO-FLUOR scintillation cocktail was added and the sample counted for radioactivity. The total amount of fecal matter for each collection period was pooled, weighed, and put in an oven at 60°C for about 12 hours until dry. The dry mass was then weighed and pulverized in a blender to a homogeneous mixture of fine powders. Two aliquots of the dried fecal matter (0.2-0.3 gm) were accurately weighed and placed in separate glass scintillation vials. To each vial 0.2 ml 70% perchloric acid and 0.4 ml hydrogen peroxide were added. The mixture was incubated for 6-8 hours at 60°C in a shaking water bath until all the fecal matter was completely solubilized. After the digestion, the vials were cooled for 30 minutes at room temperature before adding 12 ml of BIO-FLUOR scintillation cocktail. The vials were then kept in a dark place for at least 24 hours before counting in a Beckman LS 9000 scintillation counter. The data were used to calculate the amounts of radioactivity excreted daily in the urine and feces.

The *in vivo* release rate of norethindrone from Compound IX was fairly constant, though a slight but gradual decrease was observed during the five month period (Figures 4 and 5). The daily release of radioactivity expressed as μg NE/day/mg IX was 0.56 for days 1-36, 0.34 for days 37-66, 0.25 for days 67-87, 0.20 for days 88-136, and 0.11 for days 137-150.

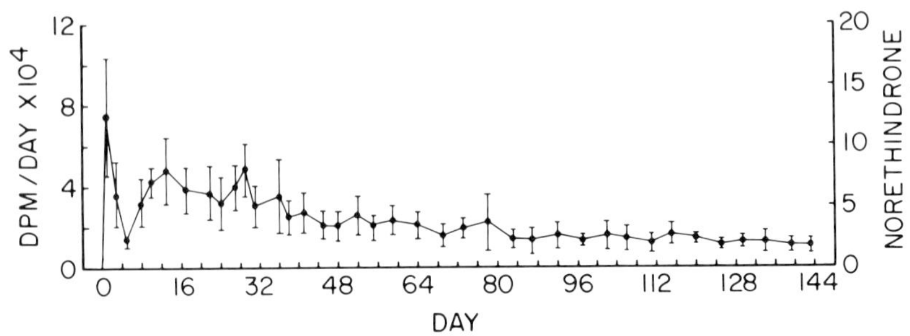


*FIGURE 4. In vivo release of NE from compound IX. 10 mg compound IX (4.9 mg NE/2.92 x $10^7$ dpm), 5 rats average ± S.D.*

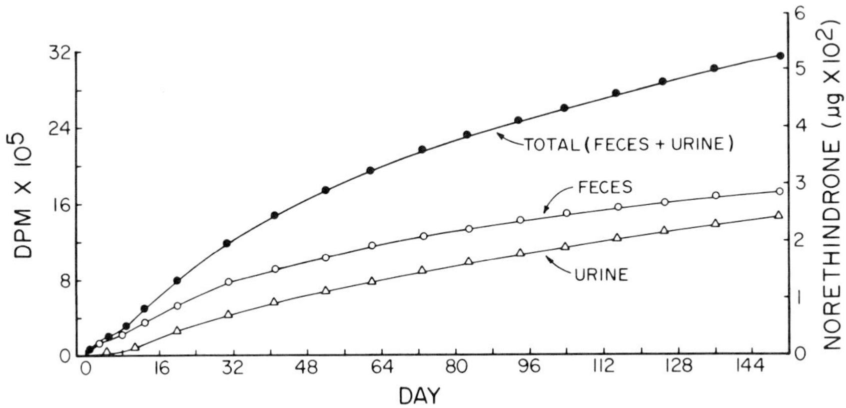


*FIGURE 5. In vivo release of NE from compound IX. 10 mg compound IX (4.9 mg NE/2.92 x $10^7$ dpm) per rat, 5 rats average.*

## V. IMMUNOGENIC AND BIOCOMPATIBILITY STUDIES

Various tests were performed using standard techniques to determine possible immunogenic and/or tissue reaction responses to Compound IX. These tests included skin tests, implantation tests, agglutination tests, precipitation tests, Agar diffusion tests, complement fixation tests, passive cutaneous anaphylaxis tests, and tissue reaction tests. Compound IX produced negative or mild responses in all of these studies.

## VI. CONCLUSIONS

Although several studies are still in progress and much work must yet be completed before the value of poly(α-amino acids) as backbones or carriers for the controlled release of drugs can be fully evaluated, it may be concluded that these systems show considerable promise. It has been demonstrated that such compounds can be successfully synthesized and that, when injected subcutaneously in rats, they release drug slowly and relatively steadily over a period of several months.

*In vitro* release data on low molecular weight 17-carbonate esters of norethindrone show that this bond is quite labile in the presence of plasma enzymes. While direct proof is not yet available, these data suggest that hydrolytic fragments attached to norethindrone through the 17-carbonate ester bond are hydrolyzed readily in plasma to release free norethindrone. These systems offer a considerable degree of flexibility in that several factors may be varied in order to influence the rate and duration of drug release. These factors include polymer molecular weight, particle size, hydrophilicity of the polymer backbone, extent of loading of drug onto available sites on the polymer, length and nature of the spacer group, type of bond between the polymer and the drug, and site of implanation or injection. When more data are available on these variants, it appears likely that valuable additions to available therapeutic alternatives will be realized.

## REFERENCES

1. S. Yolles, Acta Pharm. Suec., (supll. 32), 13 (1976).
2. Controlled Release of Biologically Active Agents, A. C. Tanquary and R. E. Lacey (Eds.), Plenum Publishers, New York (1974).
3. T. J. Roseman, J. Pharm. Sci., 61, 46 (1976).
4. G. Genagiano, M. Ermini, C. C. Chang, K. Sundaram, and F. A. Kind, Acta Endocrinology, 63, 29 (1970).
5. D. R. Kalkwarf, M. Sikov, L. Smith, and R. Gordon, Contraception, 7, 474 (1973).
6. S. W. Kim, R. V. Petersen, and J. Feijen, in Drug Design, Vol. X, A. Ariens (Ed.), Academic Press, New York, Chap. 5, 1980.
7. H. G. Batz, H. Ringsdorf, and H. Ritter, Die Makromolekulare Chemie, 175, 2229 (1974).
8. K. L. Agarwal and M. M. Dhar, Steroids, 6, 105 (1965) and reference cited.
9. C. P. Pinazzi, A. Merril, J. C. Rabadeux, and A. Pleurdeau, J. Polymer Sci. Symposium, 52, 1 (1975).
10. D. Gregonis, J. Feijen, J. Anderson, and R. V. Petersen, Polymer Preprints, 20 (2), 612 (1979).
11. J. M. Anderson, A. Hiltner, K. Sehodt, and R. Woods, J. Biomed. Mater. Res. Symp., 5, 197 (1974).
12. J. M. Anderson, D. F. Gibbons, R. L. Martin, A. Hiltner, and R. Woods, J. Biomed. Mater. Res. Symp., 5, 197 (1974).

13. S. Mitra, M. Van Dress, J. M. Anderson, and R. V. Petersen, Polymer Preprints, 20 (2), 32 (1979).
14. N. Lotan, M. Bixon, and A. Berger, Biopolymers, 8, 247 (1969).
15. M. Miyake, S. Akito, A. Teramoto, T. Norisuye, and H. Fumita, Biopolymers, 13, 1973 (1974).
16. N. Lotan, A. Yaron, A. Berger, and M. Sela, Biopolymers, 3, 625 (1965)

# APPLICATIONS OF OSMOTIC DRUG DELIVERY

*F. Theeuwes*
*B. Eckenhoff*

ALZA Research
Palo Alto, California

## I. INTRODUCTION

A class of osmotically driven systems has been developed for the controlled administration of bioactive agents. Delivery rates of these systems are:

- -High compared to the delivery rates of diffusion-controlled systems;
- -Generally independent of the properties of the active agent;
- -Independent of biological environmental influences (e.g., motility of the gastrointestinal tract, pH, or enzymatic activity) when the systems are used as implants or oral dosage forms;
- -Predictable and programmable; and
- -Equal *in vitro* and *in vivo* within specified limits.

This paper reviews some important experimental applications of osmotic systems technology in both animal and clinical studies. Two types of osmotic systems serve these applications: (1) The generic osmotic pump[1] is programmed with an osmotic agent-membrane system that is independent of and isolated from the active agent. (2) The elementary osmotic pump[2] is the simplest form of the osmotic pump. It utilizes an active agent that also serves as an osmotic (driving) agent.

[1]*ALZET® osmotic minipump systems, ALZA Corporation, Palo Alto, CA.*
[2]*ORSO® System, ALZA Corporation, Palo Alto, CA.*




ISBN 0-12-074450-3

The goals of development of these systems have been:

- To provide generic osmotic pumps as research tools for animal pharmacology; and
- To design the osmotic pumps as a dosage form to deliver drugs to the gastrointestinal tract at constant, preprogrammed rates.

## II. SYSTEMS DESCRIPTION

### A. *The Generic Osmotic Pump*

The generic osmotic pump (1), schematically represented in Figure 1, has a drug reservoir that holds liquid or semisolid formulations. A flexible partition separates this reservoir from an osmotic compartment that contains an osmotic driving agent. A semipermeable membrane separates the agent from the environment. In operation, water from the environment is imbibed across the semipermeable membrane. The influx of water, by displacing the flexible partition, displaces an equal volume of drug formulation. The semipermeable membrane and pump housing are rigidly held in place.

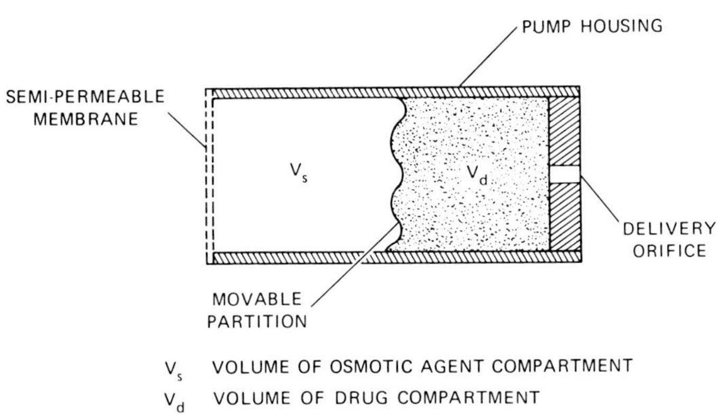


*FIGURE 1. Schematic representation of a generic osmotic pump. (Copyright 1976, ALZA Corp.)*

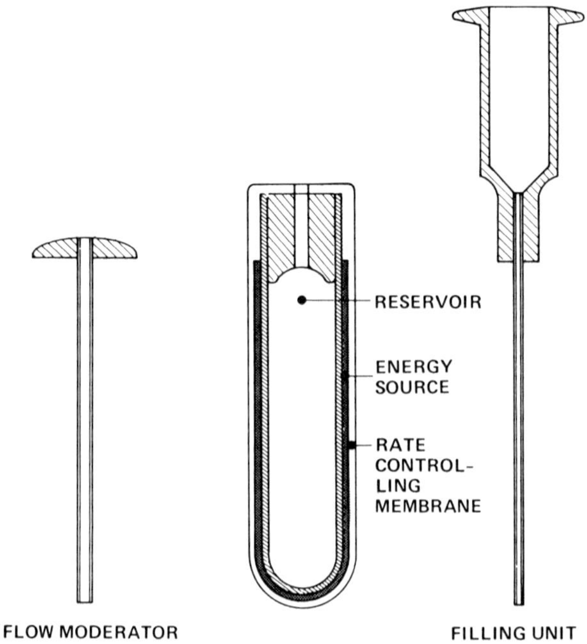


*FIGURE 2. Generic osmotic pump and components. (Copyright 1976, ALZA Corp.)*

The generic osmotic pump (2) is capsular in shape and is approximately 2.5 cm long and 0.6 cm in diameter (see Figure 2). The experimenter, using a special filling tube that is attachable to a 1 cc syringe, fills the device with a solution of an active agent through an opening located at one end. The same portal permits subsequent release of the formulation. Internally the pump consists of an inert and impermeable flexible drug reservoir with delivery portal. A thin wall of osmotic agent, enclosed in a semipermeable membrane, surrounds the drug reservoir. The osmotic agent imbibes water at a rate controlled by the rigid membrane; this rigidity causes the swelling osmotic compartment to squeeze the drug reservoir. The squeezing displaces the formulation inside the reservoir, which then flows through the delivery portal in a continuous and controlled manner.

These generic osmotic pumps have a flow moderator that is inserted through the delivery portal after filling the device. The flow moderator restricts diffusional and convective losses from the device so that the osmotic process controls the

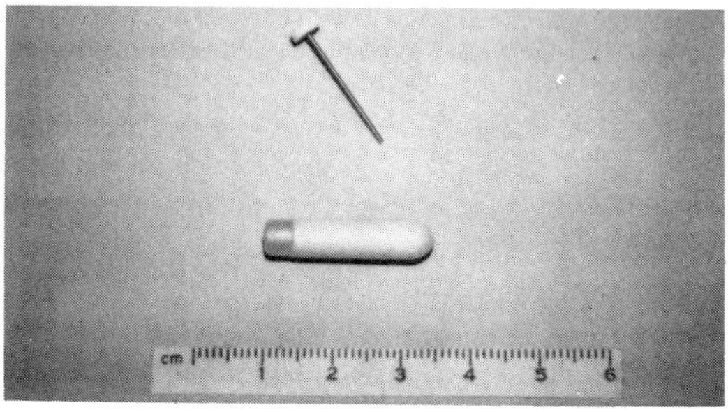


*FIGURE 3. Generic osmotic pump, with flow moderator. (Copyright 1980, ALZA Corp.)*

delivery rate. A special flow moderator is also available that permits catheter attachment for localized administration of test substances to remote areas or target organs.

Several generic osmotic pumps have been developed, differing in their delivery rates. A photograph is presented in Figure 3. Two devices have the same cross-section and internal volume, 170 μl. One unit is preprogrammed to deliver its contents at the rate of 1 μl/hr over one week, and the other unit delivers its contents at 0.5 μl/hr over two weeks. A third device has been developed which releases its contents over a 24 hour duration.

### *B. The Elementary Osmotic Pump*

The elementary osmotic pump (3) is a device in which the active agent, present alone or with an osmotic driving agent, is surrounded by a semipermeable membrane having a single delivery orifice (Figure 4). During operation (Figure 5), water from the environment is continuously imbibed across the semipermeable membrane by osmosis to produce the fluid drug formulation; the continuous membrane-controlled influx of water drives the formulation through the orifice. Several oral dosage forms based on these principles are under study.

The dosage forms are usually fabricated in tablet form (Figure 6), to be swallowed whole. While moving through the gastrointestinal tract at the speed of normal foodstuff, they dispense their contents at a controlled rate. Toward the end of the delivery period, the dosage form contains a dilute solution of drug; the membrane is excreted intact.

One such system, designed to deliver the drug acetazolamide, is described in this paper. This system's functionality has been studied *in vitro* and *in vivo* in the gastro-

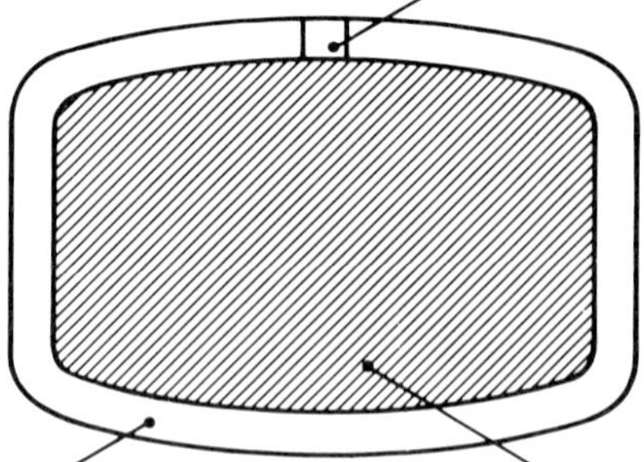


*FIGURE 4. Elementary osmotic pump cross-section. (Copyright 1975, ALZA Corp.)*

intestinal tract of dogs. Two systems, called gastrointestinal therapeutic systems (GITS), have been studied in Phase I clinical trials in normal human subjects. They contained 125 and 250 mg of the drug, and delivered 70% of these amounts at a nominal zero-order rate of 15 mg or 20 mg per hour, respectively. To distinguish the two systems, they are called, respectively, GITS 15/125 and GITS 20/250 as specified by their rate and total content. A photograph of one of these dosage forms is shown in Figure 6. The acetazolamide was present as the disodium salt, in a tablet core containing 69.3% of drug.

## III. FUNCTIONALITY OF OSMOTIC PUMPS

### A. *Methods of Measuring In Vitro Delivery*

The delivery rates of generic and elementary osmotic pumps were determined in essentially the same manner. Loosely woven mesh bags of nylon or polyethylene, each containing one system, were attached to a rod, which in turn was attached to a transfer arm that could hold up to ten rods in a row. Devices were individually immersed in test tubes containing a known amount of test fluid (e.g. water, saline, or gastric or intestinal fluid). The test tubes were held

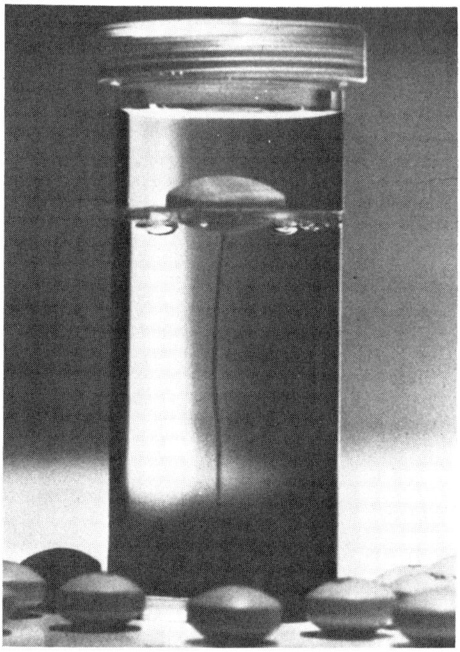

*FIGURE 5. Elementary osmotic pump in operation. The system is supported on a ledge, with its orifice downward, within a vial containing water. A dye simulating the active agent is continuously pumped downward by the system and collects at the bottom of the vial. (Copyright 1980, ALZA Corp.)*

in a rack within a temperature-controlled bath ($37^{o} \pm 1^{o}C$); the rack contained 15 rows of test tubes filled with the test solution. For the elementary osmotic pump, the transfer arm vertically agitated the systems in their separate test tubes with a stroke of 2.5 cm at a frequency of 0.5/sec. The generic osmotic pumps were tested similarly but without agitation. After a fixed period (1 to 48 hours) in the same test tubes, the systems were automatically transferred by the arm to a subsequent row of test tubes. At the end of the transfer process, the contents of each test tube contained the amount of active agent delivered during the residence time selected. For the generic pumps, the residence times were either 24 or 48 hours, depending on the release rate. The generic pumps were filled with a triphenylmethane dye solution (FD & C Blue No. 1) as a marker compound to determine the release rate pattern. The output from each pump was analyzed spectrophotometrically at a wavelength of 630 nm.

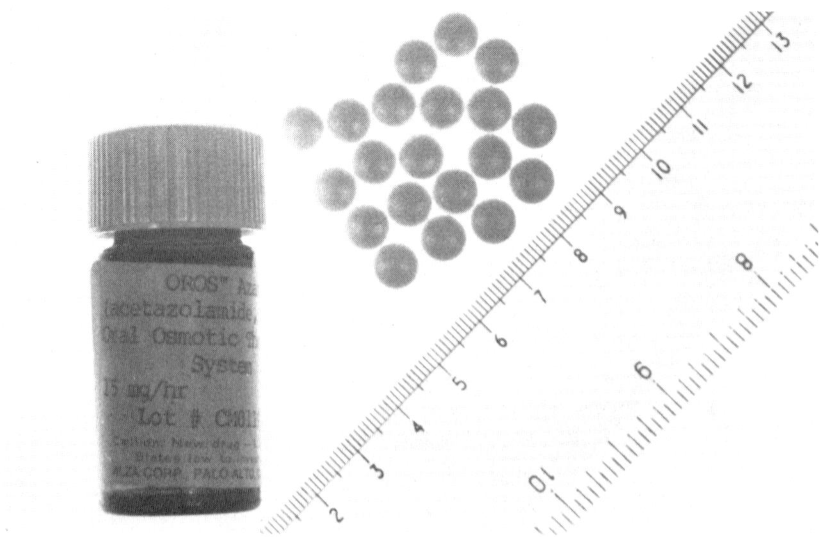

*FIGURE 6. Acetazolamide GITS 15/125. (Copyright 1980, Alza Corp.)*

For the acetazolamide elementary pumps, residence time was one hour and the test tube contents were also measured spectrophotometrically. Since the solubility of acetazolamide is strongly pH dependent, the delivery rate of the systems were tested in simulated gastric fluid[3] (without enzymes) at pH = 1.2 for the first two hours and subsequently in simulated intestinal fluid[3] (without enzymes) at pH = 7.5.

To measure the cumulative amounts of acetazolamide delivered, each system was held in a loosely woven nylon mesh bag that was totally submersed in isotonic saline at $37^{\circ}C$ and attached to the transfer arm of the equipment described above. After two-hour intervals of agitation, the systems were removed from the transfer arm and the total amount delivered was calculated. Each system was weighed individually. Its initial acetazolamide content was given by: (total initial weight - membrane weight) x 0.693. The final drug content was determined by dissolving the remaining drug in a known amount of water and measuring the concentration of solution by U.V. spectrophotometry.

[3]*U.S.P. XVIII, p. 1026-1027, The United States Pharmacopeia, 18th Revision.*

*B. Methods of Measuring In Vivo Delivery*

*In vivo* calibration experiments with the generic pumps were conducted by implanting pumps, filled with triphenylmethane dye solution, subcutaneously in rats. A short length of 60 gauge polyethylene catheter tubing, also prefilled with the dye solution, connected each minipump to an externally worn collector reservoir secured to the animal's back with wound clips. The contents of the collection reservoir were analyzed spectrophotometrically each day.

The cumulative amounts of acetazolamide delivered by the elementary pump systems *in vitro* were determined in four dogs. The animals received their normal diet of dry food 16 hours prior to administration of the systems, but no food on the day of the study. Labeled devices were orally administered to each dog at 1, 2, 4, 6, 10, and 12 hours prior to sacrifice. After each dog was sacrificed with an overdose of pentobarbital, the systems were recovered. The amount of acetazolamide remaining in each system was then determined and, knowing the original drug content, the amount delivered was the difference.

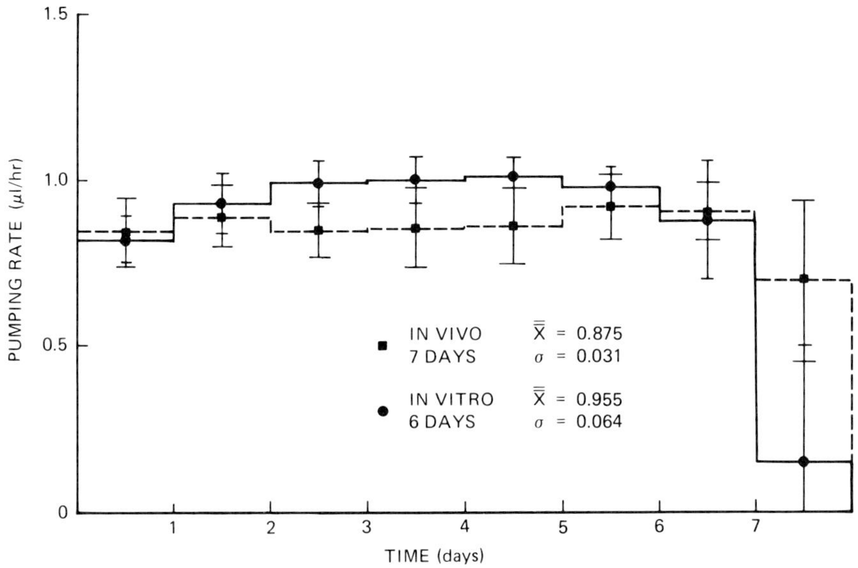


*FIGURE 7. Release rate pattern in vitro (normal saline, 37°C) and in vivo (rat) of a generic osmotic pump. (Copyright 1980, ALZA Corp.)*

## *C. Results of Functionality Measurements*

### *1. Generic Osmotic Pump*

*a. In vitro and in vivo performance.* Figure 7 compares *in vitro* and *in vivo* release rate patterns for the seven day generic osmotic pump. As Figure 7 shows, the reproducibility of delivery rate of the pump is ± 10% and does not change in different environments. However, the average pumping rates over time in the rat, mouse, and in saline were different, probably due to minor differences between temperatures of the *in vitro* medium and body temperatures. The release rate in the subcutaneous space of a rat or a mouse is constant at approximately 90% and 85%, respectively, of the rate in saline at 37°C.

*b. Performance of osmotic pump with 200 μl volume and 24 hour release rate profile.* *In vitro* release rate data for the 24 hour generic osmotic pump are presented in Figure 8. This pump has a 200 μl volume and is designed to deliver its contents over a 24-hour period. Pumping rates into isotonic saline, simulated gastric fluid, and simulated intestinal fluid at 37°C are shown. The pumping rates are independent of pH and environment.

*2. Elementary Osmotic Pump.* *In vitro* data for 10 acetazolamide systems indicate that the delivery rate is independent of pH (see Figure 9). At steady state, the delivery rate was the same in both gastric and intestinal fluid.

A comparison of the cumulative amounts of drug delivered, expressed as acetazolamide delivered as a function of time is presented in Figure 10. During the zero-order delivery period (0-6 hours), *in vitro* and *in vivo* data agree within the limit of experimental error. After a 12-hour period, the average amounts delivered *in vivo* appear to be about 10% lower than the amounts delivered *in vitro*.

# IV. APPLICATIONS

## *A. Generic Osmotic Pump*

One application of these pumps is in animal pharmacology. They are used singly in most small laboratory species: mouse, rat, guinea pig, cat, and rabbit. Multiple devices

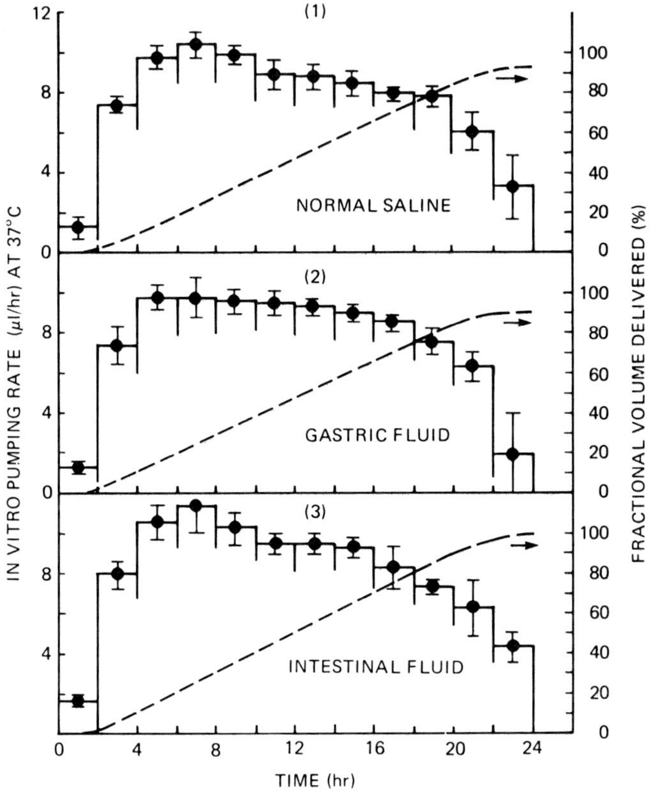


*FIGURE 8. In vitro release rate of a 200 μl osmotic pump. (Copyright 1980, ALZA Corp.)*

are useful in large animals: dogs, sheep, cows and monkeys. The pumps can be used to administer pharmacologic agents both locally and remotely. Catheter applications include microperfusion of the cerebral ventricular system, the hypothalmus, and visual cortex of the brain; the fornix and stroma of the eye; the systemic system; the gut lumen; and a number of other biological systems. In the past two years the principal interests among users have been biochemical phenomena dealing with enzymes, blood, respiration, salt and water balance, hormones, metabolism, nutrition, narcotics and opiates, growth, and cancer chemotherapy.

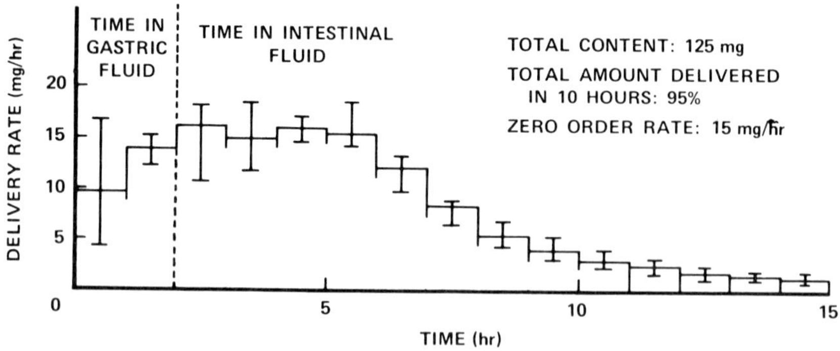


*FIGURE 9. In vitro release rates of acetazolamide from GITS 15/125 elementary osmotic pumps in simulated gastric and intestinal fluid (USP), without enzymes, at 37°C (n = 10). I indicates the range of experimental data. (Copyright 1980, ALZA Corp.)*

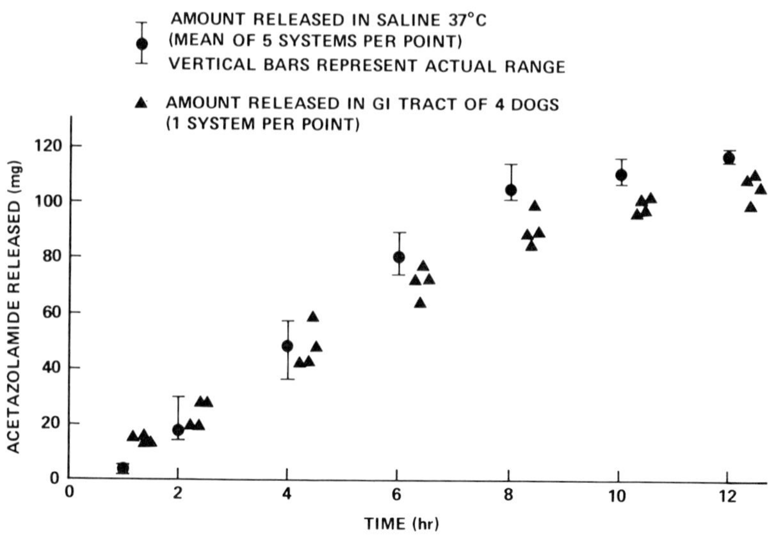


*FIGURE 10. Cumulative amounts of acetazolamide released from GITS 15/125 elementary osmotic pumps. (Copyright 1980, ALZA Corp.)*

A typical example follows. Rate with congenital diabetes insipidus (4) were treated with osmotic pumps delivering different concentrations of antidiuretic hormone (ADH) at a constant rate. During day 4 of the study, prior to treatment, the urine output of three groups of six rats was 184.1 ± 44.9 ml/day (n = 18). On day 7, ADH (lysine vasopressin) was delivered by the pump at a rate of 0.5, 5, or 10 units/kg/wk (a single week per group of rats). Changes in urine volume (Figure 11) showed a clear relation between the different rates of ADH delivery and its antidiuretic effect.

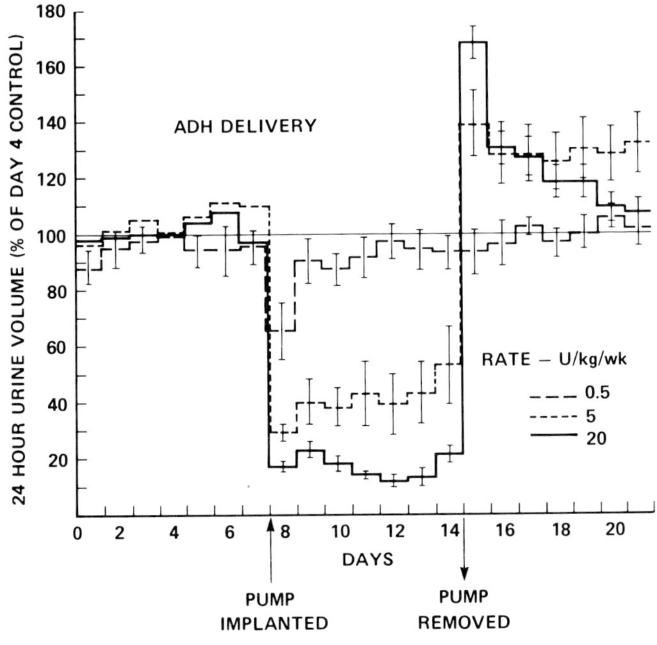


*FIGURE 11. Effect of lysine vasopressin, delivered at three different concentrations, on urine volume in rats with congenital diabetes insipidus. The 24-hour urine volumes are expressed as a percentage of the day 4 value. Each value is the mean of the data from 6 rats per group; vertical bars indicate standard errors. (Copyright 1976, ALZA Crop.)*

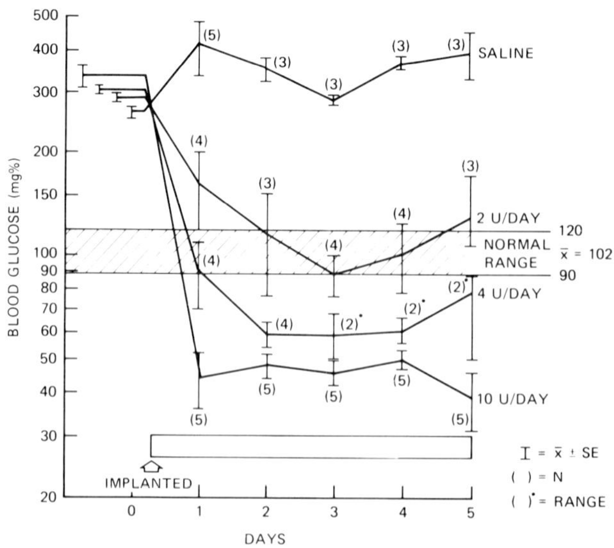


*FIGURE 12. Delivery rate-response curves of insulin in the streptozotocin-diabetic rat. (Copyright 1980, ALZA Corp.)*

Generic osmotic pumps have also served as an artificial pancreas in rats with experimentally (streptozotocin) induced diabetes. Pumps were filled with different concentrations of unmodified crystalline bovine insulin in isotonic saline and implanted subcutaneously to infuse insulin doses of 2, 4, and 10 units per 200 gram rat per day. Control rats had isotonic saline-filled pumps implanted subcutaneously. Figure 12 shows the temporal dose-response curve. Blood glucose data, monitored as a function of dosing schedule, showed that 2 units/day returned the streptozotocin-diabetic rat to normal glycemic levels.

## *B. Generic Osmotic Pumps with 24-Hour Release Rate Profile*

Experimental systems of this type can serve as precursor oral and rectal dosage forms to simulate delivery-rate-versus-time profiles and can serve in pharmacokinetic and pharmacologic studies to define optimum drug delivery rates, delivery durations, and bioavailability. Definition of rate and duration profiles permits their reproduction in oral or rectal dosage forms for clinical use, such as the elementary osmotic pump.

### C. *Elementary Osmotic Pumps in Clinical Trials*

Acetazolamide is a well-known drug, given systemically, that is effective in the treatment of ocular hypertension in glaucoma. In present dosage forms, however, the drug often elicits such effects as vomiting, tingling extremities, and confusion (5). These side effects, which appear to occur at high blood concentrations of the drug, may be due to impaired red cell activity (6). In any case, they often preclude use of acetazolamide.

The pharmacology of acetazolamide has not been completely defined, but the evidence points toward the existence of an optimal plasma concentration range. Minimum effective concentrations have been reported to range from 5 to 10 μg/ml (7), but recently the minimum has been defined as 4 μg/ml (6).

The potential toxicity of drug substances can be summarized by the therapeutic index (TI):

$$TI = \frac{C^*_{max}}{C^*_{min}}$$

where $C^*_{max}$ is the maximum tolerable concentration in plasma (above which side effects occur), and $C^*_{min}$ is the minimum effective concentration in the plasma. For acetazolamide, Lehman *et al.* (7) have determined a therapeutic index of 2.7.

Dosage form design provides a means for controlling the plasma concentration profile, as reflected by a new criterion recently devised: the dosage form index (DI) (8). This index is the ratio of maximum-to-minimum plasma concentration occurring within one dosing interval during repetitive drug administration. Engineering the dosage form to maintain the DI less than the TI permits safe and effective treatment. In line with Lehman's data, therefore, gastrointestinal therapeutic systems have been developed (8) for acetazolamide to maintain a dosage form index of less than 2.7 when the drug is dosed at a level above that providing minimum therapeutic effect.

Clinical studies were performed in normal persons to establish plasma concentration profiles for acetazolamide, and the extent of its absorption, following its repetitive administration in two types of dosage forms (8). One was a sustained release product[4] containing 500 mg of the drug; the other was an elementary osmotic pump of the type described in Section II B.

Garner *et al.* (5) reported that side effects were more severe when patients took one 250 mg tablet of acetazolamide four times daily (i.e., 1000 mg daily) than when they took a single 500 mg sustained-release tablet once daily. The ocular hypotensive effect of the two regimens, however, was comparable. Since the 500 mg, once-daily regimen appeared superior, it was compared with similar data obtained with elementary osmotic pumps delivering acetazolamide.

Two types of systems were used, one delivering 20 mg of drug/hr and containing 250 mg total drug (GITS 20/250). The other contained 125 mg total drug and delivered at 15 mg/hr (GITS 15/125). The 20 mg/hr device was administered once a day; the GITS 15/125 was administered twice a day. Both the GITS 20/250 and the sustained release dosage form were studied in 8 volunteers in a cross-over study. Following this study, five of the eight volunteers were also dosed with the GITS 15/125. Each dosage form was administered for up to ten days to allow blood sampling during successive periods at steady state. Sampling of blood at frequent intervals during days 1, 3, and 10 verified the achievement of steady state.

The average plasma concentrations for all subjects was plotted as a function of time in Figure 13. These results indicate that one GITS 15/125 dosed twice daily resulted in the most constant plasma concentration profiles at quasi-steady state. Both the sustained dosage form and the GITS 20/250 system, dosed once a day, produced more pronounced fluctuations in plasma concentration as reflected by the dosage form index.

The total urinary recovery of acetazolamide from each dosage form during the entire regimen was 63%. The average plasma concentration obtained for each dosage form was also the same per unit dose, about 13 μg/ml-gram of daily dose.

For both the sustained release dosage form and the GITS 20/250, the dosage form indices were calculated as ratios of maximum-to-minimum plasma concentrations encountered in a 24-hour dosing period during days 3 and 10. For the GITS 15/125, the index was calculated as the ratio of maximum-to-minimum plasma concentrations encountered in the 12-hour steady state interval on days 3 and 10, averaged over all subjects. Since only one cycle of detailed plasma data was

[4]*Diamox*® *Sequel*®, *Lederle.*

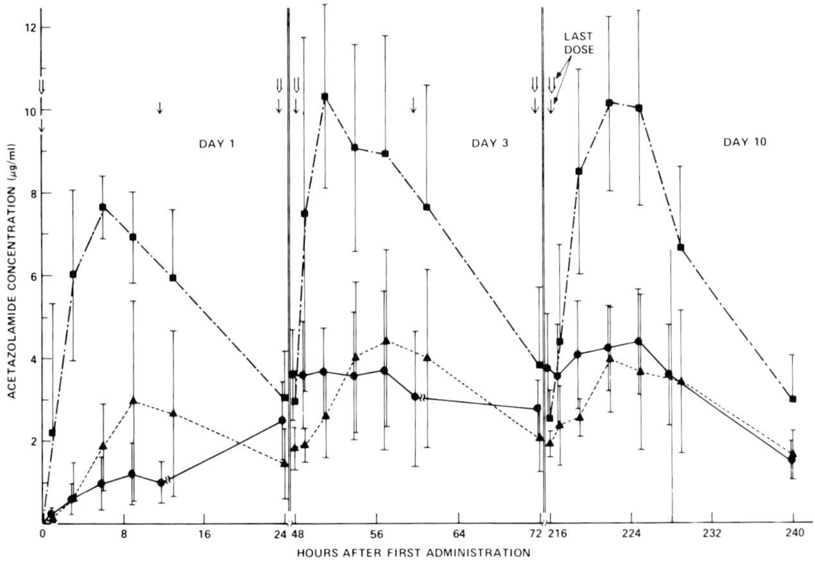


*FIGURE 13. Average acetazolamide plasma concentrations in normal human subjects following administration of: ■ Sustained-release dosage form (dosed once a day at 8 a.m., n = 8); ▲ a GITS 20/250 (dosed once a day at 8 a.m., n = 8); ● GITS 15/125 (offset 0.2 hours) (dosed twice daily at 8 a.m., 8 p.m., n = 5). (Copyright 1977, ALZA Corp.)*

taken per day and per dosage form for all three dosage forms, one measure of the DI is obtained per dosage form and per day (Figure 13).

A plot of the averages of the dosage form indices over all days and subjects (Figure 14), showed that selection of the GITS 20/250 provided an improvement, by a factor of 1.5, over the sustained-release dosage form. A further improvement, by a factor of 3.2 over the sustained-release dosage form was obtained with the GITS 15/125 administered twice a day. The dosage form index, as defined here, clearly decreases with more frequent dosing. Since increasing the frequency is inconvenient and reduces compliance, however, the overall effect on treatment may be detrimental.

Following these studies in normal volunteers, the effects of the GITS 15/125 were assessed in 7 glaucomatous patients in a double-blind study with complete cross-over design (9). The purpose was to determine whether side effects would be

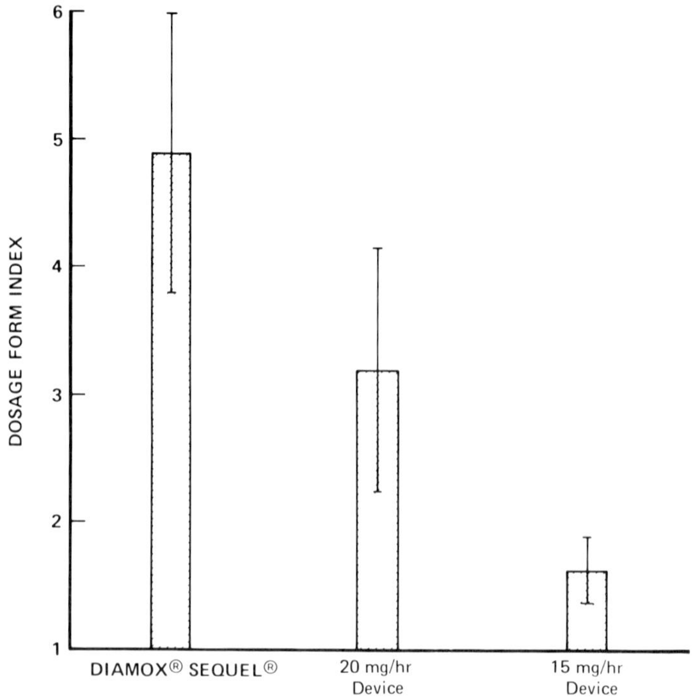


*FIGURE 14. Average ratio of maximum-to-minimum plasma concentration of acetazolamide on Day 3 to that on Day 10. Standard deviations are shown (8). (Copyright 1977, ALZA Corp.)*

reduced relative to those encountered in therapy with dosage forms having a higher dosage form index, at doses producing equal intraocular pressure lowering.

Four treatments were each administered twice daily at 8 a.m. and 8 p.m.: one GITS 15/125, two GITS 15/125, one conventional 250 mg tablet, or a placebo. Patients remained on a fixed five-day regimen per week for four weeks, starting a new regimen on each Sunday at 8 a.m., and ending it on Thursday at 8 p.m. Each dosage form was administered in a size 00 gelatin capsule (which disintegrates completely in artificial gastric fluid within 5 minutes). The scored 250 mg tablets were broken in half and inserted into gelatin capsules. The placebo capsules contained lactose.

Throughout the study, intraocular pressures (IOP) were measured for each regimen on four consecutive days, Monday through Thursday. The average treated IOP, plotted against

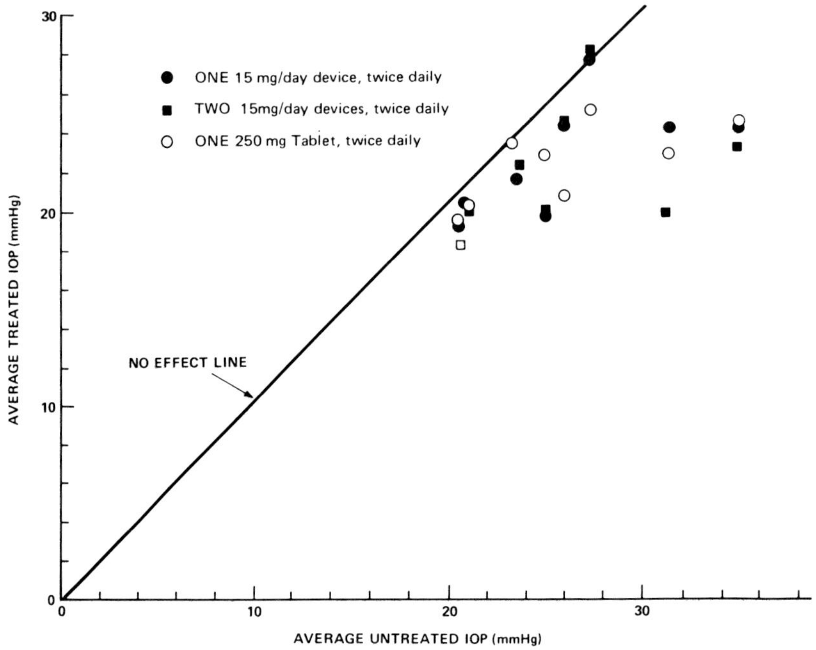


*FIGURE 15. Average treated vs. average untreated intraocular pressure for each patient for all regimens. The data for individual subjects fall on a vertical line. On the diagonal, the untreated IOP equals the treated IOP; thus, points cluster on the diagonal if treatment exerts no effect (9). (Reproduced by permission of the American Medical Association.)*

the average untreated IOP on a scattergram (Figure 15), indicated that all three drug regimens were about equally effective.

On the fifth day of each regimen, during the quasi-steady state, plasma concentrations were measured at four time periods when extreme values were expected. The average normalized plasma concentrations during each regimen (Figure 16) and the average dosage form indices (Figure 17) confirmed the findings obtained for normal subjects. The data indicated that the DI for conventional tablets dosed twice daily is comparable to that of the sustained-release product dosed once a day in the study on normal subjects. The dosage form index (DI) for the conventional products dosed on these two schedules is larger by a factor of 3 than the DI for the GITS 15/125 when dosed at the preferred schedule previously identified.

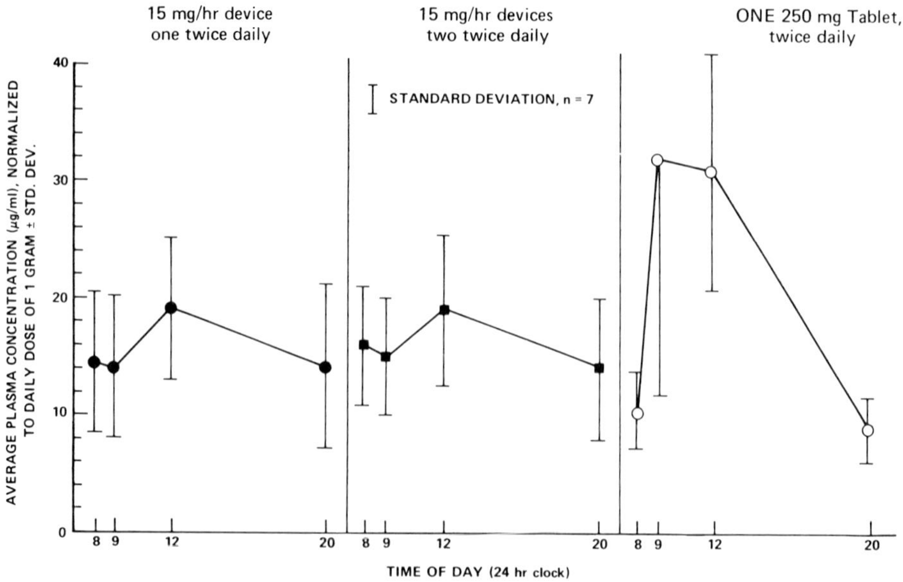


*FIGURE 16. Average normalized acetazolamide plasma concentrations at steady state. Dosing times for all regimens were 8 and 20 hours (9). (Reproduced by permission of the American Medical Association.)*

For each 12-hour interval throughout each treatment period, patients were asked to score specific side effects. A total of nine reports were produced per patient per regimen. The effects scored were: nausea, stomach distress, vomiting, drowsiness, loss of appetite, loss of sexual desire, tingling hands, tingling feet, blurred vision, diarrhea, frequent urination, rash, and confusion. The occurrence of all side effects was totaled for individual treatments and expressed as a percent of total responses (9 scores times seven patients = 63). The results (Table I) indicate that the incidence of side effects with the regimens using GITS 15/125 was significantly less than with the conventional tablets ($P < 0.05$).

Of the 13 side effects tabulated for the seven patients, four (drowsiness, tingling hands, tingling feet, and confusion) showed statistically significant differences in incidence ($P < 0.05$) with different treatments. With both regimens involving the use of the GITS 15/125, all four side effects were significantly less frequent ($P < 0.05$) than with the sustained-release tablets; the reduction for each was at least 50%.

TABLE I. Percent of Side Effects (Responses) Marked as Either Mild, Moderate, or Severe

| Side effect | Dosage form and frequency of administration | | | |
|---|---|---|---|---|
| | Placebo twice daily | One GITS 15/125 twice daily | Two GITS 15/125 twice daily | One 250 mg tablet twice daily |
| Any side effect | 16.7 | 22.2 | 17.5 | 35.5 |
| Drowsiness | 0.0 | 7.9 | 7.9 | 25.4 |
| Tingling hands | 0.0 | 11.1 | 11.1 | 22.6 |
| Tingling feet | 0.0 | 0.0 | 11.1 | 22.2 |
| Confusion | 0.0 | 0.0 | 1.6 | 9.7 |

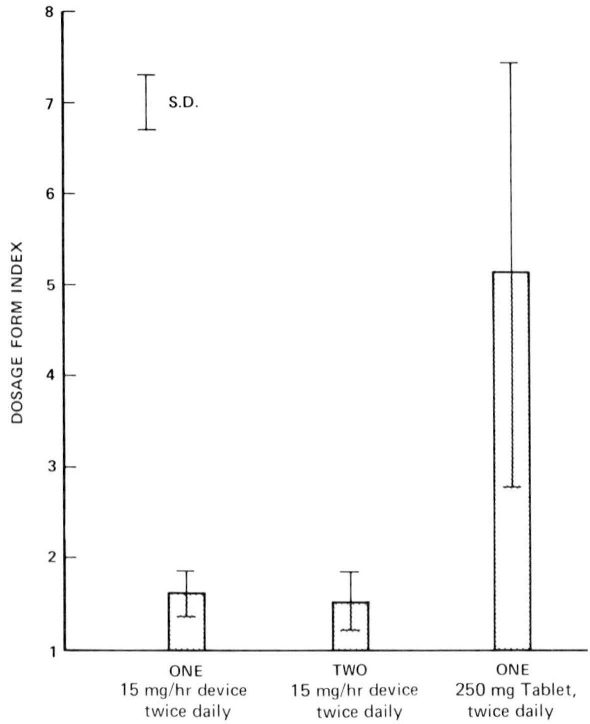


*FIGURE 17. Ratio of maximum-to-minimum acetazolamide plasma concentrations during one dosing interval at the steady state, averaged over all patients (9). (Reproduced by permission of the American Medical Association.)*

ACKNOWLEDGMENTS

The authors thank Mr. F. Landrau for his support in developing the generic osmotic pump; Mr. N. Damani and Mr. A. Ayer for their support in developing the acetazolamide elementary osmotic pump system; and Dr. J. Langston, Mr. E. Willis, Ms. L. Sendlebeck, and Mr. D. Swisher for the *in vitro* and *in vivo* experiments.

REFERENCES

1. F. Theeuwes and S. I. Yum, Ann. Biomed. Eng., 4, 343 (1976).

2. R. Capozza, B. Eckenhoff, and S. I. Yum, J. Med. Eng. Technol., 1, 281 (1977).
3. F. Theeuwes, J. Pharm. Sci., 64, 1987 (1975).
4. S. K. Chandrasekaran, F. Theeuwes, and S. I. Lum, Drug Design, Vol. 8, E. Ariëns (Ed.), Academic Press, New York, p. 133 (1979).
5. L. L. Garner, E. F. Carl, and J. R. Ferwerda, Am. J. Ophthalmol., 55, 323 (1963).
6. B. R. Friedland, J. Mallonee, and D. R. Anderson, Arch. Ophthalmol., 95, 1809 (1977).
7. B. Lehmann, E. Linner, and P. J. Wistrand, Adv. Biosci., 5, 197 (1969).
8. F. Theeuwes and W. Bayne, J. Pharm. Sci., 66, 1388 (1977).
9. F. Theeuwes, W. Bayne, and J. McGuire, Arch. Ophthalmol., 96, 2219 (1978).
10. N. Damani and F. Theeuwes, Indian J. Pharm., 39, 1 (1977).
11. S. I. Yum, S. A. Tillson, and F. Theeuwes, Proc. Fifth Int. Congr. Endocrinology, Federal Republic of Germany, July 18-24, 1976, p. 366.

# POLYMERS FOR THE SUSTAINED RELEASE OF MACROMOLECULES: APPLICATIONS AND CONTROL OF RELEASE KINETICS

*Robert S. Langer*
*William D. Rhine*
*Dean S. T. Hsieh*
*Rajan S. Bawa*

Department of Nutrition and Food Science
Massachusetts Institute of Technology
Cambridge, Massachusetts
and
Department of Surgery
Children's Hospital
Boston, Massachusetts

*In an earlier study, we described the first biocompatible polymeric delivery systems capable of continuously delivering macromolecules (M.W. > 1000), such as proteins, for prolonged periods (1). These systems were formed by mixing macromolecules in a polymer solution, pouring the mixture into small, conical molds, and evaporating the solvent. The resulting pellets, when exposed to aqueous medium, released macromolecules in biochemically active form for over 100 days (2).*

*Since their development, these sustained-release systems have been valuable in several biological studies. For example, they have provided the basis for bioassays of growth factors derived from tumors (2), cartilage (3), the vitreous body (4), and macrophages (5). They have also been used in studies of vascular proliferation (6) and regression (7). In this report, we discuss several new potential applications of these systems, including possible clinical uses, such as continuous insulin delivery or as an immunization method, and potential biological applications, such as creating chemical concentration gradients in chemotaxis studies. We also report investigations aimed at increasing our understanding*




ISBN 0-12-074450-3

*of the mechanism of macromolecular release from these polymeric systems and the factors controlling release kinetics. The data presented here suggest that macromolecules are continuously released from polymeric matrices via migration through channels and that critical parameters in determining release rates include drug loading, particle size, and solubility.*

## I. POTENTIAL APPLICATIONS

### *A. Insulin Delivery*

Diabetics normally inject insulin once or twice a day. This is not only inconvenient but, more importantly, causes fluctuations in blood glucose levels. This poor control may be, in part, responsible for diabetic complications such as blindness and heart and kidney disease (8).

The ability of sustained-release polymers containing insulin to maintain normoglycemia was tested in diabetic rats. Male rats (21 days old, CD strain; Charles River Breeding Laboratories) weighing 150-250 g were made diabetic with streptozotocin. Ethylene-vinyl acetate copolymer discs (1.3 cm diameter, 1 mm high) containing 100 mg of bovine pancreatic insulin were implanted subcutaneously in the lower abdomen. Implant sites were closed with a Michel wound clip.

The results are summarized in Figure 1. Control animals implanted with empty polymers or no polymer showed consistently high blood glucose levels. In contrast, rats with insulin-containing polymers had normal glucose levels for one month. In addition, the rats in the experimental group gained weight at a normal rate, whereas growth of the controls was depressed as a result of the disease (Figure 2).

Additional experiments monitored variations in blood glucose levels during a one-day period. Figure 3 shows variations in blood glucose concentrations on day 16 of treatment. During the 31-hour study, glucose levels for treated animals were 95 ± 13 mg%; in contrast, the controls maintained levels of 338 ± 61 mg%. Furthermore, treated animals voided approximately 40 ml of urine in 31 hours; urine glucose was not detected in the treated rats. Untreated controls voided nearly 200 ml of urine during 31 hours, and urine glucose concentrations exceeded 500 mg%.

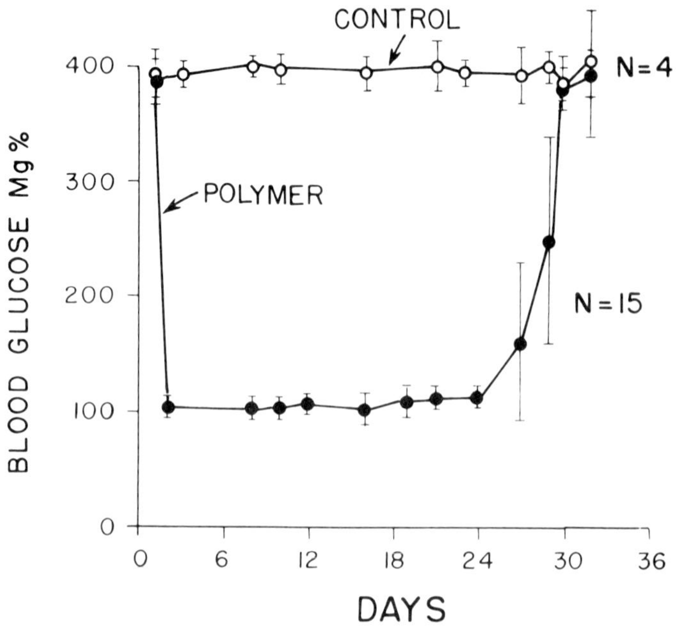


*FIGURE 1. Implantation of single insulin-containing polymers into diabetic rats. Controls were diabetic rats receiving empty polymers or no polymer.*

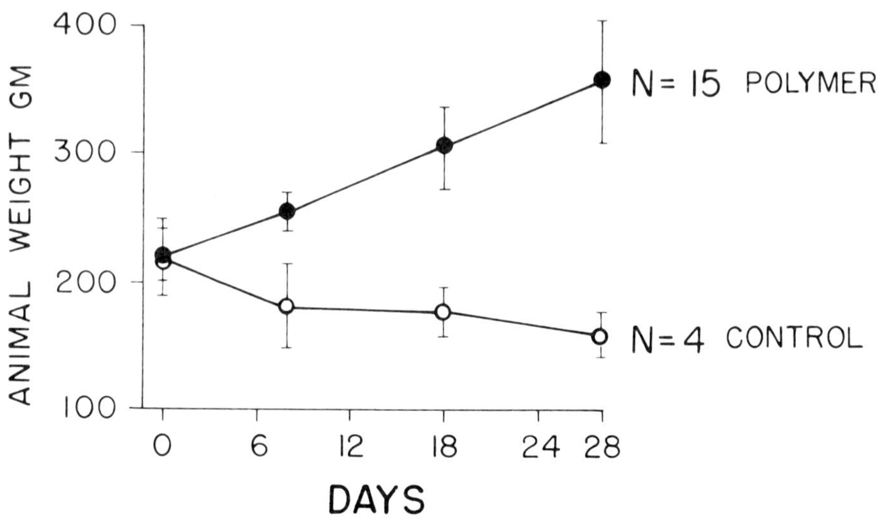


*FIGURE 2. Body weights of control and polymer-treated rats.*

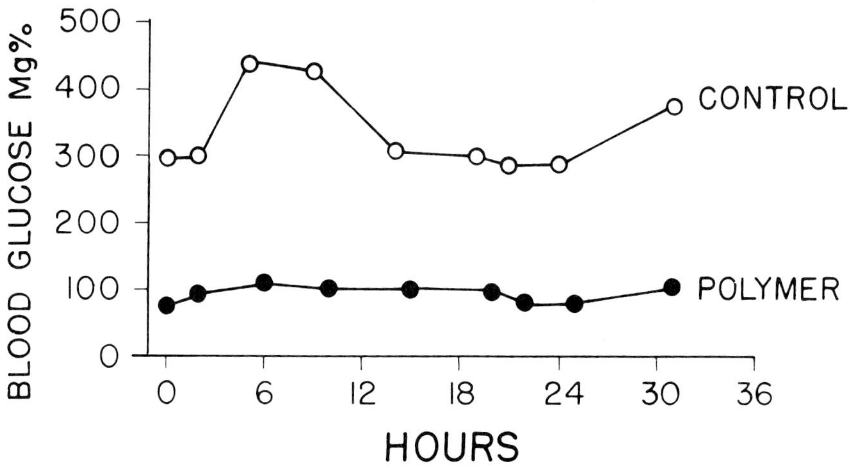


*FIGURE 3. Blood glucose determinations made every few hours around the clock for two treated animals and one control animal.*

In another experiment, the same diabetic rat was used as its own control (Figure 4). Polymer implantation lowered blood glucose from diabetic to normal levels for a one-week period. The polymer was then removed from the rat for one week and stored at 4°C. During this time period, the blood sugar level returned to a diabetic value. The same polymer was then reimplanted in the rat and blood glucose levels again returned to approximately 100 mg%.

### B. *Immunization Procedures*

Conventional methods for immunizing both animals and humans generally require multiple injections and often cause tissue irritation (9). In a recent study, we considered the possibility that sustained-release polymers for macromolecules might provide a simple, safe, and effective means of immunization by acting as a series of continuous minishots (10). To test this, ethylene-vinyl acetate copolymer pellets containing 100 μg of bovine serum albumin were fabricated according to previously described methods (1) and implanted in C57 black mice. The immune response stimulated by sustained antigen delivery from one pellet was compared to the response to a conventional method of immunization, i.e., two injections of bovine serum albumin emulsified in

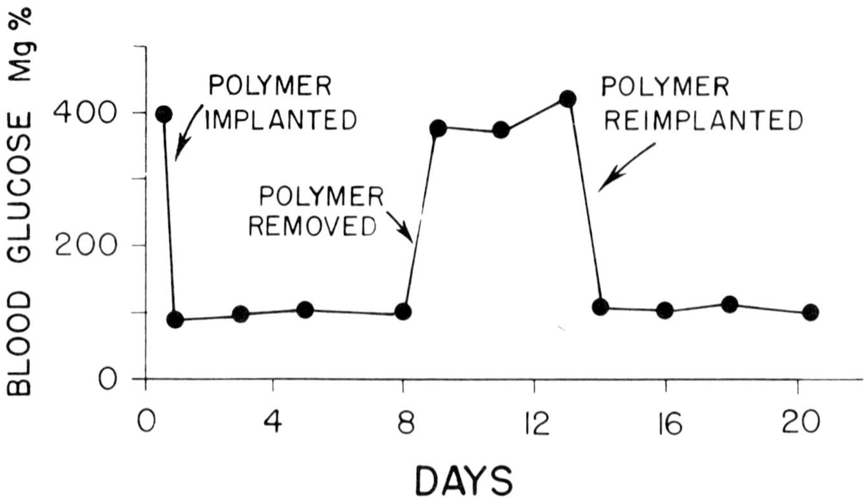


*FIGURE 4. Experiment in which a single rat was used as its own control. Glucose was at a diabetic level without the polymer, and at a normal with the polymer.*

complete Freund's adjuvant (CFA) (11). The immune response was measured by indirect haemagglutination. The results, summarized in Figure 5, indicate that the immune response induced by antigen in polymer was comparable or superior to that induced by the same total dose of antigen using the conventional technique. Furthermore, a single polymer caused a sustained immune response for half a year, and no tissue irritation was observed with the sustained-release system. We have also shown that this immunization procedure is effective for antigens other than albumin and over a wide molecular weight range (Figure 6).

## C. *Chemotaxis Studies*

Chemotaxis (movement toward or away from the source of a chemical) has been observed in a wide variety of organisms. Although numerous techniques have been developed to create concentration gradients to study this phenomenon, these methods are often too intricate and lack versatility (12-14). We believed that sustained-release polymers might provide a simple and inexpensive means of creating concentration gradients to study chemotaxis. To examine this possibility, we

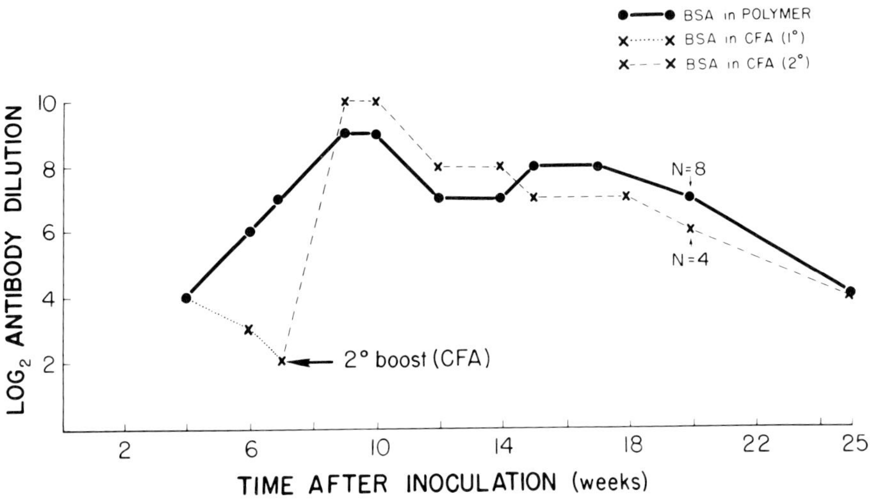


*FIGURE 5. Antibody titre in C57 black mice as a function of time, as determined by haemagglutination. The standard deviation on a given point is 1.16.*

studied the response of worms (planaria, *Dugesia dorotocephala)* to chemical gradients in an unstabilized aqueous medium. For this purpose, we adapted the T-maze system (Figure 7) of Coward and Johannes (15), who had previously demonstrated chemotaxis of the worms toward L-lysine. In 70 tests, we found that the planaria consistently migrated to the arm of the maze containing polymers impregnated with L-lysine ($p < 0.001$), whereas the worms migrated equally often to either arm of the maze if the polymers were empty or contained a control substance (e.g., L-glutamate). We have also found that these polymeric systems are useful in determining chemotaxis of bacteria (*E. coli*) toward a variety of substances (16).

## II. MICROSTRUCTURAL AND KINETIC STUDIES

The above studies suggest some of the potential applications of macromolecular polymeric delivery systems; however, the doses and release-kinetic profiles selected were purely arbitrary. It is likely that improved systems could greatly prolong both insulin and antigen delivery.

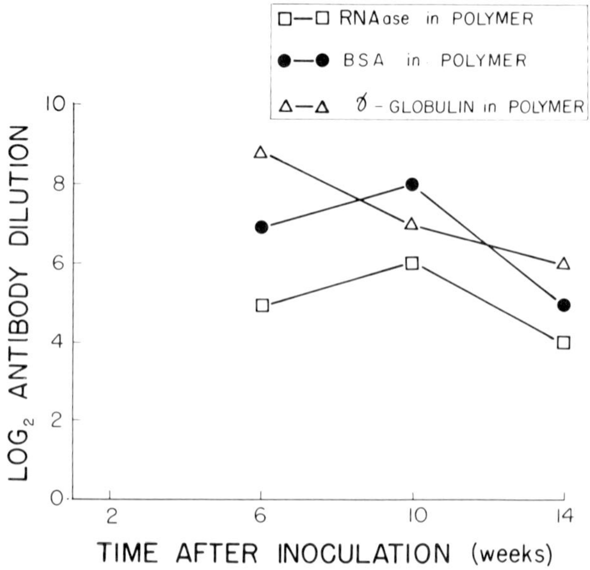


*FIGURE 6. Antibody titres of three different antigens in polymer matrices. The proteins were ribonuclease A (M.W. 14,000), bovine serum albumin (M.W. 68,000), and γ-globulin (M.W. 166,000). Each point is the mean value for five different mice.*

To obtain maximum efficiency of these systems, it is essential to understand the mechanism by which the macromolecules are released and what factors control release kinetics. To this end, we have conducted both microstructural and kinetic studies. The sustained-release polymers used in these studies were fabricated using a modification (17,18) of our original procedure (1). The critical alteration in the procedure was that both casting and drying of the polymer solution-powdered macromolecule mixture were performed at low temperature (casting at -80°C; drying in two stages, at -20°C and 20°C). This modification prevented drug settling and ensured uniform drug distribution and reproducible release kinetics.

*A. Microstructural Studies*

To examine the polymer's microstructure, 5 μ-thick sections of the polymeric systems were obtained using cryomicrotomy. When these thin sections were viewed under an

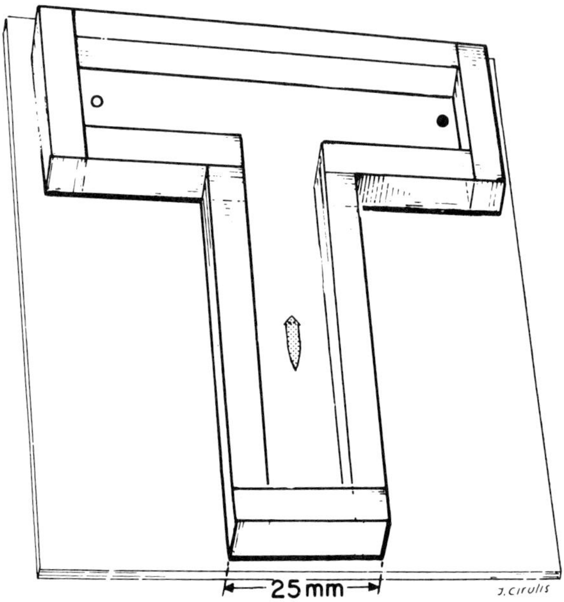


*FIGURE 7. T-maze used to assay chemotaxis of planaria. Black dots represent polymers containing a chemotactic factor; white dots indicate empty polymers. The number of times a planaria swam to a particular arm of the T-maze was counted.*

optical microscope at high power, polymers cast without protein appeared as transparent, nonporous sheets (Figure 8). This is not suprising, since molecules with molecular weights > 300 daltons generally cannot diffuse through pure ethylene-vinyl acetate copolymer films (1). In contrast, when macromolecular drugs were incorporated into the polymer films, molecules as large as 2 million daltons could be continuously released from the polymers (1).

We also obtained 5 μ-thick sections of polymers containing proteins both before any release had occurred and at the termination of release. Before release, the proteins existed as discrete aggregates within the polymer matrix (Figure 9). After all protein had been released, channels large enough to permit macromolecular diffusion could be discerned in the polymer matrix (Figure 10). All sections were viewed under 125 power magnification.

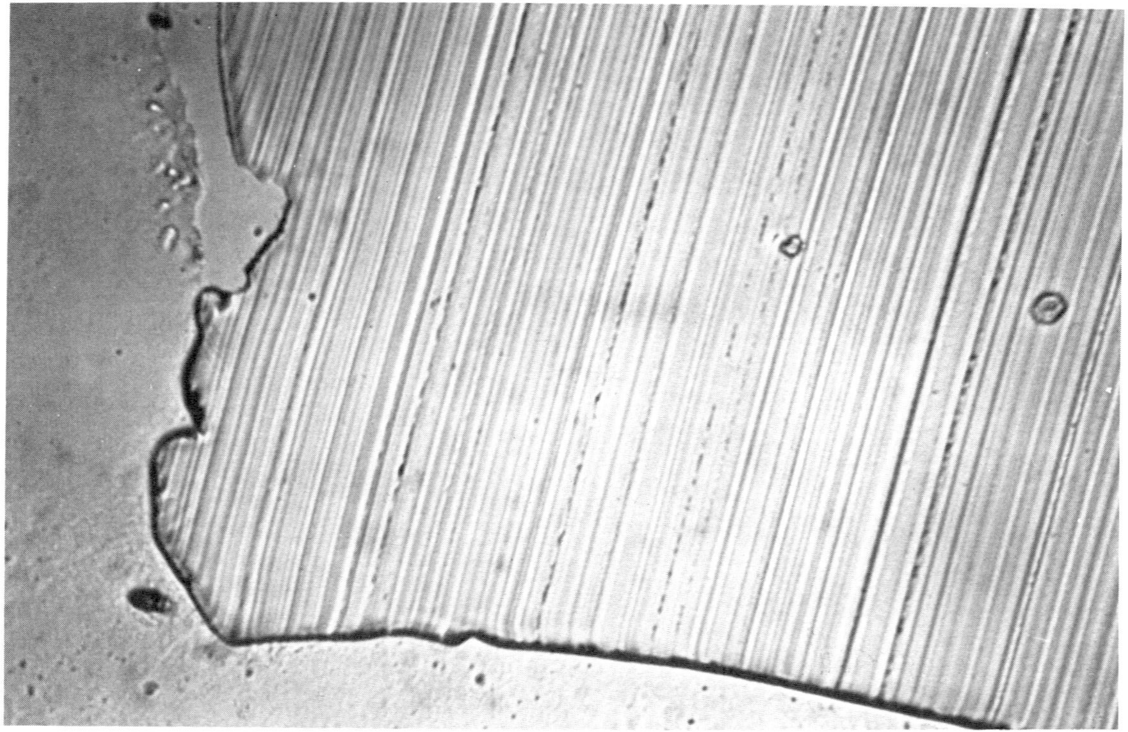

*FIGURE 8.* A 5 μ-*thick section of ethylene-vinyl acetate copolymer cast without drug.*

*FIGURE 9. A 5 μ-thick section of ethylene-vinyl acetate copolymer cast with myoglobin. Photograph was taken before release.*

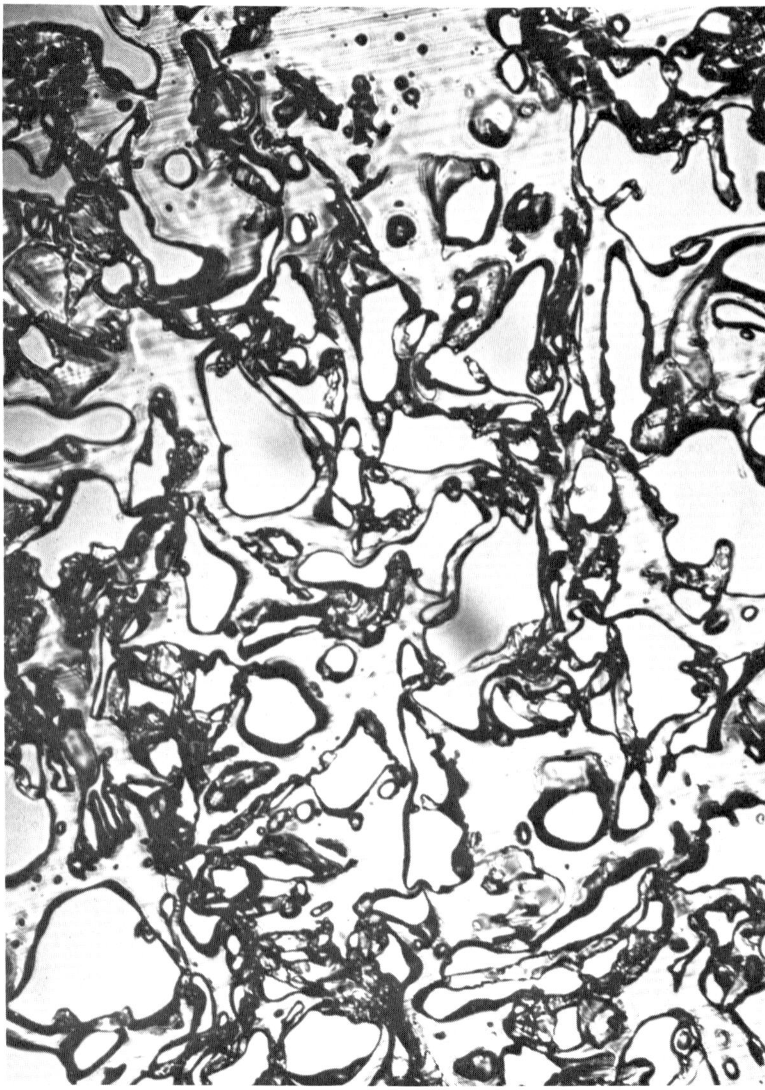

*FIGURE 10.* A 5 μ-thick section of depleted polymer initially cast with bovine serum albumin.

### B. *Kinetic Studies*

In an earlier study, we demonstrated that both drug loading and particle size could markedly alter release kinetics (18). Examples of these findings are shown in Figures 11 and 12. Changing drug particle size altered release rates by a factor of 6; increasing drug loading from 10 to 50% by weight caused a 50-fold increase in the slopes of the release curves (Figure 12).

We have also begun to examine the effects of the release environment (e.g., temperature, ionic strength, pH) on macromolecular release. The effect of pH is illustrated in Figure 13; bovine serum albumin (M.W. 68,000) was the test protein. It is interesting that the release rate was slowest at pH 5, which is very close to the isoelectric point (4.9) of albumin.

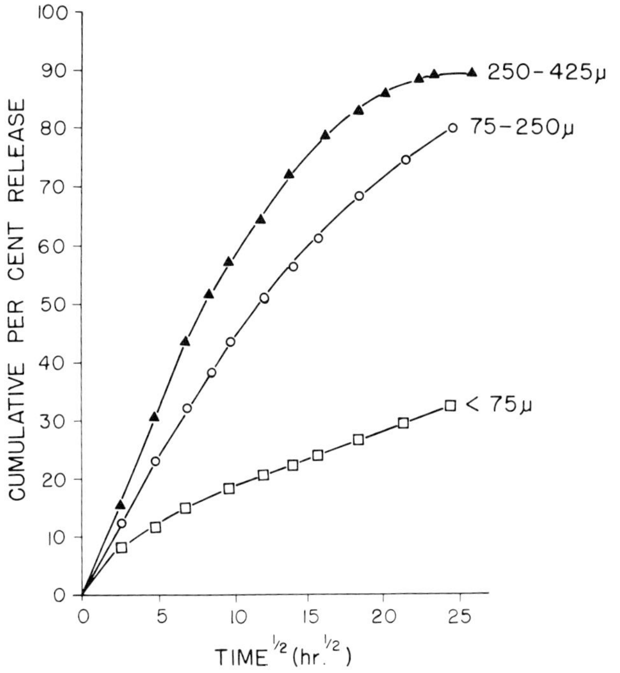


*FIGURE 11. Effect of particle size on the cumulative release of bovine serum albumin. The loading is 37.5%. Each point represents the mean of at least eight samples. Standard deviations are all less than 10% of the respective mean values.*

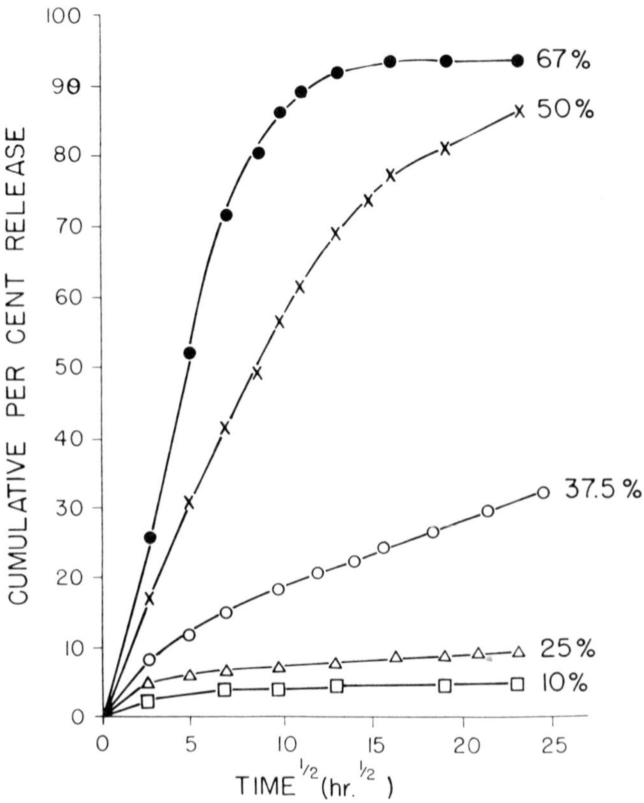


*FIGURE 12. Effect of loading on the cumulative release of bovine serum albumin. The particle size was less than 75 μ. Each point represents the mean of at least eight samples. Standard deviations are all less than 10% of the respective mean values.*

A protein's isoelectric point generally corresponds to the pH of minimum solubility. This correlation of solubility with release rate was consistent with diffusion-controlled models of release (19).

## III. DISCUSSION

These studies demonstrate several important applications of macromolecular delivery systems and further elucidate their release properties. Data obtained in the applications

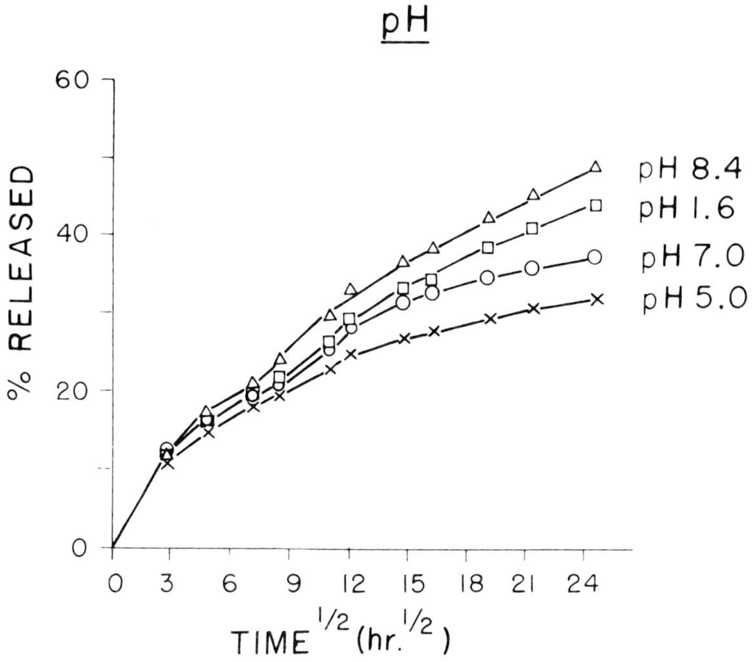


*FIGURE 13. Effect of release media pH on the cumulative release of bovine serum albumin. The particle size was 75-250 μ and the loading was 25%. Each point represents the mean of 10 samples. Standard deviations are all less than 10% of the respective means.*

experiments illustrate that sustained-release systems may be useful in blood glucose control, immunization procedures, bioassays, and chemotaxis studies. Designs of these specific release systems were not optimized and it is anticipated that improved systems will yield even more impressive results.

The ability of sustained-release polymers to provide effective alternatives to conventional modes of insulin therapy and immunization procedures in animals indicates the potential clinical importance of these systems. Drugs remain protected in the polymer and are then released in unaltered form. This may be particularly important for the long-term administration of molecules that are rapidly degraded *in vivo* or are available only in small quantities. There are numerous other macromolecules that may eventually be candidates for sustained release, including heparin, a widely used anticoagulant, and interferon, a new anti-viral compound. While the ultimate form the slow-release systems may take remains to be seen, our studies suggest the potential value of sustained macromolecular release for a variety of clinical applications.

Sustained-release polymers also provide a basis for bioassays and chemotaxis experiments, which may lead to a range of applications in basic biological studies. These polymers have many advantages over earlier methods of testing informational macromolecules *in vivo* or creating chemical concentration gradients *in vitro* and *in vivo*. Advantages include low cost, ease of fabrication, versatility, ability to localize substances in target areas, and protection from biodegradation.

The kinetic studies suggest that there are three phases of release: 1) a "burst" or initial period of rapid release, 2) a period when release was approximately linear with respect to $time^{\frac{1}{2}}$, and 3) a final period when release tapered off. The burst effect presumably was due to the dissolution of proteins on the surface and cut edges of the matrix. The intermediate phase of release that displayed $time^{\frac{1}{2}}$ kinetics (Figures 11 and 12) resembled earlier models of release of low molecular weight drugs from granular matrices (19). The correlation of protein solubility with release rate was also consistent with such models. Because macromolecules are too large to diffuse through a pure polymer film (1), it is possible that sustained release occurs *via* diffusion through channels in the matrix. The microstructural studies showed that incorporation of the macromolecules during polymer casting introduces channels that are large enough to permit macromolecular diffusion.

Our studies demonstrate the marked effect of drug particle size and loading on release kinetics. Release-rate increases caused by increases in particle size may result from the formation of larger channels or pores in the polymer matrix. Similarly, increased loadings may provide simpler pathways (lower tortuosity) and greater porosity for diffusion, both of which would facilitate the movement of water into, and proteins out of, the matrix. Studies in our laboratory are currently focused on confirming the release mechanism, establishing mathematical models to describe release kinetics, and on further exploring the potential applications of these novel drug delivery systems.

ACKNOWLEDGMENTS

This work was supported by National Institutes of Health grant 1 RO1 GM 26698-01.

REFERENCES


1. R. Langer and J. Folkman, Nature, 263, 797 (1976).
2. R. Langer and J. Folkman in Polymeric Delivery Systems, Midland Macromolecular Monograph Vol. 5, R. J. Kostelnik, (Ed.), 1978, pp. 175-196.
3. R. Langer, H. Brem, K. Falterman, M. Klein, and J. Folkman, Science, 193, 70 (1976).
4. S. Brem, I. Preis, R. Langer, H. Brem, J. Folkman, and A. Patz, Am. J. Ophthalmol., 84, 323 (1977).
5. P. Polverini, R. Cotran, M. Gimbrone, and E. Unanue, Nature, 269, 804 (1977).
6. H. Conn, M. Berman, K. Kenyon, R. Langer, and J. Gage, Invest. Ophthalmol., 19, 362 (1980).
7. D. Ausprunk, K. Falterman, and J. Folkman, J. Lab. Invest., 38, 284 (1978).
8. D. C. Job, M. D. Eschwege, C. Guyot-Argenron, J. P. Aubry, and G. Tchobroutsky, Diabetes, 25, 463 (1976).
9. O. R. Hilleman, Am. J. Med., 38, 751 (1965).
10. I. Preis and R. Langer, J. Immunol. Methods, 28, 193 (1979).
11. J. Freund, Am. J. Clin. Pathol., 21, 645 (1951).
12. S. Boyden, J. Exp. Med., 115, 453 (1962).
13. W.-W. Tso and J. Adler, J. Bacteriol., 118, 560 (1974).
14. S. J. Coward, R. C. Taylor, and R. E. Johannes, J. Biol. Physiol., 9, 32 (1967).
15. S. J. Coward and R. E. Johannes, Comp. Biochem. Physiol., 29, 475 (1969).
16. R. Langer, M. Fefferman, P. Gryska, and K. Bergman, Can. J. Microbiol., 26, 274 (1980).
17. W. Rhine, D. S. T. Hsieh, and R. Langer, Polymer Preprints, 20, 596 (1979).
18. W. Rhine, D. S. T. Hsieh, and R. Langer, J. Pharm. Sci., 69, 265 (1980).
19. T. Higuchi, J. Pharm. Sci., 52, 1145 (1963).

# CONTROLLED TRANSDERMAL DELIVERY

*S. K. Chandrasekaran*
*J. E. Shaw*

ALZA Research
Palo Alto, California

## I. INTRODUCTION

For the vast majority of therapeutic agents, the therapist's two main goals have been to attain constancy of drug action and minimization of side effects. Accordingly, the aims of pharmaceutical development have been directed toward dosage forms whose drug release rate kinetics are defined by the therapist's goals.

We have developed a new class of drug dosage forms to provide controlled administration of a drug at a predetermined rate to the surface of intact skin, from which it subsequently diffuses to the underlying capillary network and the systemic circulation. The design emphasis has been directed towards having the control of drug administration being resident in the dosage forms rather than in the skin (1,2).

In this paper, we shall attempt to describe various diffusional techniques in the design of controlled delivery systems, and the performance of a few of these systems both *in vitro* and *in vivo*.

## II. THEORY

Polymeric membranes are attractive candidate materials for the development of controlled delivery systems by virtue of the dissolution and diffusion of solute molecules in the polymer. The solute diffusion coefficient and solubility are intensive properties of the specific solute/polymer combina-




ISBN 0-12-074450-3

tion. Solubility is a thermodynamic property governed mainly by the relative chemical compositions of the polymer and solute, whereas diffusivity is a more complex kinetic parameter determined by the size and shape of the solute and the chain segmental mobility of the polymer.

For the diffusion of a solute from a polymer matrix into a fluid which is assumed to be a perfect sink, the following transport equations can be readily derived. It will be assumed that the matrix has a planar geometry with solute losses occurring through both faces.

### A. *Solute Loading Less Than Saturation ($C<C_s$)*

In this case, the polymeric matrix is equilibrated with solute, and the concentration of the solute is below the saturation concentration in the matrix, i.e., the solute is present in a totally dissolved state. The solute release rate as a function of time is given by Equations 1 and 2 (3):

$$\frac{dM_t}{dt} = 2M_\infty \left(\frac{D}{\pi \ell^2 t}\right)^{\frac{1}{2}} \tag{1}$$

for the early time approximation, and

$$\frac{dM_t}{dt} = \frac{8DM_\infty}{\ell^2} \exp\left(\frac{-\pi^2 Dt}{\ell^2}\right) \tag{2}$$

for the late time approximation, where $M_\infty$ is the total amount of solute sorbed, D is the solute diffusivity, and $\ell$ is the thickness of the matrix.

### B. *Solute Loadings Greater Than Saturation ($C>C_s$)*

This case differs from the previous one in that initially the matrix is saturated with dissolved solute, with some excess solute dispersed uniformly as small particles. Under conditions when solute diffusion in the matrix is the controlling mechanism, it has been shown that the release rate at any time 't' is given by Equations 3 and 4:

$$\frac{dM_t}{dt} = \left[\frac{D(2C_o - C_s)C_s}{t}\right]^{\frac{1}{2}} \tag{3}$$

where $C_s$ is the solute solubility in the matrix and $C_o$ is the solute loading.

Under conditions where $C_O \gg C_S$, we have

$$\frac{dM_t}{dt} = \left(\frac{2DC_OC_S}{t}\right)^{\frac{1}{2}} \tag{4}$$

If the matrix is granular, and the solute is also lost by a leaching mechanism occurring in a capillary network formed *in situ,* the solute release rate is given by Equations 5 and 6:

$$\frac{dM_t}{dt} = \left[\frac{\varepsilon D(2C_O - \varepsilon C_S)C_S}{\tau t}\right]^{\frac{1}{2}} \tag{5}$$

where $\varepsilon$ is the porosity and $\tau$ is the tortuosity of the capillary network.

Under conditions where $C_O \gg C_S$, we have

$$\frac{dM_t}{dt} = \left(\frac{2\varepsilon DC_OC_S}{\tau t}\right)^{\frac{1}{2}} \tag{6}$$

### C. *Constant Solute Activity Reservoir System*

With the solute/polymer matrix formulations presented so far, the solute release rate into a perfect sink varies with time; usually to $t^{-\frac{1}{2}}$ during the early time periods. However, if more constant solute release rates are required, the design considerations have to be modified to provide for a constant solute activity reservoir or source. If a solute is enclosed within an inert polymer membrane, the release rate of solute is given by Equation 7, provided the thermodynamic activity of the solute within the enclosure is maintained constant and perfect sink conditions are provided downstream (1,3):

$$\frac{dM_t}{dt} = DK\frac{Cs}{\ell} \tag{7}$$

where K is the solute partition coefficient and $\ell$ is the membrane thickness.

## III. EXPERIMENTAL AND RESULTS

### A. *Diffusion into a Perfect Sink In Vitro*

*1. Solute Loadings Greater than Saturation.* Monolithic systems were fabricated by incorporating scopolamine into a polymeric matrix, such that the drug loading was in excess of saturation. These systems were placed into an infinite sink and maintained at isotonic and isothermal conditions, and the *in vitro* release rate versus time profile of scopolamine is shown in Figure 1. The data points represent values experimentally measured, and the solid line represents the profile predicted using Equation 4. The system becomes devoid of excess dispersed drug after approximately 7 hours; the dotted line is the predicted release rate profile

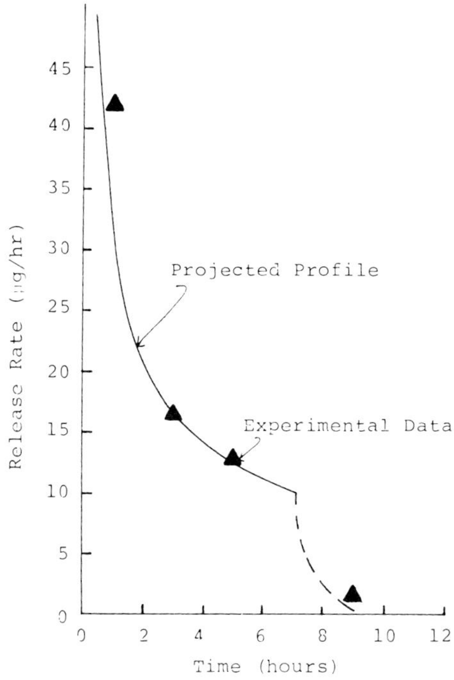


*FIGURE 1. Scopolamine release rate profile from the monolithic system in vitro.*

assuming an established linear concentration gradient in the matrix at the moment when the system becomes devoid of excess dispersed scopolamine. The agreement between experiment and theory is good.

2. *Constant Solute Activity Reservoir System.* Constant solute activity reservoir systems for transdermal applications (Transdermal Therapeutic Systems) were prepared by laminating a drug reservoir containing scopolamine in a polymeric gel between an impermeable backing membrane and a rate-controlling microporous membrane. On the dermal side of the rate-controlling membrane is an adhesive gel containing scopolamine; this gel layer serves both as an adhesive to secure the system on the skin surface and as an additional drug reservoir to provide an initial priming dose of drug prior to the controlled input of the drug to the skin surface (1).

During the priming dose period, drug diffusion from the contact adhesive layer dominates the temporal pattern of drug release, whereas during steady state delivery, rate limitation is resident in the microporous membrane. A typical *in vitro* release rate versus time profile of scopolamine from a transdermal therapeutic system into an infinite sink is shown in Figure 2. The data points represent values experimentally measured, whereas the solid line is predicted using Equations 4 and 7 (1).

## B. *Diffusion Into Human Skin In Vivo*

To follow the rate of drug input through human skin *in vivo* during application of monolithic and reservoir systems, we monitored the rate of urinary excretion of the drug, using a sensitive and specific assay (7). Following intramuscular or intravenous administration of scopolamine, only 10% of the drug is recovered in the urine in the free form, and we have assumed that a similar recovery will be obtained during transdermal administration.

The urinary excretion rates of scopolamine after the administration of a monolithic system (A-1) and reservoir systems (A-2) are shown in Figure 3. With the monolithic system, the urinary excretion rate rises during the first six hours after which the rate continuously declines attaining a very low value (<0.05 μg/hr) after 24 hours. In this case, a steady state in urinary excretion rate is not attained; the profile is indicative that human skin acts as a diffusion

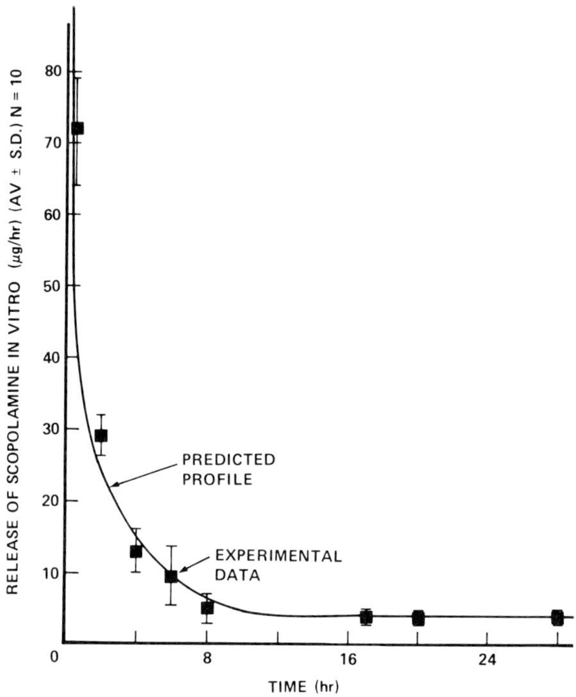


*FIGURE 2. Scopolamine release rate from the transdermal therapeutic system in vitro.*

barrier with drug transport occurring by Fickian diffusion with partition equilibrium of scopolamine being maintained at the interlayer boundaries.

Following the administration of the transdermal therapeutic system, the urinary excretion rate of drug approaches 10% of the release rate of drug from the system *in vitro*, indicating that the drug administration kinetics are governed by the transdermal system and not by patient relation factors. In this manner, the element of control has been incorporated into drug delivery.

## IV. CONCLUSIONS

Some considerations in the design of controlled drug delivery systems functioning by diffusion mechanisms have been presented. In this regard, optimal control in both rate and

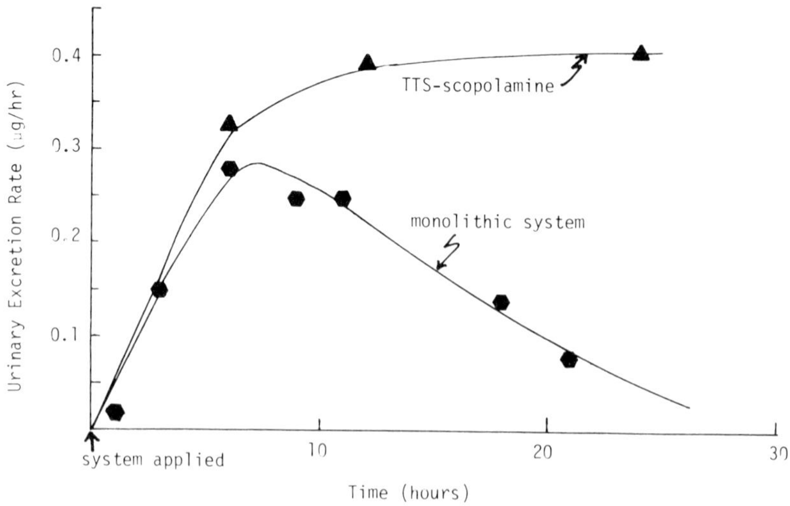


*FIGURE 3. Functionality of the scopolamine system in vivo.*

duration of drug delivery is obtained from reservoir-based systems. The transdermal therapeutic system, designed to deliver scopolamine systemically to achieve a pharmacological effect via action within the central nervous system, functions according to a specific temporally patterned program of drug delivery.

## REFERENCES

1. S. K. Chandrasekaran and J. E. Shaw, in Contemporary Topics in Polymer Science, Vol. 2, Plenum Publishing, New York, 1977.
2. S. K. Chandrasekaran, W. Bayne, and J. E. Shaw, J. Pharm. Sci., 67, 1370 (1978).
3. R. W. Baker and H. K. Lonsdale in Advances in Experimental Medicine and Biology, Vol. 47, Plenum Publishing, New York, 1974.
4. D. R. Paul and S. K. McSpadden, J. Mem. Sci., 1, 33 (1976).
5. J. Cobby, J. Biomed. Matls. Res., 12, 627 (1978).

6. Y. Samuelov, M. Donbrow, and M. Friedman, J. Pharm. Sci., 63, 325 (1979).
7. W. F. Bayne, F. T. Tao, and N. Crisologo, J. Pharm. Sci., 64, 288 (1975).

# CONTROLLED RELEASE OF ANTITHROMBOTIC AGENTS FROM POLYMER MATRICES

*Charles Ebert*
*James McRea*
*Sung Wan Kim*

Department of Pharmaceutics
University of Utah
Salt Lake City, Utah

## I. INTRODUCTION

A major concern with the use of artifical organs and biomedical devices is the untoward interactions of blood upon contacting a foreign surface, resulting in surface thrombosis. Surface thrombosis, which greatly impedes the safety and efficacy of such devices, is the end result of the complex interactions between cellular and molecular blood constituents with the polymer surface under the hemodynamic condition in which the devices operate. The subject is reviewed by Feijen (1). Although the mechanisms for thrombosis are complex and not fully elucidated at this time, progress is being made in the development of thromboresistant materials.

Current research in thromboresistant materials can be divided into two general philosophies. The first and perhaps most attractive approach is to prevent activation of the thrombogenic mechanisms at the blood/foreign surface interface. This is achieved by providing the polymer surface with the physiochemical properties to either prevent interaction between the polymer surface and the blood (2), or to selectively adsorb a nonreactive albumin layer onto the polymer surface (3). The second approach is to incorporate pharmacologically active agents with existing materials which, following controlled release at the blood/foreign surface interface, prevent thrombosis.




ISBN 0-12-074450-3

Many such controlled release, thromboresistant materials involve the release of heparin anion at blood interfaces via an ion exchange mechanism (4-6). Although these materials are thromboresistant with respect to fibrin formation, platelet adhesion onto the polymer surfaces at blood interfaces can still occur (7), which could eventually lead to thrombus formation.

It has been shown that when prostaglandin $E_1$ ($PGE_1$) was combined with polyurethanes and polyvinylchloride, pharmacologically active $PGE_1$ was released from the polymer which prevented platelet aggregation in plasma and inhibited platelet adhesion onto the polymer surfaces (8). Studies on the simultaneous administration of heparin and prostacyclin ($PGI_2$), a potent inhibitor of platelet aggregation, have shown that during extracorporeal circulation with Greyhound dogs, routine heparinization alone produced a reduction in the number of circulating platelets (indicating platelet adhesion onto foreign surfaces) and impaired normal platelet function while routine heparinization with simultaneous prostacyclin infusion produced no change in platelet plasma levels or in platelet function (9). Prostaglandin $PGE_1$ offers advantages over $PGI_2$ for purposes of simultaneous administration with heparin because: 1) $PGE_1$ is much more stable in aqueous systems than $PGI_2$, and 2) heparin antagonizes the antiaggregatory effects of $PGI_2$ (10), more than $PGE_1$ (unpublished observation).

An attractive thromboresistant material could be achieved by providing simultaneous controlled release of heparin and $PGE_1$. This requires that the matrix polymer must be capable of providing suitable simultaneous controlled release of low molecular weight solutes ($PGE_1$) and macromolecules (heparin). Hydrogels, uncrosslinked poly hydroxyethyl methacrylate poly(HEMA), were selected as the matrix polymer because hydrogels have been demonstrated to be capable of providing controlled release of a wide range of solute types (11, 12) while also possessing low interfacial free energies in aqueous systems, minimizing the tendency to adsorb biological components from blood (13).

Monolithic devices, prepared by dispersing $PGE_1$ and heparin in hydrogel matrices, were fabricated and evaluated for *in-vitro* release rates and for verification of the biological activities of the respective pharmacological agents. These studies not only investigate the feasibility of such controlled delivery devices as a thromboresistant material, but also might provide some insight into the diffusion mechanisms of solutes through hydrogels.

## II. EXPERIMENTAL

### A. *Preparation of Matrix Polymer*

Poly hydroxyethyl methacrylate poly(HEMA) was used as the matrix polymer for all controlled delivery tests. HEMA monomer (Hydron Labs) was polymerized with azobis (methyl-isobutrate) free radical initiator, prepared according to the methods of Mortimer (14), at a final concentration of 7.83 mmoles of initiator per liter of monomer in 90% (v/v) methanol for 24 hours at 60°C. The resulting polymer solution was then cooled and poly(HEMA) was precipitated out by the addition of deionized, distilled water. The poly(HEMA) was repeatedly washed with deionized, distilled water, vacuum dried and redissolved in methanol to a final concentration of 100 mg polymer/ml solution.

### B. *Preparation of Monolithic Devices*

One percent $PGE_1$ (Upjohn), 10% sodium heparin (Sigma) and a combined 1% $PGE_1$ + 10% sodium heparin polymers were prepared by mixing appropriate quantities of polymer solution, $^{14}C$ labeled $PGE_1$ (New England Nuclear) and $^{3}H$ labeled sodium heparin (New England Nuclear). The resultant mixture was solvent cast in 1-inch diameter, high density polyethylene disc shaped molds at 60°C. The molds with respective polymers were then vacuum dried. A parallel set of nonradiolabeled polymers was similarly prepared for scanning electron microscope surface analysis. Ten percent sodium heparin with 1% and 10% $PGE_2$ (Upjohn), an analog of $PGE_1$ containing two double bonds, were also prepared for additional tests regarding the heparin diffusion mechanism.

### C. *In-Vitro Release Testing*

*In-vitro* release experiments were conducted at room temperature in a phosphate buffered saline elution media (pH 7.4) contained within closed, polypropylene vessels. The respective polymers were still contained in their original casting molds to insure a constant surface area for controlled release. The release media was constantly stirred with magnetic stir bars to insure a uniform concentration profile. The solutions were sampled periodically and analyzed for $PGE_1$

and heparin levels by liquid scintillation counting of the respective radiolabeled agents using a Beckman LS9000 Liquid Scintillation Counter.

### *D. Biological Assays*

Biological assays for the released heparin were conducted in parallel with the *in-vitro* release experiments. Heparin activity in the elution media was determined by conducting activated partial thromboplastin time (APTT) tests (15) on bovine platelet poor plasma, heparinized by the addition of the elution media. Activated partial thromboplastin times were determined using a Fibrometer (Beckman, Dickinson Co.) with Ortho activated thromboplastin reagent. Heparin activities, and hence heparin concentrations, in the heparinized bovine plasma were determined by correlation of sample APTT with plasma heparin concentration vs APTT dose response profiles. From these data, heparin concentration in the elution media was calculated based on biological assay methods. Verification of biological activity is provided by correlation of biological assay results with *in-vitro* release results.

Prostaglandin $PGE_1$ biological assays were conducted using the method of platelet aggregation induced by ADP (16). The pharmacological action of $PGE_1$ is assumed to stimulate platelet membrane bound adenyl cyclase, converting ATP to cAMP and increasing intracellular levels of cAMP (17). Biological activity of $PGE_1$ can be determined by measuring the extent of inhibition of platelet aggregation exogeneously induced in bovine platelet rich plasma (PRP). Bovine plasma, collected in sodium citrate, was centrifuged at 200 g for 15 min. at room temperature. The plasma fraction was collected and adjusted to 350,000 to 450,000 platelets/$mm^3$. Aggregation was induced by the introduction of ADP (Sigma) at a final plasma concentration of $1.25 \times 10^{-4}$ M. The extent of aggregation was determined by the increased light transmittance through the PRP as aggregation occurred. Platelet aggregation is expressed as percent aggregation, using a Platelet Aggregation Profiler model 3 (Bio/Data Corp.) where nonaggregated PRP is 0% aggregation and platelet poor plasma (PPP) is 100% aggregation. Once the PRP and PPP controls have been established for a given batch of PRP, the maximum extent of aggregation is measured by the introduction of ADP, this in turn represents 0% inhibition. Incubation of PRP with excess $PGE_1$ prior to the addition of the aggregating agent ADP results in a minimum extent of aggregation or 100% inhibition. When lower concentrations of $PGE_1$ are incubated with the PRP, increasingly greater aggregation (i.e., less inhibition of aggregation) is measured. In this way a dose response curve, plotting percent inhibition as a function of the log PRP concentration of $PGE_1$, can be prepared. The following expression is used for calculating percent inhibition:

$$\text{Percent Inhibition} = 1 - \frac{\text{Aobs} - \text{Amin}}{\text{Amax} - \text{Amin}} \times 100$$

where Aobs is the percent aggregation observed for the test sample, Amin is the minimum aggretation and Amax is the maximum aggregation for a given batch of PRP, as established from the controls described above. Verification of $PGE_1$ biological activity is provided by correlation of bioassay results with *in vitro* release results.

## III. RESULTS

Figure 1 shows the effect of $PGE_1$ and $PGE_2$ on the release of heparin from 10% heparin polymers. In all cases, the amount of heparin released is proportional to the square root of time, as described by Higuchi (18); however, the type of counter solute and, more importantly, the counter solute concentration influences the heparin release profile. One percent $PGE_1$ substantially slows the release of heparin (slope = 3.17 coefficient of determination, $r^2 = 0.91$) while 1% $PGE_2$ and 10% $PGE_2$ reduces the heparin release even more so (slopes = 1.96 and 1.31, $r^2 = 0.87$ and 0.96 respectively).

Figure 2 shows the effect of heparin on $PGE_1$ release. As with heparin, the amount of $PGE_1$ released is also proportional to the square root of time. Heparin was found to increase the release of $PGE_1$ from the hydrogel matrix (slope = 0.68, $r^2$ = 1.00) as compared to PGE without heparin (slope = 0.32, $r^2$ = 0.99).

Plasma heparin concentration vs APTT calibration curves for determining heparin levels in the elution media on day #1 and day #3 of release are shown in Figures 3 and 4 respectively. APTT for plasma samples heparinized by the addition of elution media and the corresponding heparin levels in the elution media are found in Table I.

The extent of biological activity for $PGE_1$ can be determined by data described in the dose response curve of Figure 5. In a separate series of experiments, percent inhibition was plotted as a function of the log concentration of $PGE_1$ required to achieve the corresponding percent inhibition. This data fits a power curve equation where the coefficient determination $r^2 = 0.97$. By solving for the log concentration from a known percent inhibition, it is then possible

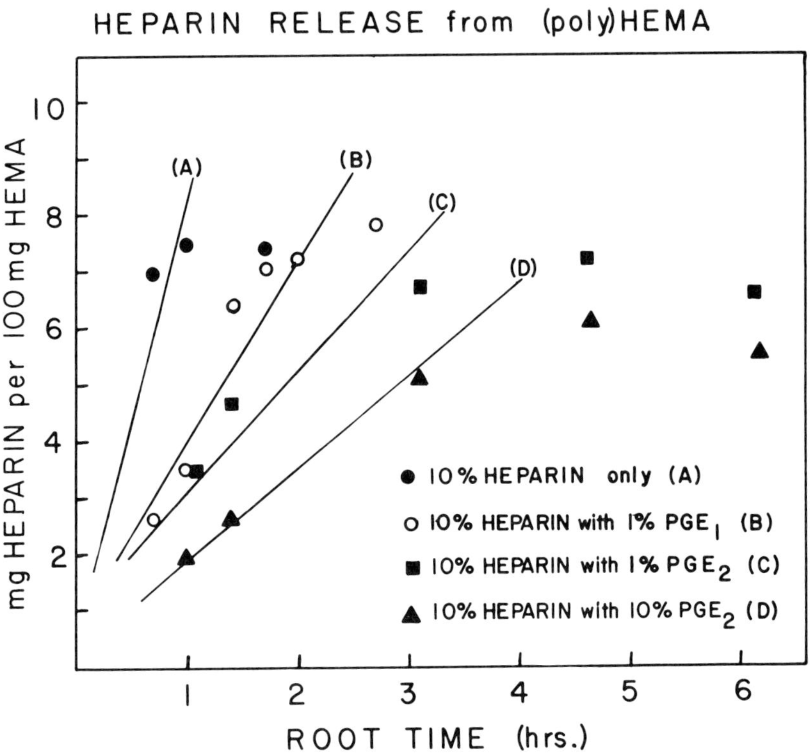


*FIGURE 1. Demonstrates relative differences in rate changes during early time heparin release with varying counter solutes and counter solute concentrations.*

to compare concentration of radiolabeled measured samples with those measured by biological activity and determine if there is any loss in biological activity.

It is important to realize in Figure 5 that as the curve approaches higher concentrations, the inhibition response rises precipitously and small changes in concentration result in significant changes in inhibition. A several fold change in concentration in this region becomes somewhat difficult to appreciate though order of magnitude changes can be easily seen.

Table II summarizes the release data of both the combined $PGE_1$ and heparin in poly(HEMA), and the $PGE_1$ in poly(HEMA) alone. This table demonstrates the corresponding values of

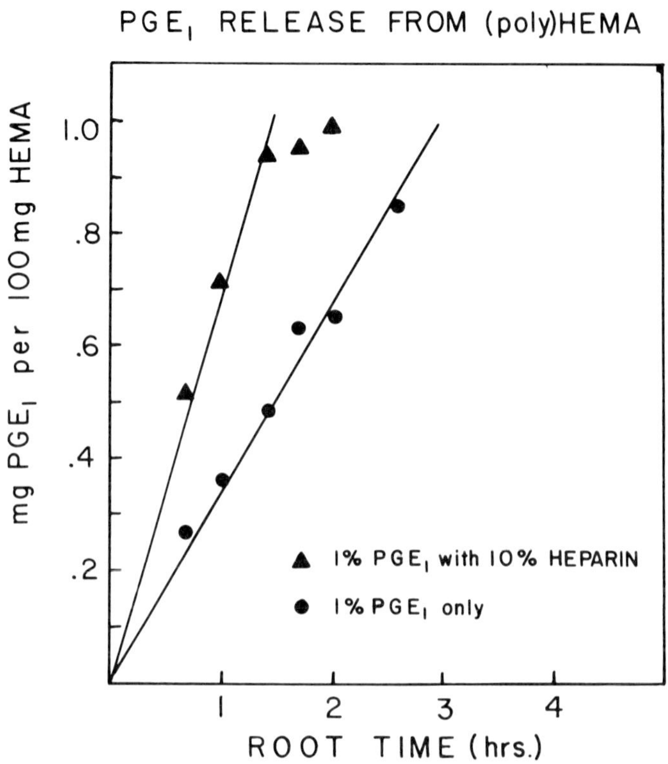


*FIGURE 2. Demonstrates relative differences in rate changes during early time release of prostaglandin $PGE_1$ in the presence of a counter solute.*

$PGE_1$ concentration as measured by carbon 14 labeled $PGE_1$ and by the biological activity assay of $PGE_1$ when calculated from the dose response curve.

No difference in the surfaces of nonhydrated (pure poly (HEMA) 1% $PGE_1$, 10% heparin and the combined 1% $PGE_1$ + 10% heparin) could be observed. Upon hydration, pure poly(HEMA), 1% $PGE_1$ and the 10% heparin polymers appeared identical, but the combined polymer demonstrated a rough, spongy like surface (See Figure 6.)

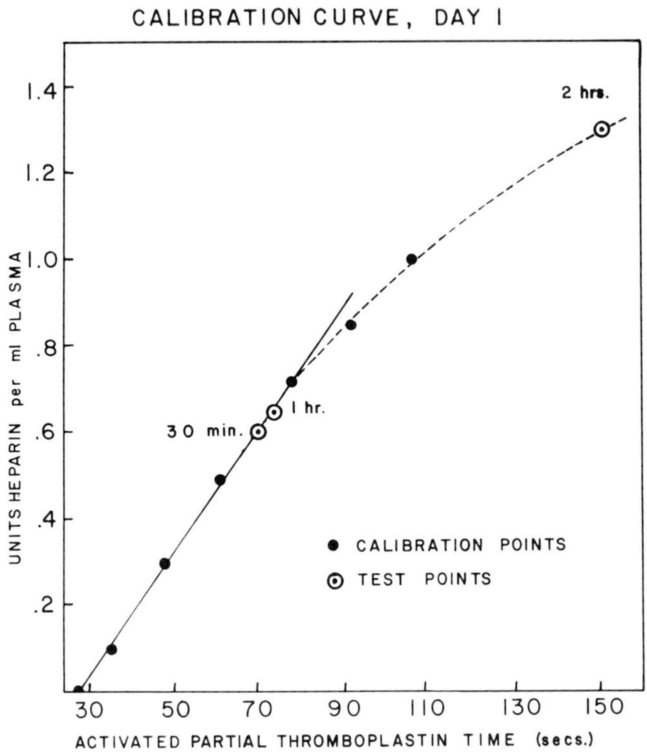


*FIGURE 3. Test points for heparin quantitation compared with calibration curve of activated partial thromboplastin time as a function of plasma heparin concentration for day one release testing.*

## IV. DISCUSSION

The role of water in hydrogels is an important factor for solute diffusion. Water in the hydrogel matrix is assumed to exist in three states: bound water, interstitial water and bulk water (19). The hydrogel can be divided into two domains (A and B) with respect to the molecular state of water and the diffusion mechanisms. The A domain consists of bound and interstitial water while the B domain consists of bulk water. There are two proposed mechanisms for solute diffusion in hydrogels: pore permeation and partition permeation. Permeation in the A domain occurs via a process where the solute dissolves into the polymer by associating with the polymer chains (including the bound and interstitial water),

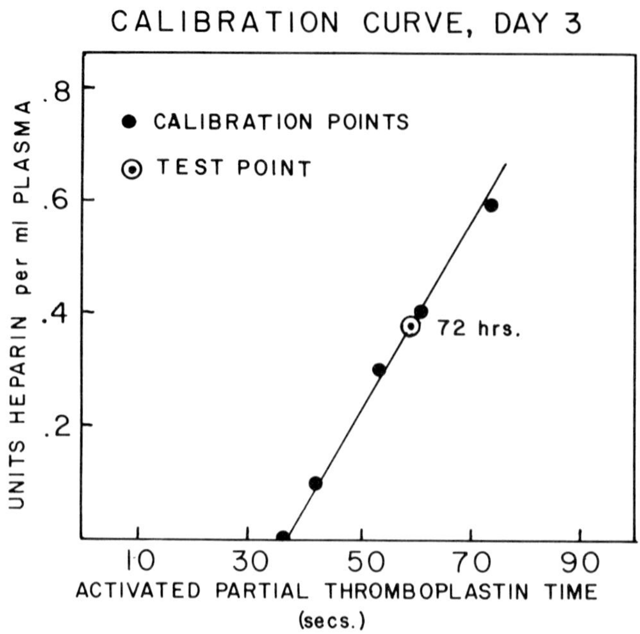


*FIGURE 4. Test point for heparin quantitation compared with calibration curve of activated partial thromboplastin time as a function of plasma heparin concentration for day three release testing.*

followed by diffusion along and between the polymer segments (20). Permeation in the B domain occurs predominantly by successive "jumps" in the fluctuating aqueous microchannels (pore mechanism) which compose the B domain (21). The presence of the hydrophilic, ionic solute heparin in the polymer matrix causes an increased internal swelling pressure with a resultant increased water fraction. Increasing the water content of the hydrogel increases the relative contribution of bulk water and is correlated with increased diffusion by the pore mechanism in the B domain (22). $PGE_1$ was demonstrated to exhibit increased diffusion rates in the presence of heparin as opposed to $PGE_1$ without heparin, indicating the $PGE_1$ diffusion is dominated by the pore mechanism in the B domain of the hydrogel. Were heparin diffusing by a pore mechanism, one would expect increased release rates for heparin in the presence of $PGE_1$ or $PGE_2$, based on the fact that $PGE_1$ and $PGE_2$ are both ionic at the pH of the release media and would therefore increase the hydration of the polymer.

TABLE I. Biological Heparin Assays

| Time | μl elution media/ml plasma | APTT | Units/ml plasma | Units in elution media | mg Heparin released from biological data | mg Heparin released from in vitro data |
|---|---|---|---|---|---|---|
| 30 min. | 50/1 | 69 | 0.60 | 630 | 2.6 | 2.1 |
| 1 hour | 50/1 | 72 | 0.64 | 670 | 2.8 | 2.8 |
| 2 hours | 50/1 | 150 | 1.32 | 136 | 5.7 | 5.1 |
| 72 hours | 50/5 | 60 | 0.38 | 166 | 6.9 | 6.4 |

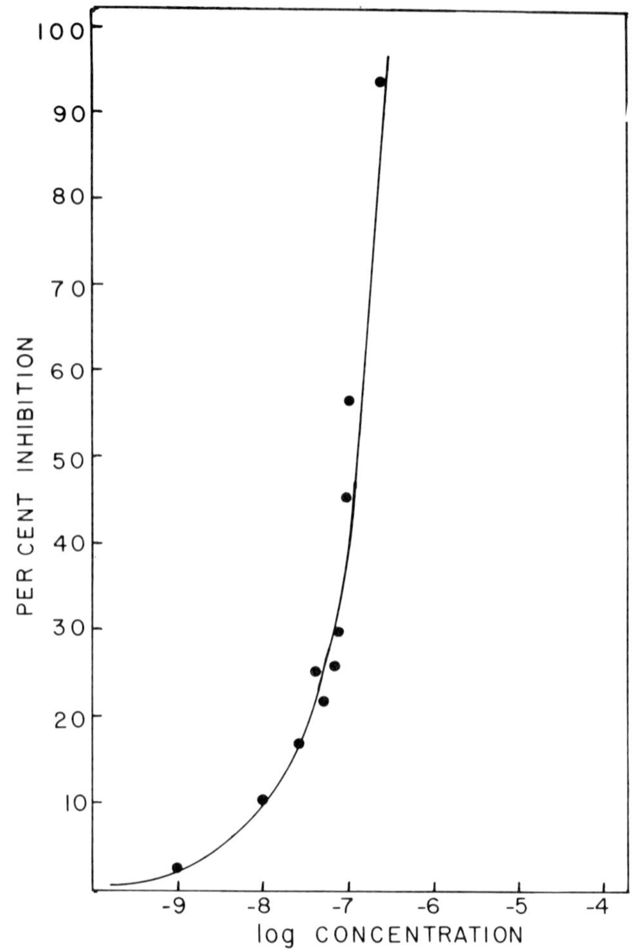


*FIGURE 5. Dose response curve for percent inhibition of ADP induced platelet aggregation as a function of log concentration of* $PGE_1$ *in platelet rich plasma.*

The converse was found to be true, heparin diffusion decreases in the presence of $PGE_1$ and $PGE_2$. It must be concluded that heparin diffusion does not occur by a pore mechanism in the B domain, but instead by a partition mechanism in the A domain.

Heparin and $PGE_1$ would then be separated on a molecular level within the hydrogel. This is demonstrated by platelet aggregation data. It has been shown that heparin antagonizes

TABLE II. Biological $PGE_1$ Assays

| Time | Dilution | % Inhibition | Log conc. from bio assay | Log conc. from in vitro assay |
|---|---|---|---|---|
| 1% $PGE_1$ | | | | |
| 4 hours | --- | 102 | -6.5 | -5.6 |
| 4 hours | 100x | 34 | -7.1 | -7.6 |
| 4 hours | 10x | 100 | -6.5 | -6.6 |
| 72 hours | --- | 95 | -6.5 | -5.4 |
| 72 hours | 100x | 8 | -8.1 | -7.4 |
| 1% PGE + 10% Heparin | | | | |
| 4 hours | --- | 101 | -6.5 | -5.4 |
| 4 hours | 10x | 102 | -6.5 | -6.4 |
| 4 hours | 50x | 67 | -6.7 | -7.1 |
| 4 hours | 200x | 14 | -7.7 | -7.7 |
| 72 hours | --- | 87 | -6.6 | -5.3 |
| 72 hours | 10x | 42 | -7.0 | -6.3 |

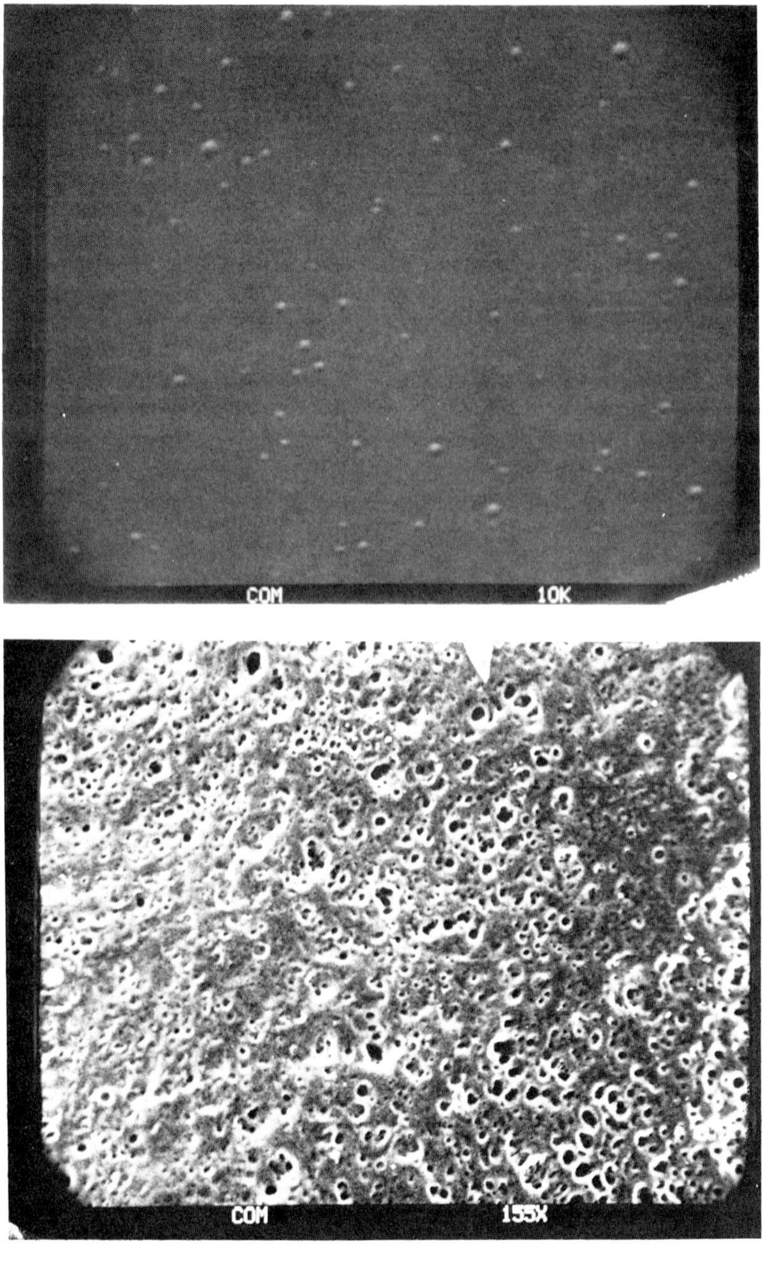


FIGURE 6. Scanning electron microscope photographs of poly(HEMA) containing 10% heparin and 1% $PGE_1$; nonhydrated sample magnified 10,000x (upper); hydrated sample magnified 155x (lower).

the antiaggregatory effect of $PGE_1$ when the two compounds are incubated together (23). Whether the antagonistic effect of heparin on $PGE_1$ is due to physical adsorption of $PGE_1$ onto heparin substrates or due to a synergistic degradation process is not known at this time (binding studies are underway in this lab to help clarify this problem). In either event, both heparin and $PGE_1$ were found to be active following release from the polymer. The fact that $PGE_1$ released from the polymer was active indicates that the $PGE_1$ and heparin were separated on a molecular level inside the polymer matrix. This conclusion is further substantiated by the observation that upon hydration, 1% $PGE_1$ polymers remain transparent (indicating that all $PGE_1$ is in solution in the bulk water comprising the B domain) while the 10% heparin polymer turns a white color and is not transparent (indicating that the heparin is not in solution but is precipitated, as would be expected for heparin in the presence of the highly organized water comprising the A domain).

It has been shown that heparin resides in the A domain and diffuses from the polymer by a partition mechanism where the heparin molecules become associated with the polymer chains and the bound and interstitial water of the A domain. The reductions in the release rates of heparin in the presence of $PGE_1$ and $PGE_2$ can be explained by assuming an electrostatic repulsive interaction exists between the negatively charged heparin anion and the hydroxide functionalities on the HEMA mers comprising the polymer chains. This repulsive interaction acts as a driving force, in addition to the predominant concentration gradient driving force, for the diffusion of heparin from the hydrogel matrix. $PGE_1$ and $PGE_2$ could retard this electrostatic repulsion by either reducing the ionic sphere of influence for the heparin anions or by altering the dielectric constant in the microenvironment between heparin and polymer molecules.

Whatever mechanisms are involved for the diffusion of heparin and $PGE_1$ from the hydrogel matrix, pharmacologically active agents are released from the polymer which could reduce surface thrombosis were the controlled delivery device employed as a thromboresistant material for biomedical devices.

ACKNOWLEDGMENT

We gratefully acknowledge Dr. J. E. Pike of the Upjohn Co. (Kalamazoo, Michigan) and Dr. S. Ronel of Hydron Medical Science, Inc. (New Brunswick, New Jersey) for generously providing the prostaglandins and purified HEMA used in this study. This work was supported by NIH Grant HL-20251. S.W. Kim is a NIH Research Career Development Awardee (HL-00272).

## REFERENCES


1. J. Feijen, in Artificial Organs, R. M. Kenedi, J. M. Courtney, J. D. S. Gaylor, and T. Gilchrist (Eds.), The Gresham Press, Old Woking, Surrey, England, 1977, p. 235.
2. J. D. Andrade, H. B. Lee, M. S. Jhon, S. W. Kim, and J. B. Hibbs, Trans. Amer. Soc. Artif. Int. Organs, 19, 1 (1973).
3. S. W. Kim, R. G. Lee, H. Oster, D. Coleman, J. D. Andrade, D. Lentz, and D. Olsen, Trans. Amer. Soc. Artif. Int. Organs, 20, 449 (1974).
4. V. L. Gott, J. D. Whiffen, and R. C. Dutton, Science, 142, 1297 (1963).
5. G. A. Grode, R. D. Falb, and J. P. Crowley, J. Biomed. Mat. Res., 6, 77 (1972).
6. H. Tanzawa, Y. Mori, N. Harumiya, H. Miyama, M. Hori, N. Ohshima, and Y. Idezuki, Trans. Amer. Soc. Artif. Int. Organs, 19, 188 (1973).
7. R. I. Leininger, J. P. Crowley, R. D. Falb, and G. A. Grode, Trans. Amer. Soc. Artif. Int. Organs, 18, 312 (1972).
8. J. C. McRea and S. W. Kim, Trans. Amer. Soc. Artif. Int. Organs, 24, 746 (1978).
9. D. B. Langmore, G. Bennett, D. Gueirrara, M. Smith, S. Bunting, S. Moncada, P. Read, N. G. Read, and J. R. Vane, Lancet, 1, 1002 (1979).
10. H. I. Saba, S. R. Saba, C. A. Blackburn, R. C. Hartman, and R. G. Mason, Science, 205, 499 (1979).
11. J. M. Anderson, T. Koinis, T. Nelson, M. Horst, and D. S. Love in Hydrogels for Medical and Related Applications, J. D. Andrade (Ed.), Symp. Series 31, American Chemical Society, Washington, D.C., 1976, p. 167.
12. D. R. Cowsar, O. R. Tarwater, and A. C. Tanquary, in Hydrogels for Medical and Related Applications, J. D. Andrade, (Ed.), Symp. Series 31, American Chemical Society, Washington, D.C., 1976, p. 180.
13. J. D. Andrade, Med. Instrum., 1, 110 (1973).
14. G. A. Mortimer, J. Org. Chem., 30, 1632 (1964).
15. I. Spector and M. Corn, J. Amer. Med. Assn., 201, 157 (1967).

16. S. F. Mohammad, R. L. Reddick, and R. G. Mason, Amer. J. Path., 79, 81 (1975).
17. J. F. Mustard and M. A. Packham, in Cardiovascular Drugs. Antithrombotic Drugs, G. S. Avery, (Ed.), University Press, Baltimore, Maryland, 1978, p. 1.
18. T. Higuchi, J. Pharm. Sci., 50, 874 (1961).
19. M.S. Jhon and J. D. Andrade, J. Biomed. Mat. Res., 7, 509 (1973).
20. D. J. Lyman and S. W. Kim, J. Polymer Sci. Symp., 41, 139 (1973).
21. L. C. Graig and W. Konigsburg, J. Phys. Chem., 65, 116 (1961).
22. G. M. Zentner, J. R. Cardinal, and S. W. Kim, J. Pharm. Sci., 67, 1352 (1978).
23. A. Reches, A. Eldore, and Y. Saloman, J. Lab. Clin. Med., 93, 638 (1979).

# HYDROGEL DEVICES FOR THE CONTROLLED RELEASE OF STEROID HORMONES

*John R. Cardinal*
*Sung-Ho Kim*
*Suk-Zu Song*

Department of Pharmaceutics
College of Pharmacy
University of Utah
Salt Lake City, Utah

*Hydrogels prepared from polyhydroxyethyl methacrylate poly(HEMA) are known as good biocompatible polymers and have potential for use in a wide variety of biomedical applications including controlled release drug delivery systems. In this study, the release of progesterone has been examined from reservoir and combined reservoir-monolithic type controlled release devices as a function of the composition of the polymers used in the preparation of the devices. It has been shown that both reservoir and combined reservoir-monolithic type devices release progesterone at zero-order rates and that the release rates can be varied by changes in the crosslinker content and the copolymer composition. The measured permeabilities obtained from release studies with reservoir devices were in good agreement with results obtained from membrane transport studies. Predicted release rates from combined reservoir-monolithic devices were in only fair agreement with the experimental values.*

## I. INTRODUCTION

In the development of controlled release drug delivery systems, it is particularly important that the release rates follow zero-order kinetics. It is only in this fashion that optimum performance and efficacy of the delivery device is




ISBN 0-12-074450-3

insured. The release rate must be sufficient to provide a constant amount of drug at the intended site of action. However, the release rate should not greatly exceed this value in order that side effects of the drugs can be reduced or eliminated. To achieve these goals, it is essential that the design of such devices be consistent with zero-order kinetics and that the devices can be prepared in a reproducible fashion.

Unfortunately, much of the work published in the field of controlled release drug delivery systems does not meet these goals. Several examples can be provided from work dealing with controlled release devices for contraceptive agents.

A great deal of effort has been devoted to the use of subdermal implants which release contraceptive agents such as norgestrel or norethindrone. Most of this work has been completed with devices prepared from silastic capsules containing crystalline drug (1-5). This design is inadequate in that as drug is depleted, the surface area of contact between drug and polymer decreases. As a consequence, it can be predicted that the release rate must decrease also. This is observed from both *in vivo* and *in vitro* results (6-8). An additional complication with such devices exists. It has been noted that *in vivo* release rates are much lower than *in vitro* values obtained with the same devices (6-8). Several authors have suggested that this is due to fibrotic encapsulation of the devices at the site of implantation (9,10). The only method available to overcome this particular problem is to utilize another polymer which does not induce or, at least, minimizes this encapsulation reaction.

Subdermal implants from nonbiodegradable devices suffer the disadvantage that they must be surgically removed. For this reason, much effort has been expended to develop biodegradable devices for contraceptive agents which release drug at zero-order rates (11-13). In most of these devices, the release rates are controlled by diffusion of drug through the biodegradable polymer. Since biodegradation must occur simultaneously with drug release, it is very difficult to achieve zero-order release rates. One alternative for biodegradable implants in which the release rate does not depend upon diffusional processes has been outlined elsewhere in these proceedings (14).

From the foregoing, it is apparent that new types of devices will be required in order to meet the desired objective of zero-order release subdermal implants. The purpose of this paper is to examine the release characteristics of a model progestational drug (progesterone) from reservoir and combined reservoir-monolithic devices prepared from hydrogels.

Such devices may serve as suitable alternatives to the silastic implant devices previously discussed. The hydrogels are known to be excellent candidate materials for biomedical implants due to their good compatibility with biological tissues and fluids (15). The results presented in this paper demonstrate that zero-order release rates can be achieved *in vitro* from suitably designed devices.

## II. EXPERIMENTAL

### *A. Materials*

Hydroxyethyl methacrylate (HEMA) was a highly purified sample obtained from Hydron Laboratories and contained the following levels of impurities: methacrylic acid, 0.06%; ethylene glycol dimethacrylate, 0.024%; and diethylene glycol methacrylate, 0.24%. Ethylene glycol dimethacrylate (EGDMA) was purified by base extraction and distillation. The preparation of the initiator, azobis(methylisobutyrate), and the monomers methoxyethoxyethyl methacrylate (MEEMA) and methoxyethyl methacrylate (MEMA) have been described previously (16).

Progesterone (Steraloids, Inc., Pauling, NY) and 1,2-$^3$H-progesterone (New England Nuclear, Boston, MA) were used as received. Both GC and TLC analyses of the unlabeled compounds indicate that these compounds were pure. TLC analyses of the radiolabeled compounds gave the same $R_f$ values as the unlabeled compounds and greater than 95% per cent of the radioactivity were associated with the primary spots.

### *B. Methods*

Rod shaped hydrogel devices were prepared in a polyethylene mold by polymerization at 60$^\circ$C for 24 hours. Azobis(methylisobutyrate) at a concentration of 7.84 mmoles/liter of monomer was used as the initiator. The dimensions of the dry devices were approximately 4 cm (height) by 0.5 cm (diameter). Subsequent to polymerization, a reservoir was created by drilling out the central portion of these rods with a drill and a 1/8" bit. The central reservoir was filled with 100 mg of a mixture of labeled and unlabeled progesterone in silicone oil. The tops of these devices were sealed by layering HEMA monomer over the silicone oil and polymerizing. Combined reservoir-monolithic devices were prepared in a fashion analogous to that described except the central reservoir was filled with a homogenous solution of labeled and unlabeled progesterone in a hydrogel monomer or monomer mixture and then polymerized. The tops of the devices were then sealed as described.

The thickness of the membrane barriers in both the reservoir and the combined devices were determined subsequent to the release studies by light microscopy. The thicknesses were determined at several points and averaged.

Progesterone release studies on the rod shaped devices were conducted at room temperature (23 ± 1.0°) in a one liter beaker containing 900 ml of deionized water. Most release experiments were performed in duplicate. To reduce boundary layer effects, the aqueous phase was stirred continuously at 1600 rpm by externally mounted constant speed synchronous motors. The devices were supported in the cell via metallic holders made from nickel-chrome wire. The medium was replaced with fresh deionized water every 24 hours. Under these conditions, the concentration of progesterone in the aqueous phase was far below saturation (<10 per cent of the saturation solubility of progesterone in water) so that sink conditions were maintained. The amount of progesterone released was determined by withdrawing approximately 0.5 ml of sample in triplicate. These samples were weighed in tared vials, 10 ml of scintillation fluid (Formula 950A, New England Nuclear, Boston, MA) was added, and the concentration was determined using a scintillation counter (Model 3385, Packard Instrument Co., Downers Grove, IL).

## III. RESULTS AND DISCUSSION

### A. *Reservoir Devices*

Figure 1 is a plot of the total mg of progesterone released versus time for a reservoir-type device prepared from poly(HEMA). Also shown in this figure is the progesterone release rate per day plotted versus time in days. Following a typical "burst" effect which lasts for about 20 days, the release rate per day is constant. From these results, it may be concluded that zero-order release rates can be achieved *in vitro* from reservoir devices prepared from hydrogels. The use of silicone oil for the suspending medium provides a condition wherein the concentration of drug is constant until the drug is depleted. This constant concentration of drug within the reservoir is a necessary condition

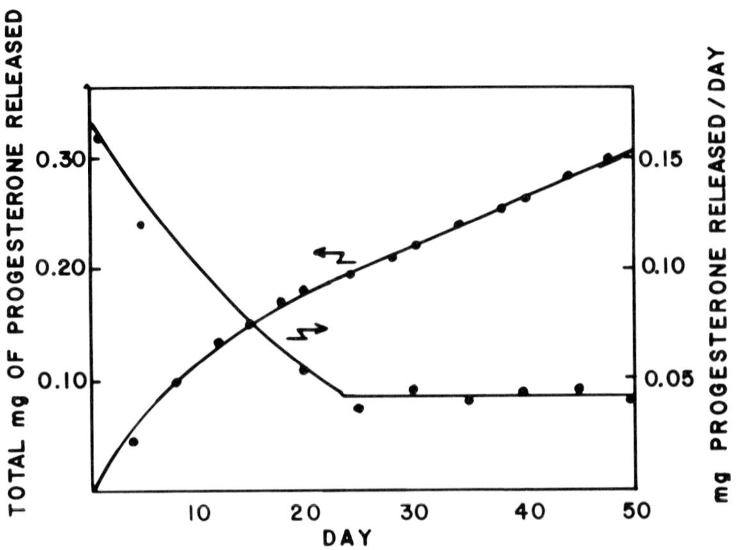


*FIGURE 1. Release of progesterone from a reservoir device prepared from poly(HEMA).*

for zero-order release rates. This conclusion is not new and has been demonstrated previously for the release of steroids from silastic capsules (17).

One advantage of the hydrogels relative to silastic is that the permeability of the membrane may be altered by changes in the composition of the monomer mixture utilized in the preparation of the membrane. Previous work from our laboratory (16) has shown that the permeability coefficients are dependent upon the water content and the crosslinker content. Permeability coefficients decrease as the water content decreases or the crosslinker content increases.

For membranes prepared with poly(HEMA) the water content can be varied by copolymerizing with varying amounts of MEMA or MEEMA. Crosslinker contents can be varied by copolymerizing with either EGDMA or tetraethylene glycol dimethacrylate (TEGDMA).

In the present study, the effects of copolymerizing HEMA with either MEMA or EGDMA on the release rate from reservoir devices was examined. In all cases, the release rates per day were constant following the initial burst effect. The measured permeabilities of progesterone in these devices are given in Table I. These permeability coefficients were determined via the following equation (18):

$$\frac{dM_t}{dt} = \frac{2\pi hPC_{aq}}{\ln(r_o/r_i)} \tag{1}$$

where $dM_t/dt$ is the release rate from the device, h is the height of the progresterone filled portion of the device (cm), P is the permeability coefficient ($cm^2$/sec), $C_{aq}$ is the saturation solubility of progesterone in water (38 μg/ml) (19), and $r_o$ and $r_i$ are the outside and inside radii of the device (cm). It is evident that copolymerizing HEMA with either MEMA or EGDMA leads to a decreased permeability of progesterone.

The permeability coefficients are in good agreement with those obtained in our previous work (Table I). From these results, it may be concluded that the presence of the silicone oil in the reservoir does not affect the permeability of progesterone in the device. Furthermore, it may be concluded that *in vitro* release rates of steroids from such reservoir

*TABLE I. Release Characteristics for Progesterone From Reservoir-Type Hydrogel Devices.*

| *Polymer* | *Length (cm)* | *Thickness (mm)* | *Permeability[a] x $10^{-7}$ $cm^2$/sec* | *Permeability[b] x $10^{-7}$ $cm^2$/sec* |
|---|---|---|---|---|
| *HEMA* | *2.85* | *1.2 ± 0.1* | *10.7* | *8.4* |
| *HEMA + 2.3 percent EGDMA* | *1.80* | *0.9 ± 0.1* | *2.3* | *2.4* |
| *MEMA-HEMA[c] (1:2)* | *3.20* | *1.0 ± 0.1* | *2.8* | *3.0* |
| *MEMA-HEMA[c] (1:4)* | *2.70* | *1.0 ± 0.1* | *3.1* | *2.7* |

[a]*Obtained from release studies.*
[b]*Obtained from membrane permeation studies (16).*
[c]*Copolymers in weight percent.*

devices can be predicted via Equation 1 given the dimensions of the device and the permeabilities of the various steroids obtained in our previous work (20).

In typical experiments performed in clinical trials with subdermal implants, silastic capsules with the dimensions of 3 cm in length and a wall thickness of about 0.09 cm are used. Such devices release about 12 μg/day of d-norgestrel and about 39 μg/day of norethindrone (8). If similar devices were prepared with poly(HEMA), *in vitro* release rates of about 18 μg/day and 32 μg/day could be expected based upon the permeabilities and aqueous solubilities of these steroids (20,21). Based on these results, it would appear that poly(HEMA) devices are sufficiently permeable to these steroids to provide adequate blood levels for contraceptive uses.

## *B. Combined Reservoir-Monolithic Devices*

Progesterone release was determined also for combined reservoir-monolithic type hydrogel devices. The release rates from such devices may be predicted from the equations developed by Roseman and Higuchi (22). These authors were concerned with the release of drugs from monolithic devices. In their theoretical development, they considered the effects of aqueous boundary diffusion layers on the release from these devices. The equations developed for this case should be suitable for release from combined reservoir-monolithic devices provided that the characteristics assigned to the aqueous boundary layers are ascribed to the outer rate controlling membrane in the combined devices. The appropriate equation is:

$$(1-F)\ln(1-F) + (1 + k')F = kt \tag{2}$$

where $k' = \dfrac{4C_sD_m}{Aa_o^2}$

and $k = \dfrac{2D_mh_aC_s}{D_aa_oC_a}$

and $F = \dfrac{a_o^2 - a^2}{a_o^2}$

F = fraction of drug released.
A = concentration of drug in matrix (mg/cm$^3$).

$D_a$ = diffusion coefficient in the boundary membrane ($cm^2$/sec).
$h_a$ = thickness of boundary membrane (cm).
$C_a$ = solubility of drug in boundary membrane ($mg/cm^3$).
$C_s$ = drug solubility in matrix ($mg/cm^3$).
$D_m$ = diffusion coefficient in matrix ($cm^2$/sec).
$a_o$ = radius of matrix cylinder (cm).
$a$ = distance from center of cylinder to receding drug boundary (cm).
$t$ = time (sec).

Equation 2 defines the fraction of drug released. The following equation describes the release rate:

$$\frac{dM_t}{dt} = \frac{2\pi h a_o C_a D_a D_m C_s h_a}{C_s h_a \left[-\frac{1}{2} D_a K a_o \ln(1-F) + D_m H_a\right]} \qquad (3)$$

where $h$ = height of cylinder (cm).
$K = C_a/C_s$

To achieve zero-order release from combined reservoir-monolithic devices, it is essential that the permeability of the outer membrane (reservoir) be less than that of the core (monolithic) portion. As mentioned, the permeabilities of hydrogels depend upon the water content and/or the nature and concentration of the crosslinker. In the present study, we have examined the release of progesterone from devices prepared with an outer membrane prepared from poly(HEMA). An inner core matrix containing progesterone was made from either poly(MEEMA) or a poly(HEMA-MEEMA) copolymer. The equilibrium water contents of these core materials are higher than that of poly(HEMA). Because of this, the core material is of higher permeability to progesterone than is the outer membrane (16).

The fraction of drug released from the combined devices is plotted versus time in Figure 2. At times greater than about 20 days, the fraction released per day is constant. For the device prepared with a poly(MEEMA) core and a poly (HEMA) coating, the release rate continuously increases at early times. This increase is probably due to an increase in the diffusivity as the water content of the device increases (19). For both devices, the release rate per day is constant at times greater than twenty days.

Also shown in Figure 2 are plots of the predicted values of the fraction released versus time, for these devices based upon Equation 2. The required parameters for this

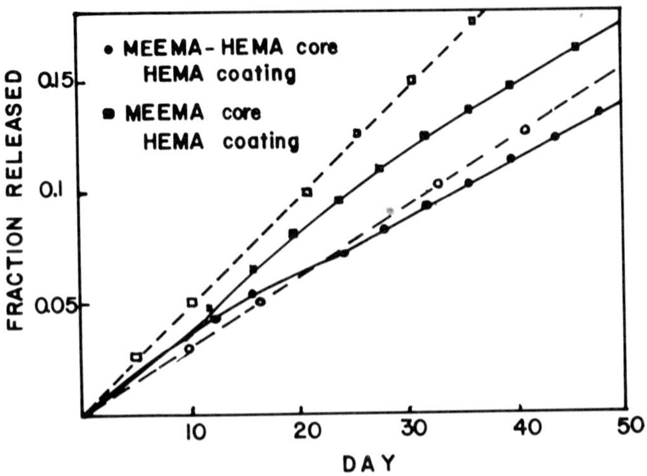


*FIGURE 2. Release of progesterone from combined reservoir-monolithic devices. Key: filled symbols are experimental, open symbols are calculated.*

calculation were taken from our previous work (16, 19). The observed and predicted release rates together with the dimensions of the devices are given in Table II.

*TABLE II. Release Characteristics for Progesterone From Combined Type Hydrogel Devices.*

| Monolithic core | Coating | Coating thickness (cm) | Core radius (cm) | Length (cm) | Release rate (µg/day) exp. | calc. |
|---|---|---|---|---|---|---|
| MEEMA-HEMA (2:1)[a] | HEMA | 0.10 ± 0.01 | 0.16 | 3.37 | 75 | 88 |
| MEEMA | HEMA | 0.10 ± 0.01 | 0.21 | 3.13 | 68 | 106 |

[a]Weight percent

Reasonable agreement between the observed and predicted release rates is obtained for the device prepared with a poly (HEMA-MEEMA) core and a poly(HEMA) coating. Rather poor agreement is obtained with poly(MEEMA). The reasons for this are uncertain, however, two likely explanations are possible. It is difficult to precisely define the dimensions of the core material and outer coatings. The calculations according to Equation 2 are quite sensitive to changes in these dimensions. Also, the permeability of poly(MEEMA) may be somewhat less than predicted due to a decreased water uptake for the poly(MEEMA) core material. The swelling of this core is probably restricted due to the outer coating of poly(HEMA).

In spite of the limitations outlined relative to our ability to predict the observed *in vitro* release rates from the combined reservoir-monolithic devices, the results obtained demonstrate that zero-order release rates can be achieved from such devices. Further refinements in the techniques of preparation will be required before such devices can be considered of practical use in controlled release applications.

## ACKNOWLEDGMENTS

The authors would like to thank Hydron Laboratories for their generous donations of HEMA and Drs. J.D. Andrade, D.E. Gregonis, and S.W. Kim for their contributions to this work. Support was provided from NIH grant HD 09791-03.

## REFERENCES

1. E. Coutinho, A. R. DaSilva, C. E. R. Matlos, N. C. Nielsen, M. Osler, J. Wiese, P. Holma, S. Diaz, H. B. Croxatto, F. A. Sanchez, A. Faundes, L. L. Williams, L. Hew, O. McDonald, S. Segal, H. Nash, D. Robertson, A. Jain, J. Stern, and I. Sivin, Contraception, 18, 319 (1978).
2. A. Faundes, I. Sivin, and J. Stern, Contraception, 18, 355 (1978).
3. S. Diaz, M. Parez, E. Quinteros, D. N. Robertson, and H. B. Croxatto, Contraception, 16, 155 (1977).
4. M. R. Bahgat and L. E. Atkinson, Contraception, 15, 355 (1978).
5. D. E. Moore, S. Roy, F. Z. Stanczyk, and D. R. Mishell, Contraception, 17, 315 (1978).

6. A. S. Lifchez and A. Scommegna, Fertil. Steril., 21, 134 (1970).
7. G. Benagiano, M. Erimini, C. C. Chang, K. Sundarum, and F. A. Kincl, Acta Endocrinologica, 63, 24 (1970).
8. H. A. Nash, D. N. Robertson, A. J. M. Young, and L. E. Atkinson, Contraception, 18, 367 (1978).
9. M. Erimini, F. Carpino, M. Russo, and G. Benagiano, Acta Endocrinologica, 73, 335 (1973).
10. D. Takkar Jegaseelan, U. N. Bhuyan, K. R. Laumas, and V. Hingorani, Contraception, 15, 39 (1977).
11. D. L. Wise, J. P. Gregory, P. M. Newberne, L. C. Bartholow, and J. B. Stanbury, in Polymeric Delivery Systems, R. J. Kostelnick, (Ed.), Gordon and Breach Science Publishers, 1978.
12. C. G. Pitt and A. Schindler in Controlled Release of Bioactive Materials, R. W. Baker (Ed.), Academic Press, New York, 1980.
13. K. R. Sidman, A. D. Schwope, W. D. Steber, S. E. Rudolph, and S. B. Poulin, in Controlled Release of Bioactive Materials, R. W. Baker (Ed.), Academic Press, New York, 1980.
14. R. V. Petersen, J. M. Anderson, C. G. Anderson, S. M. Fang, J. Feijen, D. E. Gregoins, S. Mitra, and S. W. Kim, in Controlled Release of Bioactive Materials, R. W. Baker (Ed.), Academic Press, New York, 1980.
15. B. D. Ratner and A. S. Hoffman, in Hydrogels for Medical and Related Applications, J. D. Andrade, (Ed.), ACS Symp. Series 31, American Chemical Society, Washington, D.C. 1976.
16. G. M. Zentner, J. R. Cardinal, and S. W. Kim, J. Pharm. Sci., 67, 1352 (1978).
17. R. E. Lacey and R. E. Cowsar, in Controlled Release of Biologically Active Agents, A. C. Tanquary and R. E. Lacey, (Eds.), Plenum Press, New York, 1974.
18. R. W. Baker and H. K. Lonsdale, in Controlled Release of Biologically Active Agents, A. C. Tanquary and R. E. Lacey, (Eds.), Plenum Press, New York, 1974.
19. S. Z. Song, J. R. Cardinal, S. H. Kim, and S. W. Kim, J. Pharm. Sci., (in press).
20. G. M. Zentner, J. R. Cardinal, J. Feijen, and S. Z. Song, J. Pharm. Sci., 68, 970 (1979).
21. K. Sundaram and F. A. Kincl, Steroids, 12, 517 (1968).
22. T. J. Roseman and W. I. Higuchi, J. Pharm. Sci., 59, 353 (1970).

# DETERMINATION OF DIFFUSION COEFFICIENTS BY SORPTION FROM A CONSTANT, FINITE VOLUME

*P. I. Lee*

Central Research
CIBA-GEIGY Corporation
Ardsley, New York

*Based on a refined integral method, the approximate analytical solutions for sorption from a constant, finite volume in planar, cylindrical and spherical geometries are obtained. These solutions provide a simple accurate means of determining diffusion coefficients from sorption experiments without resorting to either laborious master plots or numerical curve fitting. The accuracy and range of applicability of these solutions are established by comparison with the exact solutions. The results are also applied to absorption and desorption experiments for the determination of diffusion coefficients of dexamethasone and phenylbutazone in a poly-(hydroxyethyl methacrylate) hydrogel.*

## I. INTRODUCTION

For the design of a controlled release device, be it of the membrane-reservoir type or of the matrix-monolithic type, an accurate measure of the diffusion coefficient of active ingredient in the system is essential to predict the release rate. The permeation time-lag method has been widely used for this purpose. However, it requires elimination of boundary layer resistances and materials strong enough to be made into membranes. In addition, for penetrants with small diffusion coefficients, the permeation time-lag experiment becomes impractically long (3). Another popular method for the determination of diffusion coefficient is the direct release




ISBN 0-12-074450-3

experiment (16). It generally involves the assumption of a perfect sink or a constant external concentration. Rigorously, this can at best be an approximation since there has to be some concentration change in the external solution to make the release observable. The validity of this perfect sink assumption becomes especially questionable, however, whenever the solution to sample volume ratio is small or the partition coefficient of the active ingredient in the sample is large.

A more efficient method which appears to have been underutilized is the determination of diffusion coefficient by following the solute concentration change in a constant finite volume as the solute is absorbed or desorbed by the solid. This method has the distinct advantage that both the diffusion and distribution coefficients can be obtained accurately from just one experiment with a simple apparatus and a wide range of accessible experimental conditions (1-4). Furthermore, the sample geometry and mechanical strength are not restricted in such an experiment.

Despite the simplicity and diversity of this method, it has been used only infrequently (1-4). A large part of the obstacles to its application may be attributed to the mathematical complexity of final working equations especially for non-planar geometries. Similar complexities can be found in other diffusion systems (5). The key problem lies in the fact that in all the solutions to diffusion problems, the experimentally measurable quantity, namely the fraction absorbed or desorbed, is always expressed as an infinite series of the unknown variable, namely the diffusion coefficient D or dimensionless time $\tau$, defined as $\tau = Dt/a^2$. This obviously is quite cumbersome for routine usage.

It is therefore the purpose of this study to develop approximate analytical solutions for this sorption system based on a refined integral method (7-9). The solutions express the unknown variable $\tau$ explicitly in terms of measurable external concentration changes in such a way that they can readily be applied to general problems of sorption from a constant, finite volume in planar, cylindrical and spherical geometries. Another purpose of the study is to demonstrate the applicability of these solutions to the measurement of drug diffusion in hydrogels which are generally weak mechanically in the swollen state.

## II. ANALYSIS

For a solid sorbant with uniform initial solute concentration $C_1$ in equilibrium with bulk concentration $C_{b1}$, the general diffusion equation describing the concentration distribution C in the solid after exposing it to a constant, finite volume of well stirred solution of concentration $C_{b2}$ at time t = 0 is

$$\frac{\partial C}{\partial t} = x^{-\alpha}\frac{\partial}{\partial x}(x^{\alpha} D\frac{\partial C}{\partial x}) \quad (1)$$

where x = 0 at the center and x = a at the surface with $\alpha$ = 0, 1 and 2 for planar sheet, cylinder and sphere, respectively. Assuming equilibrium between surface and fluid at t, the initial and boundary conditions are

$$C(x,0) = C_1 \quad (2)$$

$$\left.\frac{\partial C}{\partial x}\right|_{x=0} = 0 \quad (3)$$

$$C(a,t) = kC_b(t) \quad (4)$$

$$\text{and } -(\frac{\lambda a}{a+1})\frac{\partial C}{\partial t} = D\frac{\partial C}{\partial x} \text{ at } x = a \quad (5)$$

where a is the half thickness for planar sheet and the radius for cylinder and sphere, $C_b$ is the bulk concentration with additional subscript 1,2 and $\infty$ for that at $t < 0$, $t = 0$ and $t = \infty$, respectively. D is the solute diffusion coefficient in the solid and $\lambda$ is the effective volume ratio defined as

$$\lambda = \frac{C_{b\infty}-C_{b1}}{C_{b2}-C_{b\infty}} = \frac{V}{2KAa} \text{ for planar sheet} \quad (6)$$

$$\frac{V}{k\Pi a^2 L} \text{ for cylinder}$$

$$\text{and } \frac{3V}{4k\Pi a^3} \text{ for sphere}$$

with V as the total liquid volume, k the distribution coefficient, A the area of each side of the sheet and L the length of the cylinder.

The exact solutions for this diffusion system have been obtained and applied to sorption experiments by Carman and Haul (1). These have been summarized by Crank (5). Recently, Spencer and Barrie (6) extended the solution to include a time-lag for planar geometry. Carman and Haul formulated, for each geometry, two solutions for the time dependence of concentration variable, one converging rapidly for large time t and the other converging rapidly for small t. All of these solutions require either a laborious construction of master plot of calculated values or a numerical data-fitting routine when they are applied to the analysis of experimental data. A quick method was suggested by Carman and Haul for planar geometry which involves making a semi-log plot of the leading term in the long-term solution with respect to time and calculate the diffusion coefficient from the slope of the linear portion. Unfortunately, this quick method fails in cylindrical and spherical geometries, where the convergence of the original solutions is slow.

In light of these difficulties, it is apparent that there exists a need for simple and accurate alternative solutions for the analysis of sorption from a constant, finite volume, especially for the cylinder and sphere. In this study, based on a refined version of Goodman's (7,8) heat balance integral method due to Volkov and Li-Orlov (9), approximate analytical solutions are obtained for sorption from a constant, finite volume with constant diffusion coefficient. Briefly, these solutions are derived by integrating the diffusion equation twice with respect to the space coordinate followed by an appropriate choice of trial concentration profile for the integral equation. More detailed derivations will be given elsewhere (10). Only the resulting equations are presented here.

*Planar Sheet*

$$\tau = \frac{3\lambda^2}{4}\left[\ln\left(\frac{C_b-C_{bl}}{C_{b2}-C_{bl}}\right) + \frac{1}{2}\left(\frac{C_{b2}-C_{bl}}{C_b-C_{bl}}\right)^2 - \frac{1}{2}\right] = \frac{Dt}{a^2} \tag{7}$$

*Cylinder*

$$\tau = \frac{3}{40}\lambda^3[(\frac{C_{b2}-C_{b1}}{C_b-C_{b1}})-1]^3 + \frac{3(3\lambda+5)}{160}\lambda^2[(\frac{C_{b2}-C_{b1}}{C_b-C_{b1}})-1]^2 + \frac{3\lambda^2(5-3\lambda)}{80}[(\frac{C_{b2}-C_{b1}}{C_b-C_{b1}})-1] - \frac{3\lambda^2(5-3\lambda)}{80}\ln(\frac{C_{b2}-C_{b1}}{C_b-C_{b1}}) \quad (8)$$

*Sphere*

$$\tau = -\frac{\delta}{3} - (\frac{\lambda+2}{3})\ln[\frac{4\lambda+4-(\delta-2)^2}{4\lambda}] + \frac{2}{3}(1+\lambda)^{0.5}\ln\left[\frac{[2(1+\lambda)^{0.5}+(\delta-2)][(1+\lambda)^{0.5}+1]}{[2(1+\lambda)^{0.5}-(\delta-2)][(1+\lambda)^{0.5}-1]}\right] \quad (9)$$

$$\text{with } \delta(\tau) = 2 - 2[(\lambda+1)-\lambda(\frac{C_{b2}-C_{b1}}{C_b-C_{b1}})]^{0.5} \quad (10)$$

In all the above geometries the concentration term is related to the fraction absorbed, $\frac{W}{W\infty}$, by

$$\frac{C_b-C_{b1}}{C_{b2}-C_{b1}} = 1 - \frac{1}{1+\lambda}(\frac{W}{W\infty}) \quad (11)$$

Equations (7) - (10) are the final working equations for sorption from a constant, finite volume with the dimensionless time defined by $\tau = Dt/a^2$.

## III. ACCURACY

Previously, after the effective volume ratio $\lambda$ is established from Equation (6) for a given experiment, the diffusion coefficient D of the solute in the solid can only be evaluated from experimental data by using either a curve-fitting routine or a master plot. The former requires a digital computer and a curve-fitting program. The latter requires a laborious construction of master plot of fraction absorbed vs. $\tau$ from the exact solutions for planar sheet, cylinder and sphere followed by reading off values of $\tau$ for the corresponding experimental fraction absorbed values. A plot of $\tau$ vs. t then gives a straight line, from the slope of which the diffusion coefficient is evaluated.

The quick method of Carman and Haul avoids the labor of construction of master plot. However, it is only applicable to planar geometry and to $\lambda$ values larger than 0.4, there are no corresponding labor-saving quick methods for cylindrical and spherical geometries.

In contrast to these, the solutions derived in the present study express $\tau$ explicitly in terms of the measured bulk concentration changes. A $\tau$ vs. t plot can easily be constructed for the evaluation of diffusion coefficient. The accuracy of

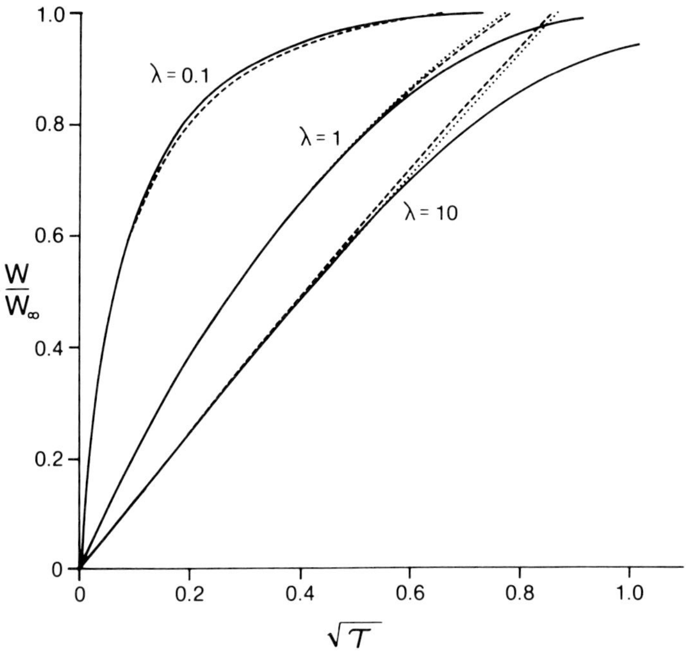


*FIGURE 1. Comparison of approximate analytical solution with existing solutions for planar sheet.*
*——— exact solution; ..... alternative solution of Carman and Haul; ----- solution obtained in this study, Equation 7.*

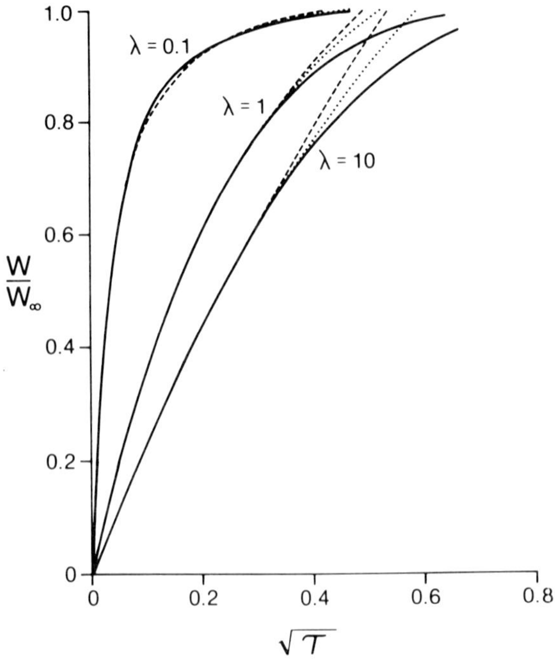


*FIGURE 2. Comparison of approximate analytical solution with existing solutions for cylinder. ——— exact solution; ..... alternative solution of Carman and Haul; ----- solution obtained in this study, Equation 8.*

these solutions can be realized from Figures 1-3 where we have compared the solutions obtained in this study, the alternative solutions of Carman and Haul as well as the exact solutions by plotting the fraction absorbed vs. $\sqrt{\tau}$ for various values of $\lambda$. The results clearly indicate that the solutions obtained here closely approximate the alternative solutions given by Carman and Haul which constitute the major part of the exact solutions. For small values of $\lambda$ our solutions almost completely coincide with the exact solutions (within 2%). As $\lambda$ gets larger, deviations start to appear. However, only values at the largest values of $\tau$ are affected, the major part of the absorption can still be approximated by our solutions with good accuracy. For $\lambda$ values greater than 10, which correspond to experimental conditions of near constant bulk concentration, the solutions for constant external concentration could be used.

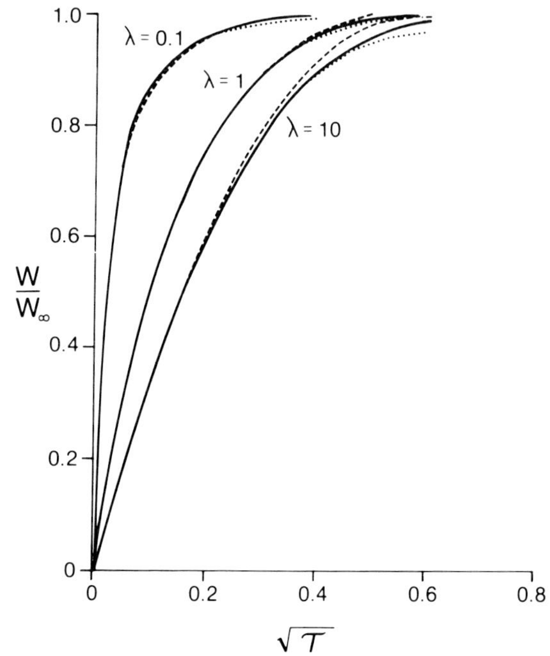


*FIGURE 3. Comparison of approximate analytical solution with existing solutions for sphere. ——— exact solution; ..... alternative solution of Carman and Haul; ----- solution obtained in this study, Equations 9 and 10.*

The applications of these solutions in the determination of diffusion coefficients from experimental data are further illustrated in Figures 4-7. Where the solutions obtained in this study have been used to analyze the data of Carman and Haul on the absorption of butane in silica gel and granules and the data of Rice and Foo (11) on the absorption of NaCl in Bio-Rad AG11A8 resin particles. The diffusion coefficients, calculated from the linear portions of the $\tau$ vs. t plots in Figure 4-7, are tabulated together with the reported values in Tables I and II. It can be seen that our solutions yield the same diffusion coefficients as that obtained by Carman and Haul from the master plot. As to the results of Rice and Foo, the validity of their calculated diffusion coefficients is questionable because they have applied the quick method to spherical geometry by using only the leading term of the solution. The quick method generally fails in cylindrical and spherical geometries where the convergence

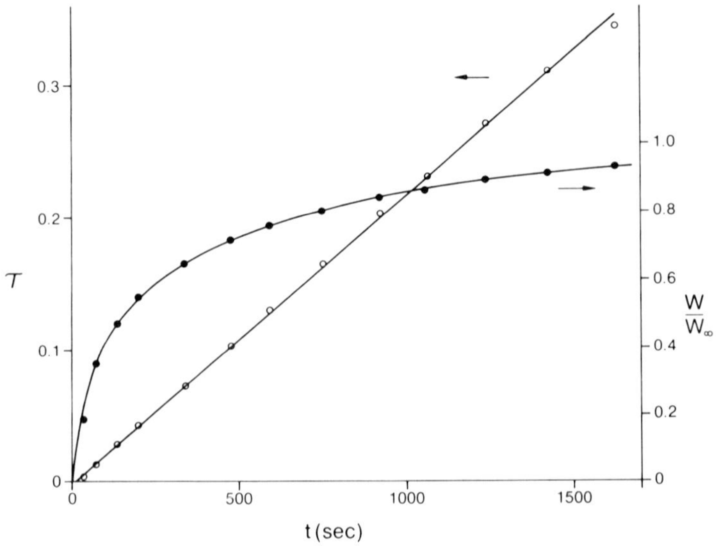


*FIGURE 4. Analysis of Carman and Haul's data on the absorption of butane at -10°C in Linde silica plug by the approximate analytical solution for planar sheet; a = 0.75 cm, λ = 0.42 giving D = 1.2 x $10^{-4}$ $cm^2$/sec.*

*TABLE I. Comparison of Diffusion Coefficients Calculated in this Work with Those Reported by Carman and Haul (1).*

| | Diffusion coefficient, D($cm^2$/sec) | |
|---|---|---|
| | Absorption of butane at -10°C in Linde silica plug | Absorption of butane at 0°C in Silica gel granules |
| Carman and Haul | 1.23 x $10^{-4}$ | 1.91 x $10^{-5}$ |
| This work | 1.2 x $10^{-4}$ | 1.9 x $10^{-5}$ |

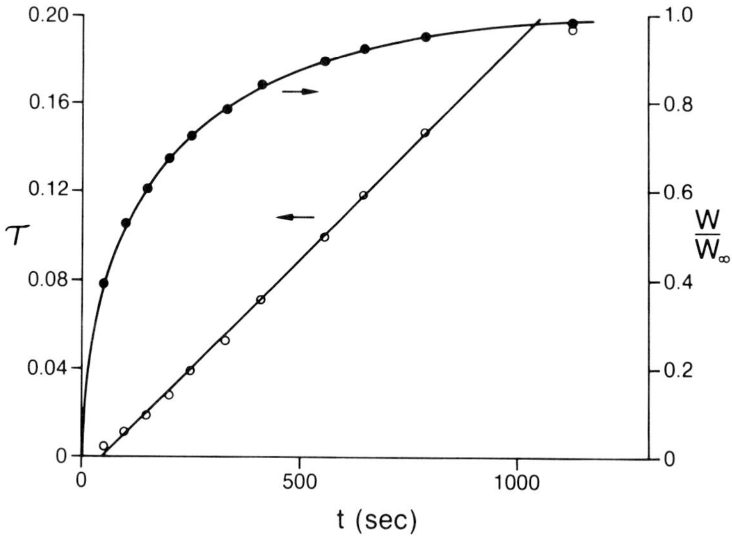


*FIGURE 5. Analysis of Carman and Haul's data on the absorption of butane at $0^{o}C$ in silica gel granules by the approximate analytical solution for sphere; a = 0.3, $\lambda = 1.24$ giving $D = 1.9 \times 10^{-5}$ $cm^2/sec$.*

*TABLE II. Comparison of Diffusion Coefficients of NaCl in Bio-Rad AF11A8 Resin at $50^{o}C$ Calculated in this Work with Those Reported by Rice and Foo (11)*

| | Diffusion coefficient, $D \times 10^7$ ($cm^2/sec$) | |
|---|---|---|
| | 40-80 mesh | 60-80 mesh |
| Rice and Foo | 2.8 | 3.7 |
| This work | 3.1 | 3.2 |

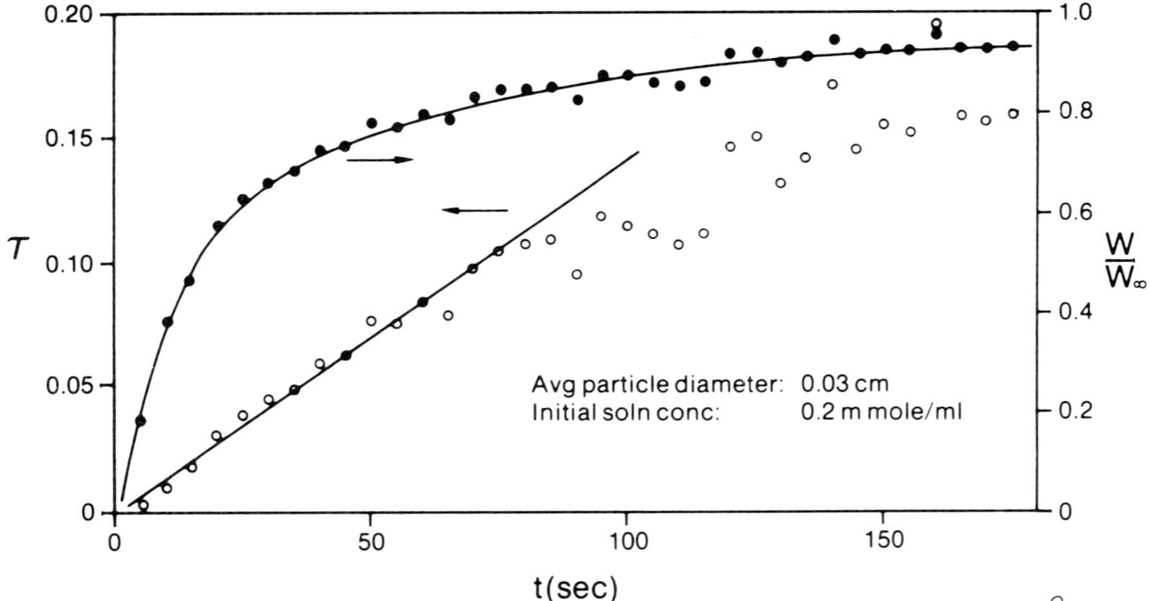


*FIGURE 6. Analysis of Rice and Foo's data on the absorption of NaCl at $50^{\circ}C$ in Bio-Rad AG11A8 resin particles of 40-80 mesh by the approximate analytical solution for sphere; $\lambda$ = 4.55 giving $D = 3.1 \times 10^{-7}$ $cm^2/sec$.*

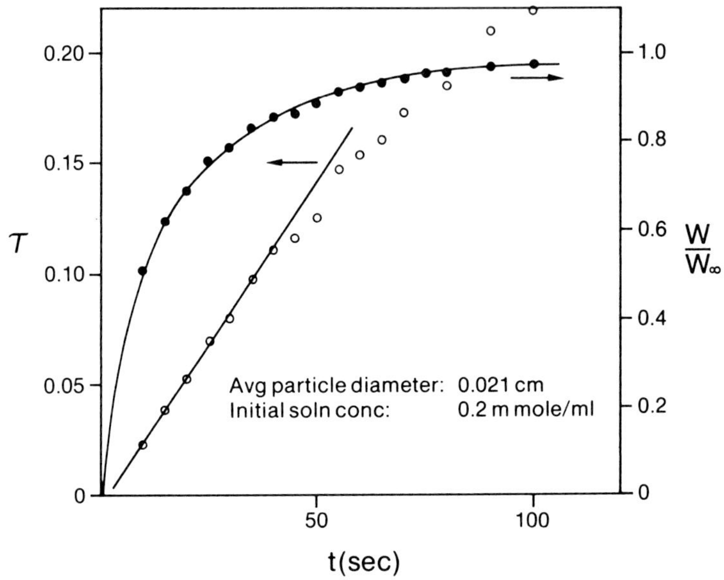


*FIGURE 7. Analysis of Rice and Foo's data on the absorption of NaCl at 50°C in Bio-Rad AG11A8 resin particles of 60-80 mesh by the approximate analytical solution for sphere; $\lambda$ = 4.55 giving D = 3.2 x $10^{-7}$ $cm^2$/sec.*

of the original solution is slow (1). In addition, the quick method utilizes only the last 20-30% of the absorption where most of the data fluctuations occur (Figures 6 and 7). The discrepancy in Rice and Foo's calculated diffusion coefficients for different size ranges of the same resin (Table II) clearly reflects the inadequacy of their data analysis. On the other hand, with the use of the solutions derived here, data up to 80-90% of the absorption can be used (Figures 4-7) in contrast to the total reliance on a limited portion of the absorption in the quick method. The closeness of the diffusion coefficients calculated from Rice and Foo's data for different mesh size ranges of the same resin (Table II) further illustrates the utility of our solutions.

## IV. APPLICATION

Good blood compatability has made hydrogels promising candidates for use as biomedical implantable devices (12-14). Their uses in the controlled release of drugs have been investigated extensively (13-15). In this section the use of poly(hydroxyethyl methacrylate) poly(HEMA) hydrogel in the delivery of dexamethasone, an anti-inflammatory steroid, and phenylbutazone, an anti-rheumatic agent, is investigated by the absorption technique described in the previous sections.

The hydrogel sheets were prepared by polymerizing, between sealed Mylar plates, distilled hydroxyethyl methacrylate monomer with 0.6 mole % tetraethylene glycol dimethacrylate as the crosslinker. The polymerization was carried out with t-butyl peroxy-2-ethylhaxanoate as the initiator at $80^{\circ}C$ for three hours and $100^{\circ}C$ for one hour. The poly(HEMA) sheets were washed in distilled water for four weeks (final equilibrium water content 39%) and then equilibrated in phosphate buffer (pH 7) prior to use.

For the absorption experiments, discs of diameters ranging from 1.118 to 1.143 cm and thicknesses ranging from 0.12 to 0.14 cm were cut out from the swollen poly(HEMA) sheets. Each absorption experiment was carried out in a rectangular silica UV cell with 1 cm optical path length. Typically, the cell was filled with 3 ml of drug solution of concentration $5.1 \times 10^{-5}$M prepared in phosphate buffer. It was then placed into a Beckman ACTA C-III UV-visible spectrophotometer and stirred with a built-in magnetic stirrer. The temperature control in the sample compartment was achieved by the use of a Haake FE constant temperature circulating bath.

After the establishment of a baseline, the poly(HEMA) disc hanging on a thin stainless steel wire was introduced to the upper half of the UV cell. The cell was then capped to prevent solvent evaporation. The change of bulk concentration as the drug was absorbed by the poly(HEMA) disc was recorded continuously at 243 nm for dexamethasone and 265 nm for phenylbutazone.

With the use of Equations (9) and (10), $\tau$ vs. t plots were constructed and the diffusion coefficients were calculated from the slopes. Both the diffusion and partition coefficients were determined from the same set of data. Typical data analyses are shown in Figures 8 and 9 for the absorption of dexamethasone in poly(HEMA) at $25^{\circ}C$ and $37^{\circ}C$. It can be seen that the linear $\tau$ vs. t plots utilize data up to 80% of the absorption. The results of the study including

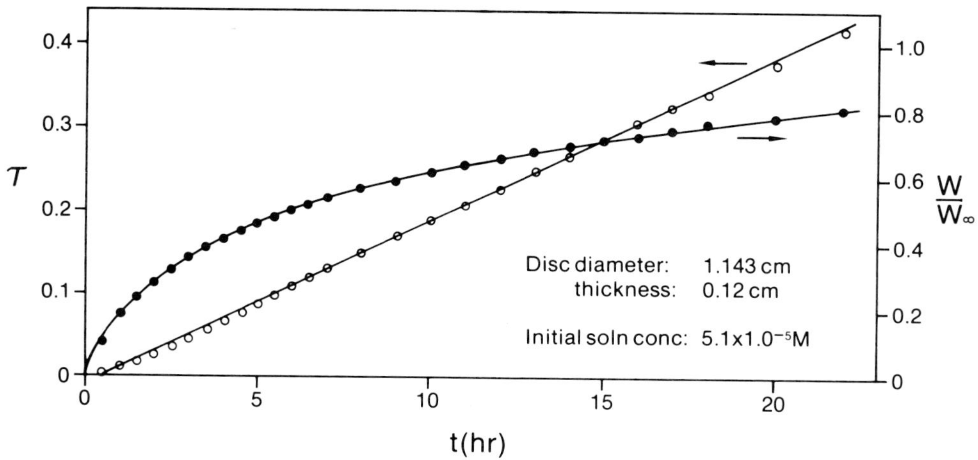


*FIGURE 8. Absorption of dexamethasone at pH 7 and 25°C in a poly(HEMA) disc from a constant, finite volume;* $k = 51.4$, $\lambda = 0.621$ *giving* $D = 0.67 \times 10^{-8}$ *cm²/sec.*

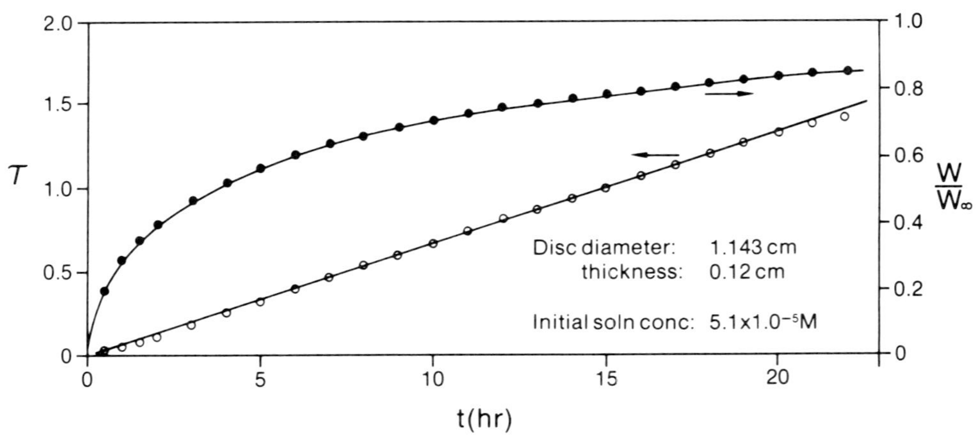


*FIGURE 9. Absorption of dexamethasone at pH 7 and 37$^{O}$C in poly(HEMA) disc from a constant, finite volume; k = 51.1, λ = 0.465 giving D = 1.18 x 10$^{-8}$ cm$^2$/sec.*

the predicted permeabilities for dexamethasone and phenylbutazone through poly(HEMA) membranes are summarized in Table III.

In order to test the applicability of absorption results to controlled release, a desorption experiment on dexamethasone was carried out at 25°C in a constant, finite volume on one of the discs equilibrated in a previous absorption run. The $\tau$ vs. t plots for the absorption and subsequent desorption experiments are shown together in Figure 10. The closeness of the absorption and desorption data not only suggests the constancy of diffusion coefficient of dexamethasone in poly(HEMA) within the concentration range of this study but also indicates the applicability of Equations (7)-(10) to both the absorption and desorption experiments.

## V. DISCUSSION

The solutions obtained in this study for sorption from a constant, finite volume enable one to calculate $\tau$ directly from simple algebraic equations [Equations (7)-(10)] based on

*TABLE III. Diffusion Coefficients, Partition Coefficients and Predicted Permeabilities of Dexamethasone and Phenylbutazone in poly(HEMA) Hydrogel at pH 7 Obtained by Absorption from a Constant, Finite Volume*[a]

| *Compound* | *Temp(°C)* | *$D \times 10^8$ ($cm^2/sec$)* | *k* | *$P \times 10^7$*[b] *($cm^2/sec$)* |
|---|---|---|---|---|
| *Dexamethasone* | *25* | *0.67* | *51.3* | *3.44* |
| | *37* | *1.18* | *51.1* | *6.03* |
| *Phenylbutazone* | *25* | *1.07* | *15.4* | *1.65* |
| | *37* | *1.88* | *17.1* | *3.21* |

[a]*Initial solution concentration: $5.1 \times 10^{-5}M$ in phosphate buffer.*
[b]*Permeability defined as $P = kD$.*

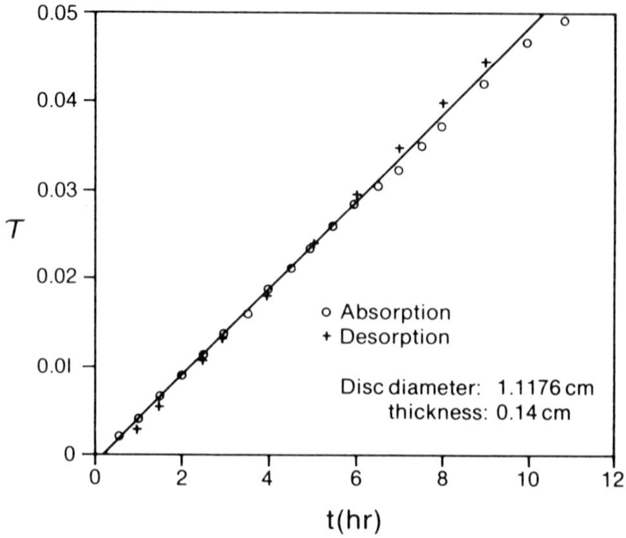


*FIGURE 10. τ vs. t plot for the absorption and subsequent desorption of dexamethasone at pH 7 and 25°C from poly(HEMA) disc in a constant, finite volume with initial concentration 6.0 x $10^{-5}$M for the absorption and 0 for the desorption; k = 51.4, λ = 0.397 giving D = 0.67 x $10^{-8}$ $cm^2$/sec from the slope of the solid line.*

absorption as well as desorption data. The diffusion coefficient can then be evaluated easily from the slope of a τ vs. t plot. The experiments on the diffusion of dexamethasone and phenylbutazone in poly(HEMA) further illustrate the utility of these results. The simplicity of the apparatus and wide range of accessible experimental conditions make this absorption technique extremely suitable for the determination of transport characteristics of drugs in polymers and other solid matrices especially for drugs with small diffusion coefficients and polymers with weak mechanical strength such as hydrogels. The refined integral method utilized in our derivation is not limited to systems with constant diffusion coefficients. In fact, general solutions including the concentration dependency of diffusion coefficients can be derived. This will be examined in a forthcoming paper (10).

## VI. NOTATION

$a$ = half thickness for planar sheet and radius for cylinder and sphere.
$A$ = area of each side of planar sheet.
$C$ = solute concentration in solid at any time.
$C_1$ = initial solute concentration in solid.
$C_b$ = bulk concentration at any time.
$C_{b1}$ = bulk concentration in equilibrium with $C_1$.
$C_{b2}$ = bulk concentration at the beginning of sorption experiment.
$C_{b\infty}$ = final bulk concentration in equilibrium with $C_\infty$.
$D$ = diffusion coefficient.
$k$ = distribution coefficient defined by Equation (4).
$L$ = length of cylinder.
$P$ = permeability defined as $P = kD$.
$t$ = sorption time.
$V$ = total external volume.
$W$ = amount absorbed at any $t$.
$W_\infty$ = total amount absorbed.
$x$ = spatial coordinates.
$\alpha$ = indices with values 0, 1 and 2 for planar sheet, cylinder and sphere, respectively.
$\tau$ = dimensionless time defined by Equation (7).

## REFERENCES


1. P. C. Carman and A. W. Haul, Proc. Roy. Soc., 222A, 109 (1954).
2. J. L. Lundberg, M. B. Wilk, and M. J. Huyett, J. Polym. Sci., 57, 275 (1962).
3. A. S. Michaels, W. R. Vieth, and H. J. Bixler, Polym. Lett., 1, 19 (1963).
4. W. Brown and K. Chitumbo, J.S.C. Faraday I, 71, 1 (1975).
5. J. Crank, Mathematics of Diffusion, 2nd edition, Clarendon Press, Oxford 1975.
6. H. G. Spencer and J. A. Barrie, J. Appl. Polym. Sci., 20, 2557 (1976).
7. T. R. Goodman, Trans. ASME, 80, 335 (1958).
8. T. R. Goodman, J. Heat Transfer, 83, 83 (1961).
9. V. N. Volkov and V. K. Li-Orlov, Heat Transfer Sov. Res., 2, 41 (1970).
10. P. I. Lee, to be submitted for publication.
11. R. G. Rice and S. C. Foo, Ind. Eng. Chem. Fundam., 18(1), 63, 68 (1979).
12. J. D. Andrade, Med. Instrum., 1, 110 (1973).

13. R. A. Abrahams and S. H. Ronel, J. Biomed. Mater. Res., 9, 355 (1975).
14. J. D. Andrade (Ed.), Hydrogens for Medical and Related Applications, ACS Symp. Series 31, American Chemical Society, Washington, D.C. 1976.
15. G. M. Zentner, J. R. Cardinal, and S. W. Kim, J. Pharm. Sci., 67(10), 1352 (1978).
16. T. J. Roseman and W. I. Higuchi, J. Pharm. Sci., 59, 353 (1970).

# A NEW FAMILY OF MONOLITHIC HYDROGELS FOR CONTROLLED RELEASE APPLICATIONS

*W. R. Good*

CIBA-GEIGY Corporation
Pharmaceuticals Division
Suffern, New York

*K. F. Mueller*

CIBA-GEIGY Corporation
Central Research
Ardsley, New York

## I. INTRODUCTION

In the field of controlled release of low molecular weight compounds from polymeric devices, we have seen a proliferation of control mechanisms from chemical degradation (1-3) to simple diffusion (4,5) to more exotic combination mechanisms (6). In this paper we are concerned with diffusion mechamisms which rely heavily on structure and thermodynamic interactions within a crosslinked polymer network. The polymer networks studied fall into the general class referred to as hydrogels. Although hydrogels have long been studied for controlled release applications (7-9), the networks described in this paper are structurally novel and versatile.

The networks described in this paper are two-component systems best described as macromer-crosslinked graft copolymers (MCGP). However, this nomenclature is not totally descriptive of this type structure. Figure 1 shows an idealized schematic representation of an MCGP structure. Structures of this type, in which a polymer chain of one type terminates at a chain of another type, were first studied by Bamford




ISBN 0-12-074450-3

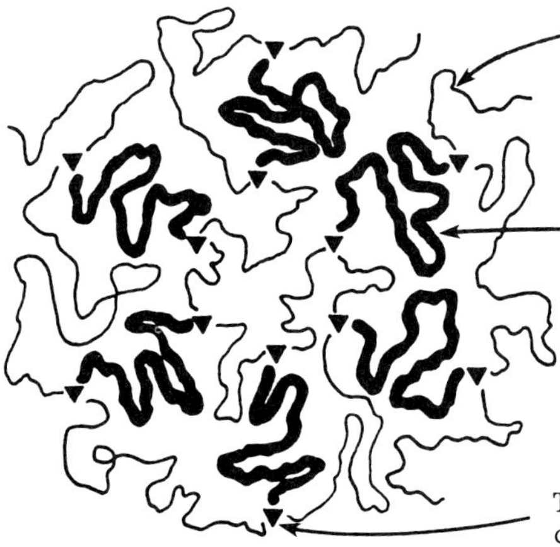


*FIGURE 1. Schematic representation of an MCGP structure.*

*et al.* (10,11). If the two types of polymer chains are thermodynamically compatible, a single-phase system results. Two incompatible types of chains result in phase separation and domains will be formed consisting of a dispersion of one type embedded in a matrix of the other type. Recently, the mixing and elastic properties of the compatible swollen two-component networks have been put on a firm theoretical foundation (12). However, incompatible systems have not yet been treated. The networks described in this paper are generally incompatible in the unswollen state. The "macromer" component is a hydrophobic oligomer of butane diol while the "main chain" is hydrophilic poly-2-hydroxyethyl methacrylate. Variation of the molecular weight of the macromer and its concentration in the finished network allows for a wide range of water swellability, drug loading and release rates in an aqueous environment.

### *A. Network Synthesis*

Preparation of MCGP was accomplished by simultaneous copolymerization and crosslinking of 2-hydroxyethyl-methacrylate (HEMA) in the presence of the macromer, polytetramethylene oxide, which contain reactive vinyl groups at its termini. Polytetramethylene oxide is extremely soluble in HEMA, the resultant solution being viscous but easily polymerized in molds or as a suspension using free radical initiation. This method is preferable to the alternative graft-crosslinking of existing chains technique because of its greater flexibility.

Polytetramethylene oxide (Polymeg®, Quaker Oats Co.) is obtained as a (α,ω) diol having number average molecular weight of 1,000 or 2,000. The diol is dissolved in an appropriate amount of a diisocyanate such that linear chain extension occurs and all chains are isocyanate terminated. The relative amounts of diisocyanate and diol determine the number average molecular weight of the product. For the present study we have prepared macromer having molecular weights given in Table I. Number average molecular weights were determined by reacting an aliquot with excess dibutylamine and back titrating with HCl.

The macromers were dissolved in HEMA where the isocyanate termini react with the monomer to produce end groups on the macromer capable of polymerization. Moreover, because the end groups have the same methacrylate structure as the monomer, nearly ideal copolymerization is expected.

Normally, the macromer is not isolated, but rather dissolved in the monomer in the desired concentration. The mixture is then mixed with a free radical initiator, 2,2$^1$-azobis-isobutyronitrile (AIBN) and either polymerized in molds to produce flat sheets or it is dispersed in a saturated salt solution to produce beads or pearls ranging in diameter from about 0.1 mm to 1.5 mm. Both types of polymers were prepared for this study. After polymerization is complete, the material is washed several days in running water and finally allowed to swell in a large volume of ethanol.

*TABLE I. Preparation of macromer from polytetramethylene Oxide and Diisocyanate*

| Mole ratio of diisocyanate to diol | $M_n$ (diol) | $M_n$ (macromer) |
|---|---|---|
| 2/1 | 1000 | 1460 |
| 2/1 | 1980 | 2450 |
| 5/3 | 1980 | 3470 |
| 3/2 | 1980 | 4770 |
| 4/3 | 1980 | 5960 |

*B. Network Characterization*

For the purpose of our study these polymer networks can be thought of as an infinitely long branched molecule. The segment between branches have a random distribution of conformations while the junction points serve only to maintain the integrity of the network. We will assume these networks obey the swelling theory first derived by Flory and Rehner (17). In reality, several factors cause non-randomness which effects properties such as tacticity, crystallinity, phase separation, and mechanical behavior. Moreover, even for networks which are thought to be random in their conformational distributions, the issues of topology, entanglements (13,14) and the effects of local fluctuations of the junction points (15,16) are all presently controversial topics in the field of network theory.

The absorption of a solute into the networks and controlled release of the solute into an aqueous environment will both be intimately related to the swelling properties of the network involved. The swelling properties of any polymer network are controlled by the combination of free energies of mixing between solvent and polymer chains and by the elastic response of the network to volume increase due to solvent absorption. At swelling equilibrium the elastic response of the network exactly offsets the difference in chemical potential of the solvent between the swollen network and external solvent phase. The equilibrium is normally expressed by,

$$\ln \phi_1 + \phi_2 + \chi_{12}\phi_2^2 + \phi_2\overline{V}_1(A\nu\phi_2^{-2/3} - B) = 0 \qquad (1)$$

where $\phi_1$ and $\phi_2$ are volume fractions of solvent and network, respectively, $\overline{V}_1$ is the solvent molar volume, A and B are related to network topology (18), $\nu$ is the number of elastic chains (related to crosslink density) and $\chi_{12}$ is the Flory-Huggins interaction parameter between solvent and network. Equation (1) establishes that the elastic and mixing effects can be separated, the mixing effect being characterized by the $\chi_{12}$ parameter and the elastic effect being related to $\nu$, the number of chains in the network. Further, it has been shown that the well-known solubility parameter (19), $\delta$, of the network is related to $\chi_{12}$ by,

$$\chi_{12} = \overline{V}_1(\delta_1 - \delta_2)/RT \qquad (2)$$

where $\delta_1$ and $\delta_2$ are solubility parameters of solvent and network, respectively. Since the coefficient of $\chi_{12}$ is quadratic in the volume fraction of network at equilibrium swelling, a plot of volume fraction of network (or, alternatively, degree of swelling) will show an extremum when $\delta_1 = \delta_2$. Moreover, since the position of the extremum is related only to solvent/chain interaction, it should be independent of crosslink density and, therefore, a measure of network composition. The magnitude of the degree of swelling is then related to crosslink density.

The solubility parameter $\delta$ is found by measuring the degree of swelling q of the polymer for a series of solvents; q is defined as the inverse of the weight fraction of network at equilibrium swelling. If we swell specimens in a homologous series of solvents having a range of solubility parameters and plot q versus solubility parameter of the solvent, the solubility parameter of the polymer network is found at the point of maximum swelling. In this work a series of ethanol/water mixtures were used as solvents. Typical swelling curves are shown in Figure 2. The five curves shown here are results of swelling experiments performed on networks having equivalent compositions of 65 parts HEMA and 35 parts macromer, the macromers having molecular weights indicated in Table I. Maximum degree of swelling occurs at the same solvent compositions for all of these networks. These results demonstrate the independence of the solubility parameter from crosslink density. The maximum degree of swelling, on the other hand, increases with increasing molecular weight of the macromer. Increasing molecular weight of the macromer at constant composition has the effect of reducing crosslink density, hence, degree of swelling is increased.

The networks studied here are two-component systems as are the solvent systems. To be perfectly rigorous would require the solution of the equations of state for a four-component system, a tedious task at best. Instead, we have taken a more empirical approach and expressed the degree of swelling as a power series in $\delta_1$,

$$q = \sum_{i=0}^{\infty} a_i \delta_1^i \qquad (3)$$

and truncated the series when the experimental data fits the calculated data within the limits of experimental error. We have found that a cubic relation is almost always sufficient to fit the data within experimental error. Hence, the apparent solubility parameter can be found by differentiating equation (3) and setting the differential to zero. We use the term "apparent solubility parameter" because the actual solvent composition within the network is not known. If

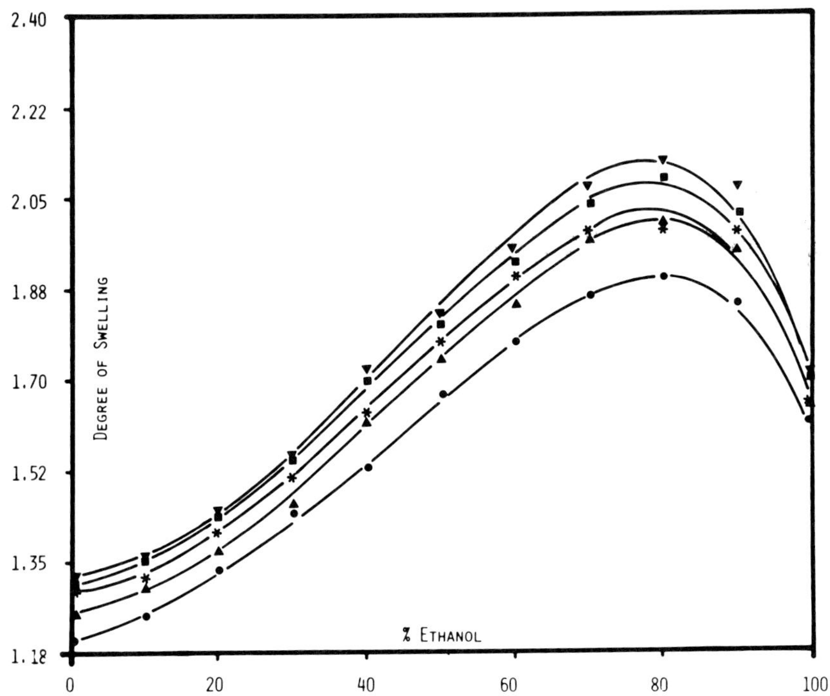


*FIGURE 2. Degree of swelling in ethanol/water mixtures for two-component networks having composition of 65 parts HEMA, 35 parts macromer. $M_n$ of macromers are 1460 (●), 2450 (▲), 3470 (✱), 4770 (■), and 5960 (▼).*

preferential absorption of one solvent component occurs, the internal solvent composition will not be the same as the solution phase. This possibility in no way compromises our conclusions as long as we do not change the components of the solvent. Therefore, the apparent solubility parameter of the polymer is given by,

$$\delta p = -a_2/3a_3 \pm \sqrt{a_2^2/3a_3^2 - a_1/3a_3} \tag{4}$$

where the $a_i$ are the power series coefficients of equation (3). The maximum degree of swelling is then given by,

$$q_{max} = a_0 + a_1\delta_p + a_2\delta_p^2 + a_3\delta_p^3 \tag{5}$$

*TABLE II. Compositions of Two-Component Networks*

| $M_n$ (macromer) | HEMA (% w/w) | Macromer (% w/w) | Crosslink density |
|---|---|---|---|
| 72 | 65 | 35 | 0.174 |
| 1460 | 65 | 35 | 0.088 |
| 2450 | 65 | 35 | 0.053 |
| 3470 | 65 | 35 | 0.039 |
| 4770 | 65 | 35 | 0.028 |
| 5960 | 65 | 35 | 0.023 |
| 72 | 97 | 3 | 0.023 |
| 1460 | 87 | 13 | 0.023 |
| 2450 | 80 | 20 | 0.023 |
| 3470 | 75 | 25 | 0.023 |
| 4770 | 69 | 31 | 0.023 |
| 5960 | 65 | 35 | 0.023 |
| 2450 | 90 | 10 | 0.012 |
| 2450 | 80 | 20 | 0.026 |
| 2450 | 70 | 30 | 0.044 |
| 2450 | 60 | 40 | 0.066 |

To study the effects of composition and crosslink density independently, we have prepared networks at constant crosslink density as well as networks at constant composition. Table II contains the compositions and crosslink densities of the networks studied, including an additional series in which the molecular weight of the macromer is held constant. All of these compositions were prepared in sheet form via bulk polymerization. Swelling measurements were performed for each of these networks using ethanol/water solutions for which solubility parameters were calculated by,

$$\delta_1 = \{(\Delta\overline{H}_v - RT)/(X_1\overline{V}_1 + X_2\overline{V}_2)\}^{\frac{1}{2}} \qquad (6)$$

where $\Delta\overline{H}_v$ is the average latent heat of vaporization of ethanol and water, $X_1$ and $X_2$ are mole fractions of ethanol and water, respectively and $\overline{V}_1$ and $\overline{V}_2$ are molar volumes of each component, respectively. The use of equation (5) implies that the mixed solvent may be considered, at least approximately, as a single solvent. Since the latent heats of vaporization of ethanol and water are nearly the same, this approximation should be reasonable. In Figure 3 we present

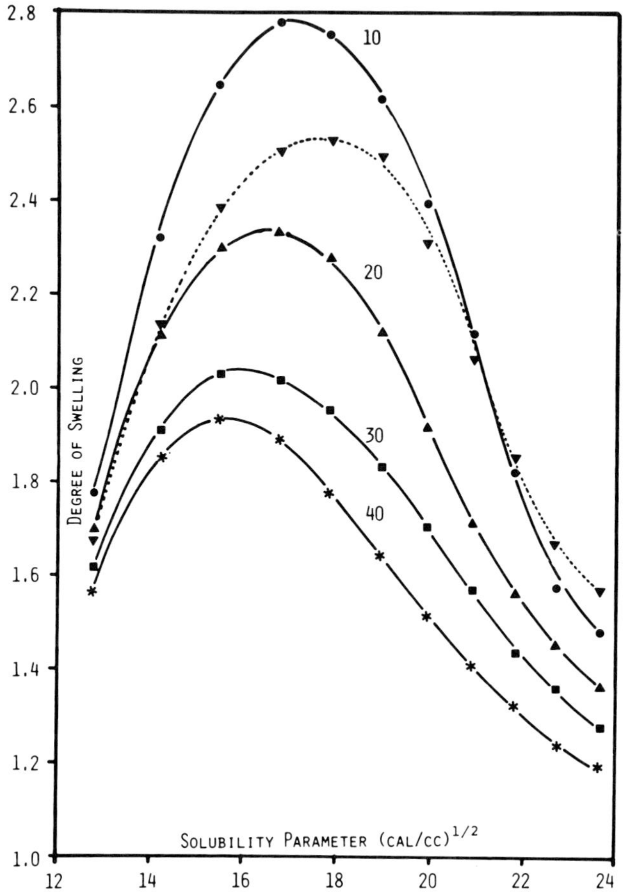


*FIGURE 3. Degree of swelling in ethanol/water mixtures for two-component networks having compositions of 10, 20, 30 and 40 parts of macromer of $M_n$ 2450. The dotted line represents the swelling curve for a conventional poly HEMA network having 1.5% TEGDMA as cross-linking agent.*

the results of swelling experiments for the series of networks in which the macromer molecular weight is held constant and the composition and crosslink densities are both allowed to vary. The shift in the position of the maximum towards lower solubility parameter as macromer content is increased is quite marked and is consistent with an increase in the hydrophobic component of the composition. Reduction in the degree of swelling for all solvent mixtures with increasing macromer content is indicative of the increasing crosslink density. The dotted curve in this figure is the swelling curve for poly(HEMA) crosslinked with tetraethylene glycol dimethacrylate, (TEGDMA) the monomeric analogue of our macromer. It is noteworthy that this curve crosses the 90:10::HEMA:macromer composition on one side and the 80:20:: HEMA:macromer composition on the other. This crossing effect illustrates the structural differences which exist between the two types of networks.

## II. DISCUSSION

In order to develop a viable controlled delivery system it is necessary to be able to achieve a sufficient drug concentration in the polymer network while simultaneously providing for diffusion rates consistent with the time frame required by the *in vivo* environment. For an oral dosage form, these requirements are defined by dosage form size and gastrointestinal residence time. In order that the polymer be acceptable from a safety point of view it is necessary to wash out all extractable by-products and residual monomers left after polymerization. Hence, the only method of loading the active ingredient into the polymer is by absorption from concentrated solution. Therefore, high degrees of swelling are required to achieve high drug loadings while low to moderate degrees of swelling in water are desired in order to control drug release rate. A correlation of $q_{max}$ (the maximum swelling in ethanol-water mixtures), $q_{H_2O}$ (the swelling in water), and $\delta p$ to macromer molecular weight should provide us with information about the effects of crosslink density and composition. In Figure 4 we plot apparent solubility parameter of the polymer versus molecular weight of macromer where we have added the molecular weight of the crosslink junction point to that of the isocyanate terminated macromer. The two curves represent data taken for constant crosslink density (✱ data points) and for constant composition (■ data points). The effect of macromer content, as seen in the upper curve, is quite substantial, while the effect of crosslink density, as seen in the lower curve, is nearly insignificant. One would conclude then that selection of a drug loading solvent is important only as function of composition. Figure 5, on the other hand, shows that maximum degree of swelling is dependant on both composition and crosslink density, although it seems crosslink density has somewhat more effect here. The upper curve, at constant composition, illustrate an interesting point. As macromer molecular weight is increased, the maximum degree of swelling

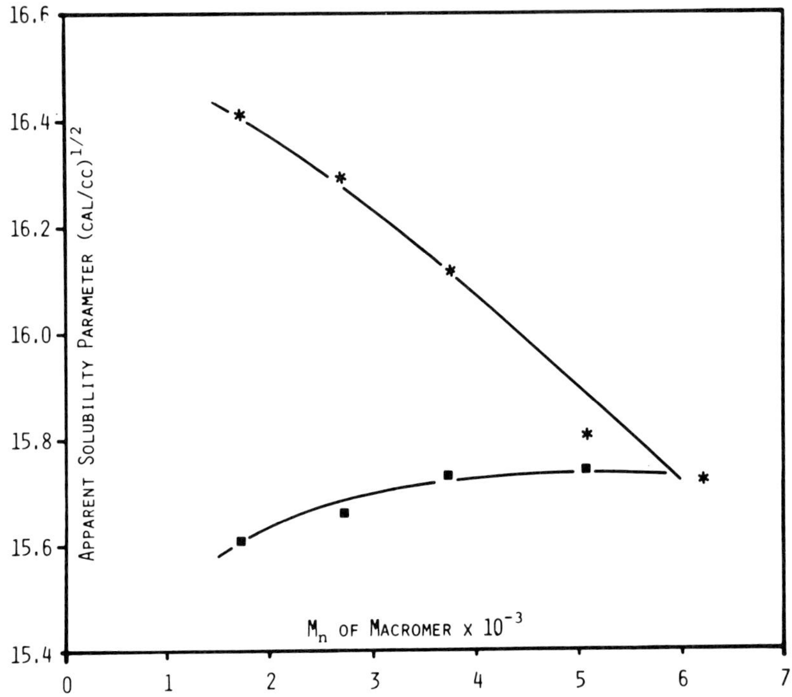


*FIGURE 4. Dependance of apparent polymer solubility parameter on macromer $M_n$ for networks of equal crosslink density (✱) and those having equivalent composition (■).*

itself goes through a maximum. This implies that at some molecular weight of macromer, at constant crosslink density, the degree of swelling is equivalent to a hypothetical network having volumeless crosslinks (i.e. $M_n = 0$) and therefore molecular weight between crosslinks along the poly(HEMA) chain is equal to the macromer molecular weight. This phenomenon has been shown to be discernable in other two-component networks as well (12). The same behavior in water degree of swelling for these specimens is observed in Figure 6.

### *A. Drug Loading*

For the present study we have chosen to illustrate drug loading and the effects on drug loading of composition and crosslink density using Phenformin-HCl, a highly water soluble compound that has found use as an oral anti-hyperglycemic.

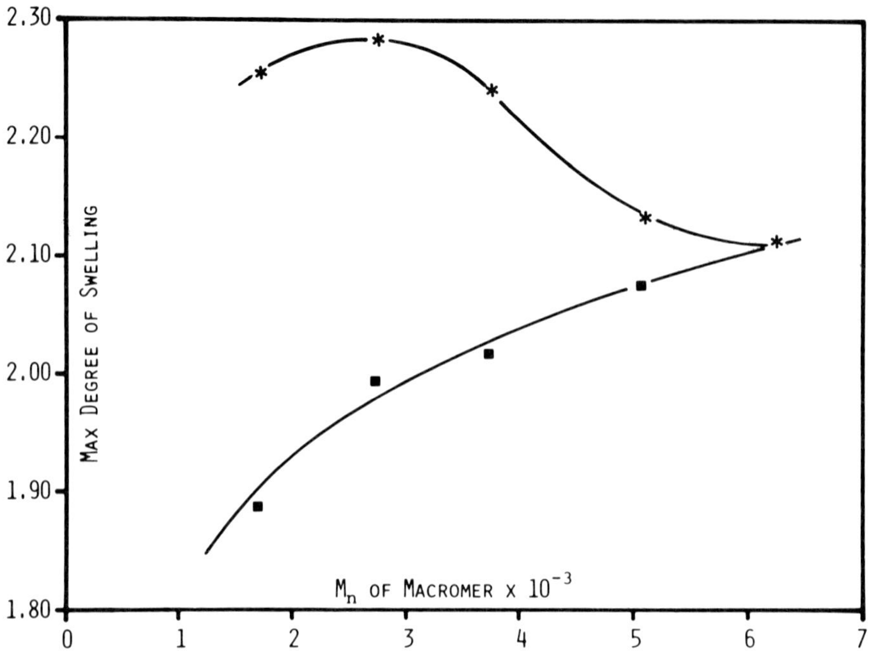


*FIGURE 5. Dependance of maximum degree of swelling on $M_n$ of macromer for networks of equal crosslink density (✱) and those having equivalent composition (■).*

The networks were allowed to equilibrate in drug solution for a period of 48 hours at room temperature. The initial concentration of drug was twenty percent (w/w) and was used in large excess such that the external concentration remains constant, even if a large partition effect exists. The specimens, which were in the form of discs having diameter of 2 cm and thickness of approximately 1 mm, were removed from the solution, rinsed and vacuum dried. Rough estimates of the diffusion coefficient for the drug indicated that equilibrium distribution through the sheet should be obtained within 36 hours. The solvent system utilized was 70 percent ethanol in water, a composition having solubility parameter near those of the networks. The drug loadings were calculated from ultraviolet spectra determined on solutions which were used to exhaustively extract the discs. Results of drug loadings are given in Table III for the series having constant composition and also for those having constant crosslink density. These data run parallel to the $q_{max}$ versus macromer $M_n$ relationship of Figure 5 quite closely. However, it is at

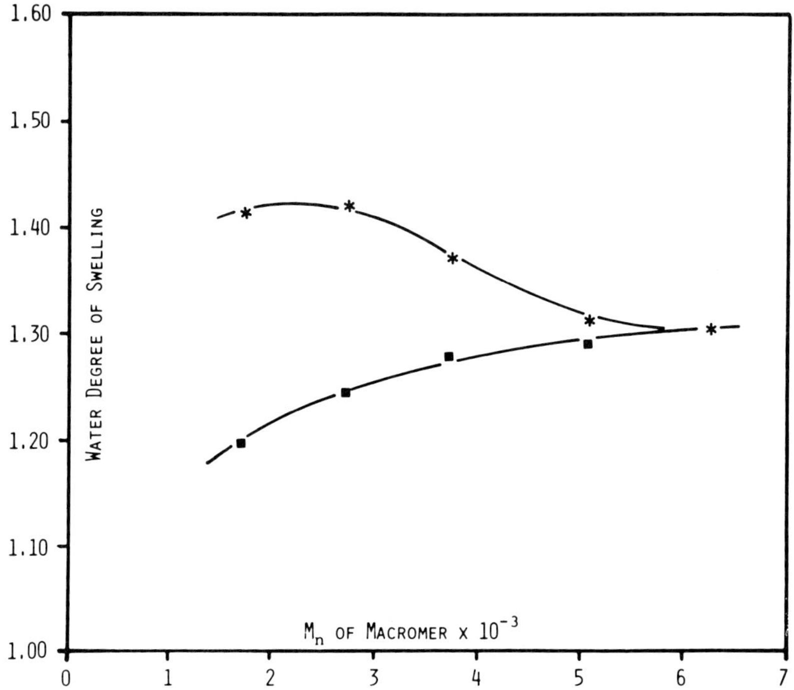


*FIGURE 6. Dependance of degrees of swelling in water on $M_n$ of macromer for networks of equal crosslink density (✱) and those having equivalent composition (■).*

once apparent that variation of crosslink density at constant composition has a smaller effect on loading than does compositions at constant crosslink density. The net effect of varying both crosslink density and composition shows the compositional variation to have more effect on final concentration of drug than the crosslink density, hence, increasing macromer content of constant molecular weight will have an accelerated effect on final drug loadings. Regardless of composition or crosslink density, $q_{max}$ correlates well with final drug loading as can be readily seen in Figure 7.

*B. Diffusion*

Release of Phenformin-HCl from the networks described in the drug loading section above was studied in a flow through system using water as the extraction medium at 37°C. Detection of the amount of drug released was accomplished by

*TABLE III. Results of Loading Phenformin-HCl in Two-Component Networks from Solutions of the Drug (20% w/w) in a 70 Per Cent Ethanol in Water Solvent System*

| $M_n$ (macromer) | HEMA (%w/w) | Crosslink density | Percent drug |
|---|---|---|---|
| 1460 | 65 | 0.088 | 19 |
| 2450 | 65 | 0.053 | 20 |
| 3470 | 65 | 0.038 | 19 |
| 4770 | 65 | 0.028 | 22 |
| 5960 | 65 | 0.023 | 22 |
| 1460 | 87 | 0.023 | 25 |
| 2450 | 80 | 0.023 | 28 |
| 3470 | 75 | 0.023 | 25 |
| 4770 | 69 | 0.023 | 22 |
| 2450 | 90 | 0.012 | 21 |
| 2450 | 80 | 0.026 | 25 |
| 2450 | 70 | 0.044 | 18 |
| 2450 | 60 | 0.067 | 14 |

monitoring the strong U.V. absorption band of this compound at 235 nm in a Perkin Elmer Model 554 UV/Visible Spectrophotometer. The disc-shaped specimens are allowed to wander throughout the extraction flask which is stirred fast enough to assure quick homogeneity of the solution and prevent concentration polarization near the specimen surface. Release curves are typical for monolithic devices, resembling the one marked as "sheet" in Figure 8.

The release data were treated in a way which reflects the fact that water is absorbed at the same time that drug is released (20). The drug diffusion coefficient is taken to be a function of the amount of water absorbed. The assumption is then made that water is absorbed by the loaded network via pure Fickian diffusion with constant diffusion coefficient. For cartesian coordinates with diffusion in one dimension, if $Q_t$ is the amount of drug diffused at time t, $Q_\infty$ is the amount diffused at infinite time, then we have

$$Q_t = Q_\infty\{1 - 8\pi^{-2} \sum_{n=0}^{\infty} (2n + 1)^{-2}\exp[-(2n+1)^2\pi^2 T(t)]\} \quad (7)$$

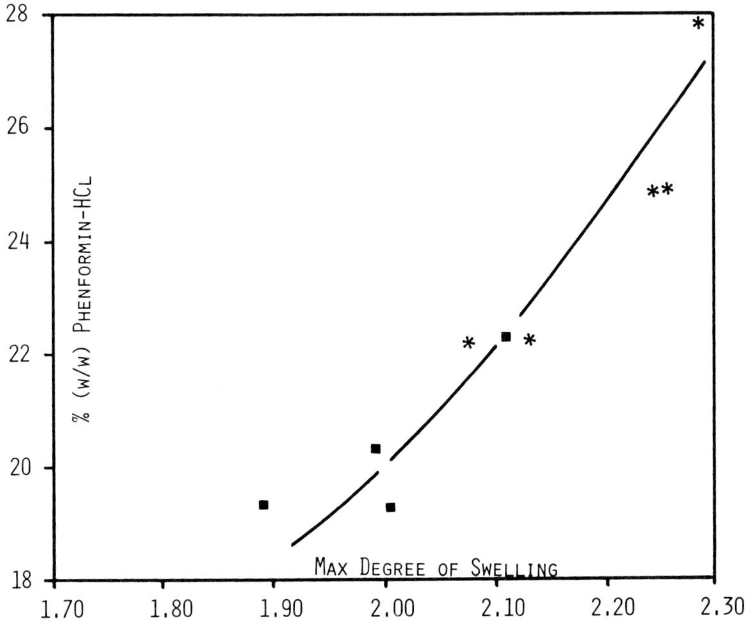


*FIGURE 7. Concentration of Phenformin-HCl in the dry networks as a function of max degree of swelling. All networks were loaded under equivalent conditions (✱) indicates networks of equal crosslink density and (■) indicates networks of equivalent composition.*

where T is defined as

$$T(t) = \int_0^t \{D(t)/a^2\}dt \tag{8}$$

where D(t) is the time dependent solute diffusion coefficient and a is the swollen sheet thickness. If D(t) is proportional to the fraction of water absorbed at time t and has the steady state value $D_s$ at equilibrium swelling, then,

$$D(t) = D_s \{1 - 8\pi^{-2} \sum_{n=0}^{\infty} (2n + 1)^{-2} \exp[-(2n+1)^2\pi^2 D^* t/a^2]\} \tag{9}$$

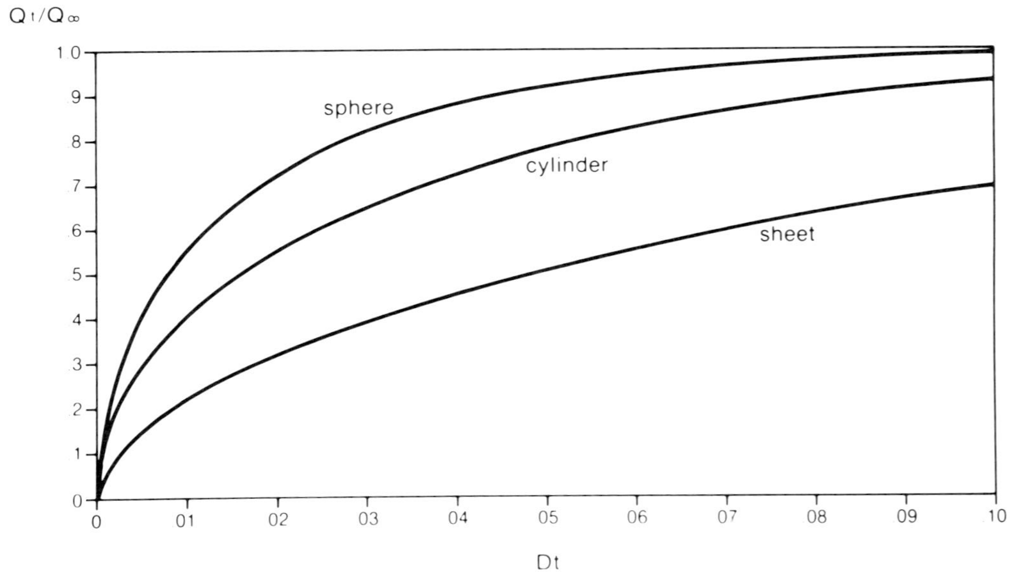


*FIGURE 8. Diffusion curves for different geometries using solutions to Fick's second law with constant diffusion coefficient.*

where D* is the solvent absorption coefficient. The function T becomes,

$$T(t) = \frac{D_s t}{a^2} - \frac{D_s}{12D^*}\left\{1 - \frac{96}{\pi^4}\sum_{n=0}^{\infty}\frac{\exp[-(2n+1)^2\pi^2 D^* t/a^2]}{(2n+1)^4}\right\} \quad (10)$$

At equilibrium swelling the infinite series vanishes and a plot of T versus time gives,

$$T \simeq (D_s t/a^2) - (D_s/12D^*) \quad (11)$$

a straight line having slope proportional to the steady state drug diffusion coefficient and intercept related to the assumed constant water absorption coefficient. Using arbitrary values for $D_s$, D* and a, we illustrate the function T(t) versus time in Figure 9. In practice, one constructs a master curve for values of T in the desired range, selects

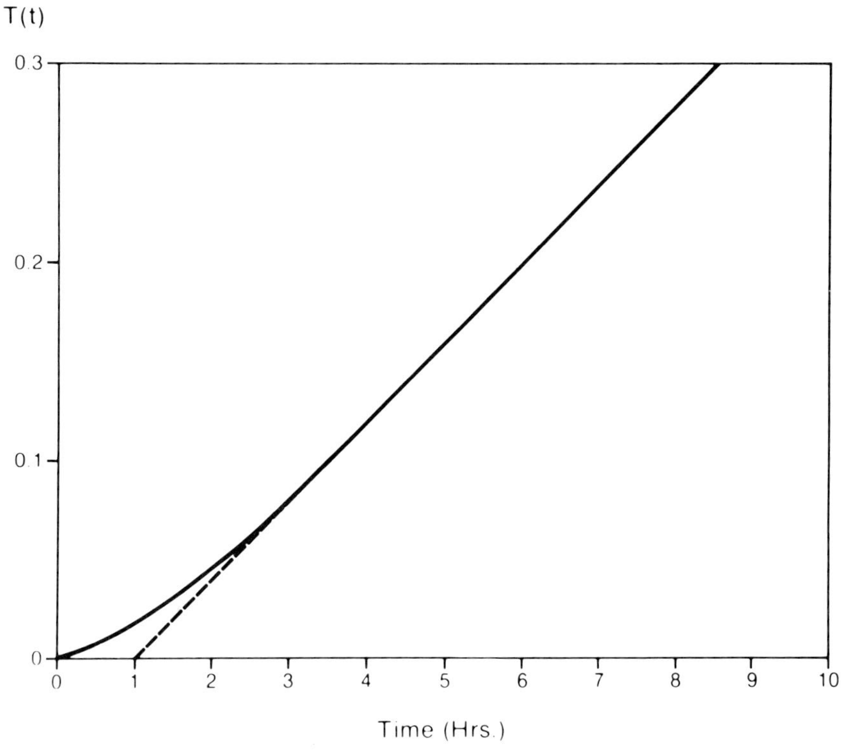


*FIGURE 9. Behavior of equation (10) for a hypothetical case of simultaneous water absorption and drug diffusion.*

the time values from the experimental release curve which have the appropriate fraction released for that value of T. The linear portion of the curve is fitted by a least squares method and $D_S$ is determined from the slope while D* is found from the time lag value for T(t) = 0. The effects of crosslink density and network composition can be seen in Figure 10, where we have plotted the steady state solute diffusion coefficient versus number average molecular weight of macromer. The upper curve was obtained from specimens at constant crosslink density, hence, the effect of network composition is illustrated. The lower line is obtained at constant composition, hence, the effect of crosslink density is seen. The effects of crosslink density and composition on the steady state diffusion coefficient are in the same direction, as expected. However, it is apparent that increasing the macromer content has more effect than does increasing the

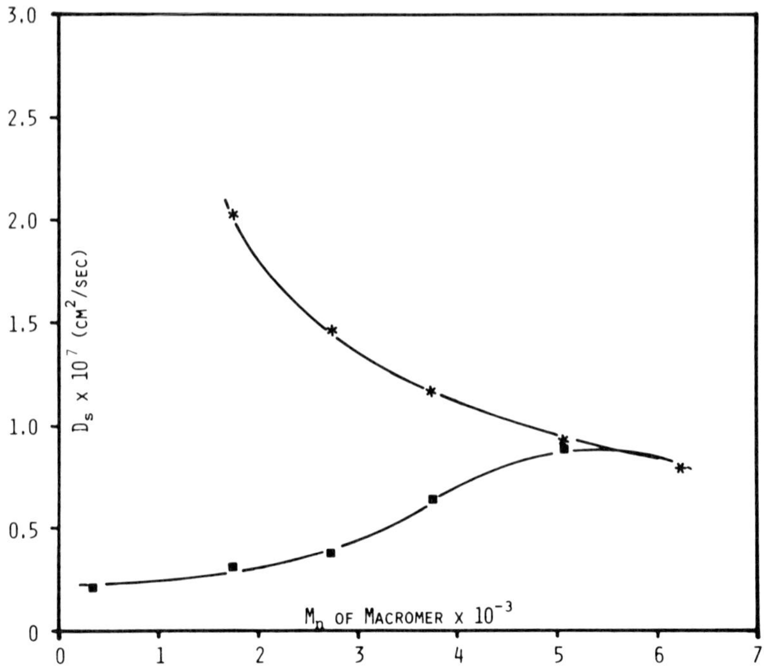


*FIGURE 10. Dependance of the steady state drug diffusion coefficient on $M_n$ of macromer for constant cross-link density networks (✱) and constant composition networks (■).*

crosslink density alone. This phenomenon is more clearly demonstrated in Figure 11, where the steady state diffusion coefficient for solute is plotted versus macromer content, regardless of whether crosslink density is constant (lower curve) or allowed to vary (upper curve). The upper curve is data calculated from specimens in which macromer molecular weight is maintained at 2450 and composition is varied as indicated. Therefore, both crosslink density and composition are increasing with increasing macromer content. The lower curve, specimens at constant crosslink density, crosses over the curve from variation of both crosslink density and composition. This crossover corresponds exactly to an inversion in relative crosslink densities between the two sets of specimens.

On the other hand, the water diffusion coefficient as defined in equation (10) appears to be almost exclusively

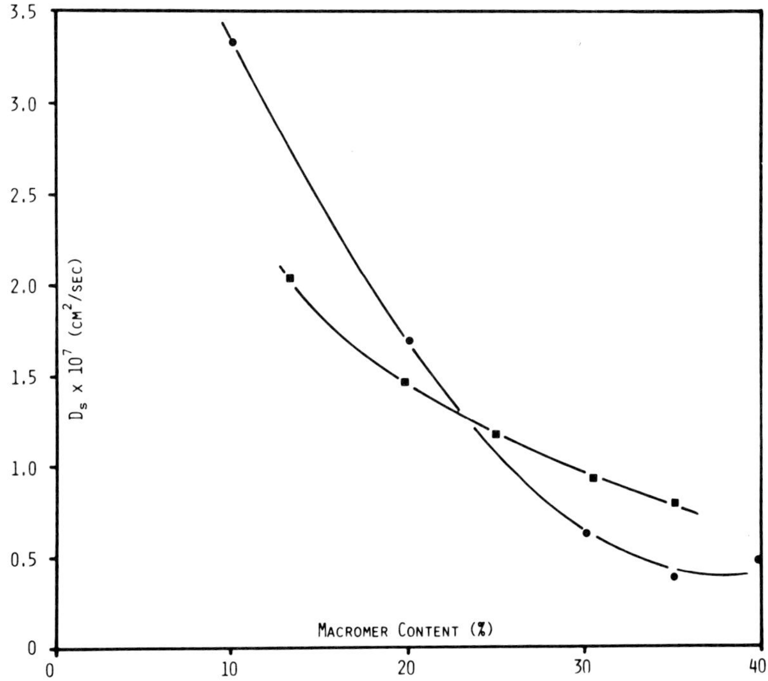


*FIGURE 11. Dependance of the steady state drug diffusion coefficient on the network macromer content. Circles indicate networks in which both crosslink density and composition are changing while squares are data for networks of constant crosslink density.*

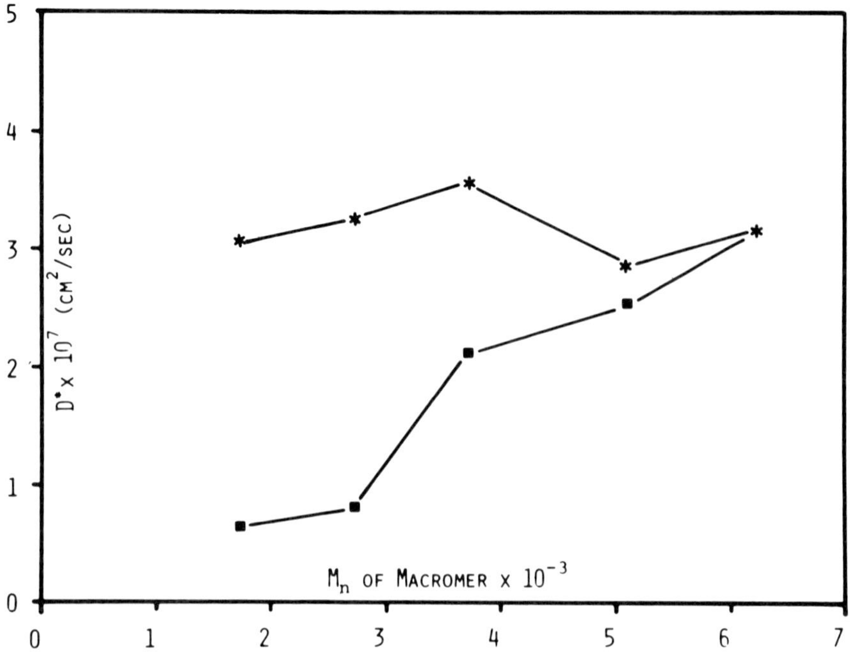


*FIGURE 12. Dependance of the water absorption coefficient on $M_n$ of macromer for networks of constant crosslink density (✱) and networks of constant composition (■).*

dependant upon crosslink density, as illustrated in Figure 12. In this figure we see an increase in D* with increasing macromer molecular weight at constant composition (i.e. decreasing crosslink density) while D* remains relatively constant with macromer molecular weight at constant crosslink density (i.e. increasing macromer content). This fact indicates that water transport occurs almost exclusively in the hydrophilic poly(HEMA) fraction of the 2-phase network system.

## III. CONCLUSION

Hydrogels can be useful as controlled release systems for pharmaceuticals, not only as implants, but also as oral dosage forms. The ability to wash the devices totally clean of all extractable impurities prior to loading the drug provides an attractive means of purification which not only renders the device completely safe for consumption, but also eliminates unwanted binders and excipients found in classical dosage forms which are absorbed in the G.I. tract along with the active ingredient.

We have shown in this paper that the usefulness of hydrogels can be expanded by the construction of 2-component systems in which the hydrophilic balance is adjusted to provide for the desired release rates for an enclosed solute. The effects of crosslink density and composition on solute diffusion coefficient are separable and complementary, providing a versatile method of controlling the release rate. Due to the 2-phase nature of the MCGP structure, solvent diffusion coefficient is unaffected by the composition, implying that transport occurs only in the hydrophilic regions. Although the present work only reports the release of a water soluble drug, we believe these systems can be used with virtually any reasonably low molecular weight pharmaceutical.

## ACKNOWLEDGMENTS

The authors would like to recognize the fine work of Mr. Walter Plankl, who prepared the macromer for this study, and Ms. Zienat Tawfik, who prepared the polymers, dosage forms, and performed the release rate measurements.

## REFERENCES

1. R. C. Capozza, L. Sendelbeck, W. J. Balkenhol, Midland Macromolecular Monographs, 5, 59 (1978).
2. T. M. Jackaniscq, H. A. Nash, D. L. Wise, J. B. Gregory, Contraception, 8, 227 (1973).
3. A. D. Schwope, D. L. Wise, and J. F. Howes, Life Sci., 17, 1877 (1975).
4. J. Folkman and D. M. Long Jr., J. Surg. Res., 4, 139 (1964).
5. R. W. Baker and H. K. Lonsdale, Adv. Exp. Med. and Bio., A. D. Tanguary and R. E. Lacy, (Eds.), 47, 15 (1972).
6. R. C. Capozza, Midland Macromolecular Monograph, 5, 201 (1978).
7. O. Wichterle, Encyclopedia of Pol. Sci. and Tech., 15, 273 (1971).
8. S. M. Podos, B. Becker, and J. Hartstein, Soft Contact Lens, C. V. Mosby, St. Louis (1972).
9. M. Aden, J. Scherrer, J. Curierman, F. Nagin, and B. S. Levowitz, Arch. Surg., 105, 100 (1972).

10. C. H. Bamford, R. W. Dyson, and G. C. Eastmond, J. Pol. Sci., 16C, 2425 (1967).
11. C. H. Bamford, G. C. Eastmond, and D. Whittle, Polymers, 12, 247 (1971).
12. W. R. Good, Ph.D. Thesis, Univ. Freiburg, W. Germany (1979).
13. O. Kramer, R. L. Carpenter, V. Ty, and J. D. Feng, Macromolecules, 7, 79 (1974).
14. R. L. Carpenter, O. Kramer, and J. D. Feng, Macromolecules, 10, 117 (1977).
15. P. J. Flory, J. Chem. Phys., 66, 5720 (1977).
16. B. Erman and P. J. Flory, J. Chem. Phys., 68, 5363 (1978).
17. P. J. Flory and J. Rehner Jr., J. Chem. Phys., 11, 512, 521 (1943).
18. K. Dusek and W. Prins, Adv. in Pol. Sci., 6, 1 (1969).
19. J. Hildebrand and R. Scott, The Solubility of Nonelectrolytes, 3rd ed., Reinhold, New York, (1949).
20. W. R. Good, Midland Macromolecular Monographs, 5, 139 (1978).

# A NEW APPROACH TO ACHIEVE ZERO-ORDER RELEASE KINETICS FROM DIFFUSION-CONTROLLED POLYMER MATRIX SYSTEMS

*William D. Rhine*
*Vikas Sukhatme*
*Dean S. T. Hsieh*
*Robert S. Langer*

Department of Nutrition and Food Science
Massachusetts Institute of Technology
Cambridge, Massachusetts
and the
Department of Surgery
Children's Hospital Medical Center
Boston, Massachusetts

## I. INTRODUCTION

Matrix devices as vehicles for drug delivery have advantages in their ease of fabrication when compared to other sustained release systems such as membrane-encapsulated reservoir devices (1). Furthermore, for certain applications such as the sustained release of macromolecules (2), the use of a matrix system is essential to achieve release. One major drawback, however, to the use of matrix systems is that they generally do not display zero-order release kinetics. Most studies have used matrix devices in the form of a rectangular slab and it has been observed that the rate of release is proportional to the square root of time (3). This is because the release rate is inversely proportional to the distance the drug must travel from within the matrix to the matrix surface. Since this diffusional distance increases with time (as drug is released) the release rate decreases. We considered that if a matrix system was constructed in such a way as to compensate for this increasing diffusion distance, zero-order release could theoretically be achieved. We have




ISBN 0-12-074450-3

therefore considered the effect of matrix geometry on the rate of the release and have found that an inwardly releasing hemisphere can provide an essentially zero-order release rate. In this paper, we present the theoretical basis for these kinetics as well as a comparison of kinetics from matrix devices of other geometries.

## II. THEORETICAL SECTION

The following analysis concerns the rate of release from an inwardly releasing hemisphere. In this case release would only occur through a concavity in the flat surface; all other surfaces would be laminated with an impermeable coating (Figure 1). Infinite sink conditions are assumed. Consider:

- $a_i$ = inner radius or radius of concavity
- $a_o$ = outer radius of device
- $r$ = any radial distance
- $R$ = radial distance to interface between dissolved and dispersed drug within the matrix (see Figure 1).
- $J$ = flux
- $C(x)$ = drug concentration (wt/vol) within the matrix at radial distance x
- $C_s$ = drug solubility in the release media (this also

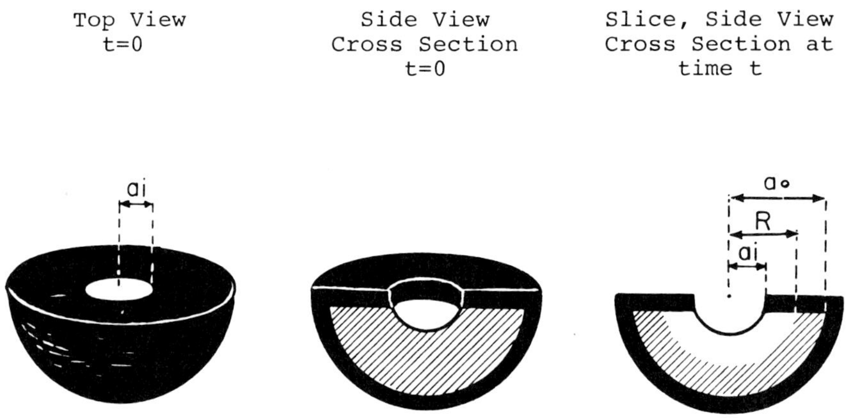


*FIGURE 1. Schematic diagram of an inwardly releasing hemisphere. $a_i$ = inner radius, $a_o$ = outer radius, R = distance to the interface between the dissolved region (white area) and the dispersed zone (dark grey). Black represents laminated regions through which release can not occur.*

equals the concentration of drug at the interface between dissolved and dispersed drug).

$C_o$ = drug concentration in the matrix at any point at t=o (this also equals the concentration of drug in the dispersed phase).

$Q(t)$ = drug released (wt) from the pellet at time t

$M(t)$ = drug retained (wt) in the pellet at time t

D = diffusion coefficient of drug

Then applying an analysis of the type originally proposed by Higuchi (4), we can write:

At steady state $$J\,2\pi r^2 = \frac{dQ}{dt} \quad (1)$$

By Fick's law $$J = \frac{dC}{dr} \quad (2)$$

$$\therefore\ dC = \left(\frac{dQ/dt}{2\pi D r^2}\right) dr \quad (3)$$

Integrating from $a_i$ to R

$$C\,(R) - C\,(a_i) = \frac{(dQ/dt)}{2\pi D}\left(\frac{1}{a_i} - \frac{1}{R}\right) \quad (4)$$

$$C(R) = C_s,\ C(a_i) = 0$$

So $$C_s = \frac{(dQ/dt)}{2\pi D}\left(\frac{1}{a_i} - \frac{1}{R}\right) \quad (5)$$

Integrating (3) from $a_i$ to r gives:

$$C(r) = \frac{(dQ/dt)}{2\pi D}\left(\frac{1}{a_i} - \frac{1}{r}\right) \quad (6)$$

Eliminating dQ/dt using (5) gives:

$$C(r) = C_s\left(\frac{R}{R-a_i}\right)\left(1 - \frac{a_i}{r}\right) \quad (7)$$

But from equation (3), $$\frac{dQ}{dt} = 2\pi r^2 D\,\frac{dC}{dr} \quad (8)$$

Combining (7) and (8)

$$\frac{dQ}{dt} = 2\pi D C_s\left(\frac{R}{R - a_i}\right) a_i \quad (9)$$

This equation describes the release rate as a function of constants and the value R. To describe the release rate as a function of time, an expression for the variation of R with time can be derived by the following mass balance.

$$\text{Drug in pellet at time } t = 0,\ M_o = C_o\frac{2}{3}\pi(a_o{}^3 - a_i{}^3)$$

Drug in pellet at time t when front is advanced to distance R is the sum of drug in undissolved region plus dissolved region:

$$M(t) = C_o\frac{2}{3}\pi(a_o{}^3 - R^3) + \int_{ai}^{R} 2\pi r^2 C(r)\,dr$$

When $C_o >> C_s$, we may neglect the second term, giving

$$M(t) \sim C_o\frac{2}{3}\pi(a_o{}^3 - R^3)$$

Thus,

$$Q(t) = M_o - M(t) = C_o\frac{2}{3}\pi(a_o{}^3 - a_i{}^3) - C_o\frac{2}{3}\pi(a_o{}^3 - R^3) = C_o\frac{2}{3}\pi(R^3 - a_i{}^3) \tag{10}$$

Differentiating the above expression gives,

$$\frac{dQ}{dt} = 2\pi R^2 C_o\frac{dR}{dt} \tag{11}$$

To get R as a function of time, using (8) and (11),

$$R(R - a_i)\,dR = a_i\left(\frac{DC_s}{C_o}\right) dt \tag{12}$$

Integrating (12) from t=0 to t and R=$a_i$ to R gives:

$$\left(\frac{DC_s}{C_o a_i{}^2}\right) t = \frac{1}{3}\left(\frac{R}{a_i}\right)^3 - \frac{1}{2}\left(\frac{R}{a_i}\right)^2 + \frac{1}{6} \tag{13}$$

## III. RESULTS

From equation (9), the release rate, dQ/dt, can be expressed as a function of the ratio ($R/a_i$). Similarly, equation (10) can be used to describe total release as a function of the same ratio. With these two equations and equation (13), which describes the time to reach any value of

$(R/a_i)$, both release and release rate as a function of time can be calculated. Figure 2 shows total release as a function of time for a hemisphere. The units of time and release are simply a combination of constants D, $C_o$, $C_s$, $\pi$, and $a_i$. Except for a brief period initially, release is nearly linear or zero-order. Figure 3 is a graph of release rate versus time. Note the rapid approach to zero-order when $R > 3a_i$. This can also be seen in equation (9); when $R >> a_i$, $dQ/dt \sim 2\pi D C_s a_i$.

The first studies exploring the use of matrix shape to achieve zero-order kinetics were by Brooke and Washkuhn (4) and Lipper and Higuchi (5). They derived equations for the

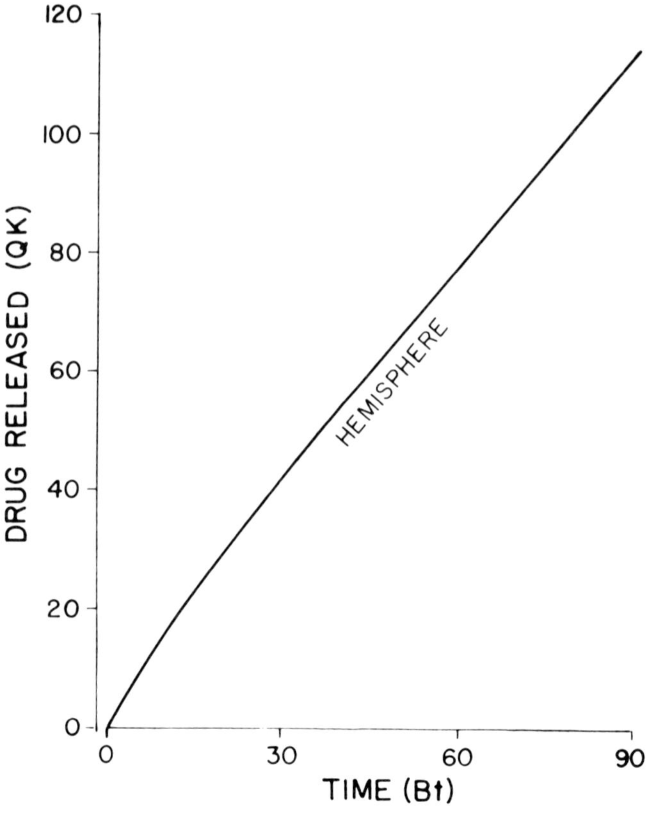


*FIGURE 2. Release versus time for an inwardly releasing hemisphere. Quantity released is plotted as QK, where* $K = 2\pi C_o a_i^3$. *Time is plotted as Bt, where* $B = DC_s/C_o a_i^2$.

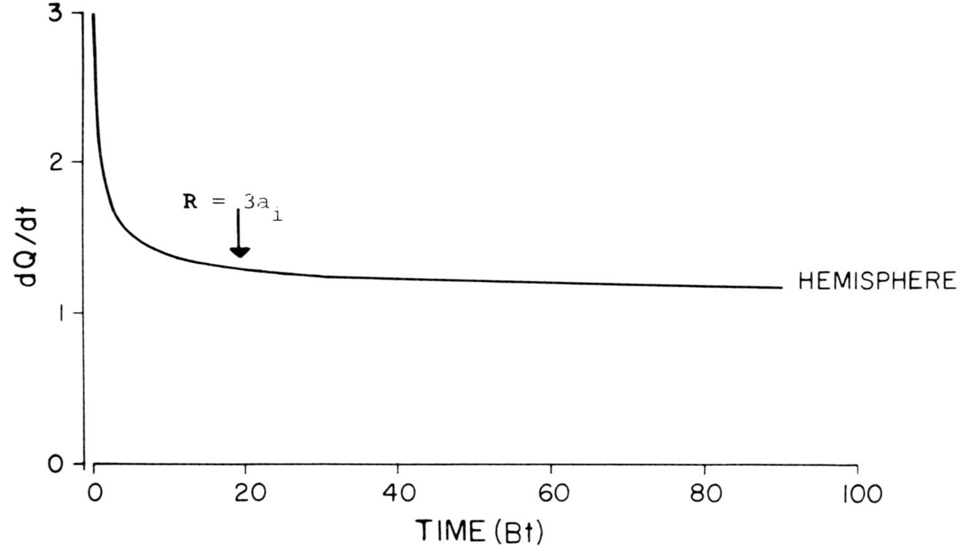


*FIGURE 3. Release rate versus time for a hemisphere. Release rate is plotted as* $(dQ/dt)Z$ *where* $Z = 2\pi C_s D a_i$. *Time is plotted as* $Bt$, *where* $B = DC_s/C_o a_i^2$.

release of drugs from an inwardly releasing sectioned cylinder. Its release kinetics are compared to the hemisphere's and to a slab's in Table I. As in the above derivation, equations for either Q, dQ/dt, or time can be listed as a function of R.

To exemplify the differences in the release kinetics of these three shapes, a hypothetical case was derived so that all devices would deliver the same total amount of drug (this amount is 114K where $K = 2\pi a_i^3 C_o$) over a set period of time (90 B where $B = C_s D / C_o a_i^2$). Both the cylinder and the hemisphere had an inner radius, $a_i$, and the area of the cylinder was set at two times that of the hemisphere (i.e., $L = 4a_i$). The total release versus time is graphed in Figure 4 for three shapes for this hypothetical case. It is clear that the hemisphere best approximates zero order release. The release *rates versus* time for these three shapes are compared in Figure 5. In the last 90% of the duration of release there is only a 16% decrease in the release rate from a hemisphere. For a cylinder there is a 35% decrease in this same time period, and for a slab, a 70% decrease occurs.

## IV. DISCUSSION

These results demonstrate the effectiveness of utilizing geometric means to achieve near zero-order kinetics using matrix delivery systems. An inwardly releasing hemisphere is a simple geometry, which closely approximates constant release, more so than either a slab or an inwardly-releasing cylinder. To take advantage of the hemispherical geometry, a device would have to be designed so that the outer diameter is greater than three times the inner diameter. For devices possessing such geometry, zero-order kinetics will essentially be achieved after a short burst and maintained for the duration of release. While we have assumed perfect sink conditions, there may be a significant drug concentration gradient between the matrix release surface and the surrounding media. The resultant bulk flow has been shown to bring release kinetics even closer to zero-order (5). It is also possible to pre-release the hemisphere to avoid the most non-linear portion of drug release which would waste only a minimal quantity of drug. In addition, it may be possible to combine non-uniform drug concentration profiles within a matrix with these geometric methods to approach zero-order kinetics even further.

*TABLE I. Comparison of Release Kinetics for the Three Figures as Shown; a Slab, an Inwardly Releasing Cylinder Section, and an Inwardly Releasing Hemisphere. In all Cases, $x = R/a_i$, where $a_i$ is the Appropriate Dimension Depicted and R Equals the Distance to the Interface Between the Dissolved Region and the Dispersed Region. Graphs of Release or Release Rate Versus Time are Calculated by Determining the Time and the Release Quantity Associated with Various Values of X. For Time Units, $B = DC_s/C_o a_i^2$.*

| | AREA | TIME BT | RELEASE RATE | TOTAL RELEASE |
|---|---|---|---|---|
| SLAB 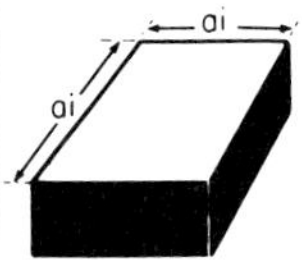  $x \geq 0$ | $a_i^2$ | $\frac{1}{2}x^2$ | $\frac{C_s D a_i}{x}$ | $C_o a_i^3 x$ |
| CYLINDER   $x \geq 1$ | $\pi L a_i$ | $\frac{1}{2}x^2 \ln x - \frac{1}{4}(x^2-1)$ | $L\pi C_s D(\frac{1}{\ln x})$ | $\frac{1}{2}L\pi C_o a_i^2 (x^2-1)$ |
| HEMISPHERE 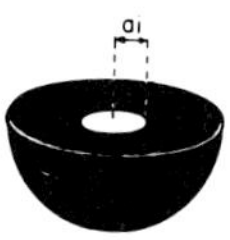  $x \geq 1$ | $2\pi a_i^2$ | $\frac{1}{3}x^3-\frac{1}{2}x^2+\frac{1}{6}$ | $2\pi C_s D a_i(\frac{x}{x-1})$ | $\frac{2}{3}\pi C_o a_i^3 (x^3-1)$ |

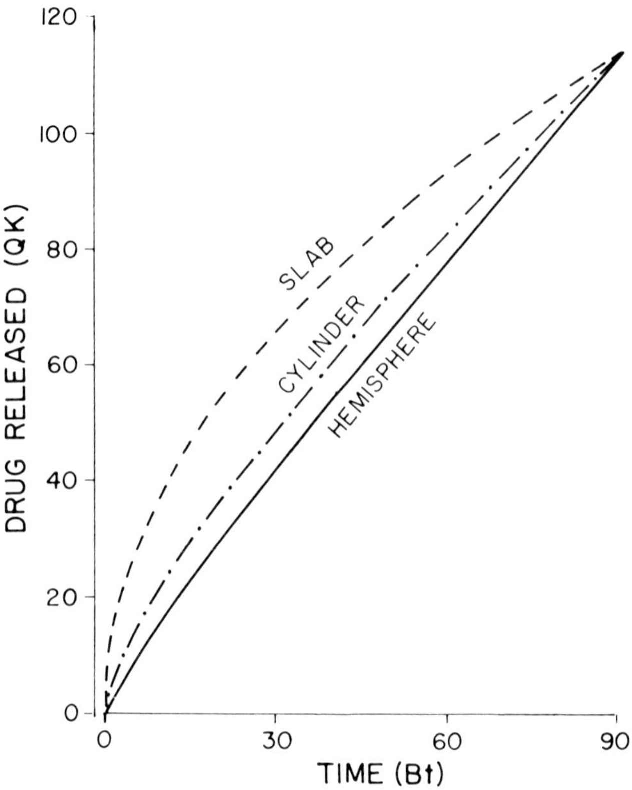


*FIGURE 4. Release versus time for three different shapes designed to give the same total release after time ∿ 90B. Both the cylinder and the hemisphere have an inner radius equal to $a_i$. The length of the cylinder is 2.3 $a_i$; the width of the slab is 2.57 $a_i$. Axis have same units as Figure 2.*

Thus, geometry provides a useful means by which to affect release rates of matrix devices. In particular, a hemisphere which releases drug only from a small concavity in its planar surface displays essentially zero-order release kinetics. Current efforts are being directed toward developing simple procedures to fabricate these devices and to experimentally validate the above theoretical treatment.

## ACKNOWLEDGMENTS

This work was supported by National Institutes of Health Grant 1 R01 GM 26698-01.

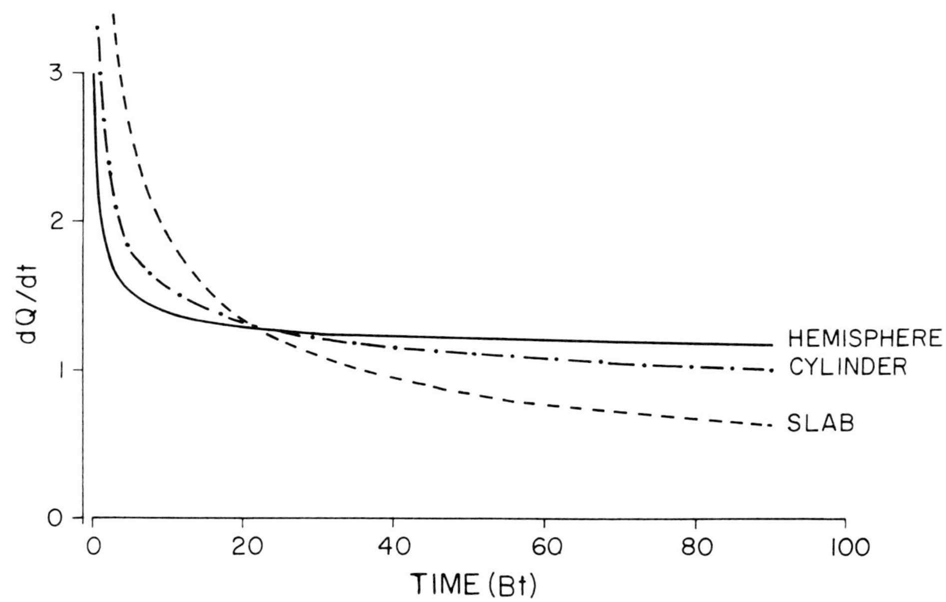


*FIGURE 5. Release rate versus time for the case shown in Figure 4. Axis have same units as Figure 3.*

## REFERENCES

1. D. R. Paul in Controlled Release Polymeric Formulations, D. R. Paul and F. W. Harris, (Eds.), American Chemical Society, Washington, D.C., 1976, p. 1.
2. R. Langer and J. Folkman, Nature, 263, 797 (1976).
3. T. Higuchi, J. Pharm. Sci., 52, 1145 (1963).
4. D. Brooke and F. J. Washkuhn, J. Pharm. Sci., 66, 159 (1977).
5. A. Lipper and W. I. Higuchi, J. Pharm. Sci., 66, 163 (1977).

# MATHEMATICAL MODEL FOR PROGESTERONE RELEASE FROM INJECTIBLE POLY(LACTIC ACID) MICROCAPSULES *IN VITRO*

*D. V. Satyanarayana Gupta*[1]
*R. E. Sparks*

Biological Transport Laboratory
Department of Chemical Engineering
Washington University
St. Louis, Missouri

## I. INTRODUCTION

The observation that progestational agents induce a change in the physical chemical properties of the cervical mucus and consequently inhibit sperm migration in humans (1-3) led to the use of low doses of progestins, locally administered to the cervix, as an effective means of contraception (3). It was shown by us in a previous paper (4) that injectible microcapsules of progesterone in poly(lactic acid) can be developed for controlled release of the drug in the cervix. The present paper deals with the development of a mathematical model to predict the release behavior of the drug from microcapsules in *in vitro* situations. The model can also be extended to *in vivo* cases where external boundary layer resistances affect release.

[1]*Present address: Polymer Department, Gulf South Research Institute, New Orleans, Louisiana.*




ISBN 0-12-074450-3

This model considers a diffusion boundary layer and a time-variant concentration on the outside of the particle as compared to Higuchi's model (5) where a perfect sink condition is assumed.

## II. THEORETICAL

### A. *System Description*

The microcapsule skeleton is an inert, homogeneous matrix which is insoluble in the surrounding medium and remains intact throughout the period of drug release. Unlike active matrix systems, which disintegrate or dissolve to release the drug, drug release in this system is dependent only on passive diffusion. Inertness also implies a high degree of biocompatibility.

### B. *Diffusion*

Fick's law relates diffusional flux to the concentration gradient as

$$J = D_e \vec{\nabla} C_A \qquad (1)$$

where J = flux, $g/cm^2$-sec, $D_e$ = effective diffusion coefficient, $cm^2/sec$, and $\vec{\nabla} C_A$ = concentration gradient of species A, $g/cm^4$.

The proportionality constant, $D_e$, the effective diffusion coefficient in the medium, takes into account the porosity of the matrix, the tortuosity factor for the diffusional path, and all other factors causing diffusion to differ from diffusion in pure solution.

In a spherical, porous matrix in which drug particles are dissolving during a diffusional process, it is likely that the diffusion coefficient will be different from the value which would be measured in a flat membrane of the polymer. The geometry of the two systems is different. Also, the capsule is a matrix of drug and polymer whose porosity would change as the drug dissolves. The flat membrane does not contain any drug.

### *C. Assumptions*

It is assumed that the matrix is solid and homogeneous with drug uniformly distributed throughout. The surrounding fluid is able to enter the matrix phase, and the drug dissolves in the permeating fluid phase. The drug-containing core shrinks with time while the overall particle diameter remains constant. The dissolved drug diffuses through the exhausted matrix. The concentration at the core matrix interface is the solubility of the drug in the polymer. On the external surface of the capsule, a liquid mass transfer resistance is assumed at the interface. The concentration of the drug in the liquid is related to the concentration of the drug in the capsular side through the partition coefficient. The drug diffuses through the surface liquid layer into the solution. As the concentration of the solution increases, the driving force across this liquid decreases and the liquid becomes a nonperfect sink. Figure 1 shows the concentration profile for the problem. The assumptions for

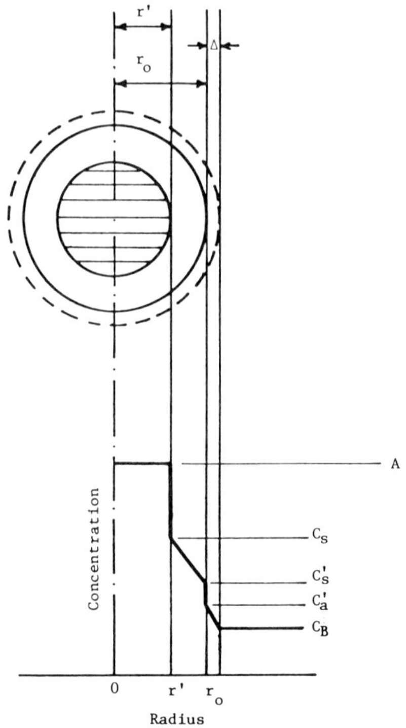


*FIGURE 1. Concentration distribution for the mathematical model.*

mathematical analysis are:

1. Pseudo-steady state exists
2. The shrinking-core model applies
3. Diffusion coefficients are constant
4. Diffusion rather than dissolution is the rate-controlling step
5. The liquid surface layer is thinner than the particle dimension
6. Diffusion is through the matrix phase rather than through pores or channels within the matrix.

### D. *Mathematical Analysis*

According to Fick's law:

$$\frac{dq}{dt} = 4\pi r^2 D_e \frac{dC}{dr} \tag{2}$$

where

$q$ = amount of drug diffusing through shell at radius $r$, g
$r$ = radius under consideration, cm
$D_e$ = effective diffusion coefficient, $cm^2/sec$
$C$ = concentration, $g/cm^3$

Because of pseudo-steady state assumption, integrating equation (1) from $r'$ to $r_o$ gives:

$$\frac{dq}{dt} = \frac{4\pi D_e (C_s' - C_s)}{\left[\frac{1}{r_o} - \frac{1}{r'}\right]} = \frac{4\pi D_e (C_s - C_s')}{\left[\frac{1}{r'} - \frac{1}{r_o}\right]} \tag{3}$$

where

Diffusional matrix layer (zone of depletion) = $r_o - r'$, cm
At $r = r' \rightarrow C = C_s$ the solubility of the drug in the matrix, $g/cm^3$.
At $r = r_o \rightarrow C = C_s'$ the concentration in the matrix at the capsular boundary, $g/cm^3$.

$$K = \frac{C_a}{C_s} = \frac{C_a'}{C_s'} = \text{partition coefficient} \tag{4}$$

The rate of diffusion across the boundary layer:

$$\frac{dq}{dt} = \frac{4\pi D_a (C_a' - C_B)}{\left[\frac{1}{r_o} - \frac{1}{r_o + \Delta}\right]} \tag{5}$$

where

$\Delta$ is the thickness of the diffusion boundary layer, cm

At $r = r_o \rightarrow C_a' =$ concentration at liquid boundary, $g/cm^3$

At $r \geq r_o + \Delta \rightarrow C_B =$ concentration in the bulk liquid, $g/cm^3$

$D_a =$ diffusion-coefficient in the boundary layer, $cm^2/sec.$

Since $\Delta << r_o$

$$\frac{dq}{dt} = \frac{4\pi r_o^2 D_a (C_a' - C_B)}{\Delta} \tag{6}$$

For *in vitro* testing of release behavior of capsules, the drug is released into a fixed amount of liquid medium and so the concentration in the bulk of the liquid, $C_B$, increases with time, i.e.

$$C_B = \frac{Q}{v} \tag{7}$$

where $Q$ = the total amount of drug released by the total dose of microcapsules, g

$v$ = volume of elution medium, $cm^3$

$$Q = \frac{qL}{\frac{4\pi}{3} r_o^3 \rho} \tag{8}$$

where $q$ = amount of drug released per capsule, g

$L$ = weight of capsules used in release experiments, g

$r_o$ = initial radius of capsule, cm

$\rho$ = density of capsule, $g/cm^2$

Using steady-state assumption, equations (3) and (6) are equal;

$$\frac{4\pi D_e (C_s - C_s')}{\left[\frac{1}{r'} - \frac{1}{r_o}\right]} = \frac{4\pi r_o^2 D_a (C_a' - C_B)}{\Delta}$$

$$= \frac{4\pi r_o^2 D_a}{\Delta}\left(C_a' - \frac{Q}{v}\right) = \frac{4\pi r_o^2 D_a}{\Delta}\left[KC_s' - \frac{3qL}{4\pi r_o^3 \rho v}\right] \tag{9}$$

After rearrangement,

$$C_s'\left\{\frac{D_e}{\left[\frac{1}{r'} - \frac{1}{r_o}\right]} + \frac{Kr_o^2 D_e}{\Delta}\right\} = \left\{\frac{D_e C_s}{\left[\frac{1}{r'} - \frac{1}{r_o}\right]} + \frac{3qLr_o^2 D_a}{4\pi r_o^3 \Delta\rho v}\right\}$$

$$= \left\{\frac{D_e C_s}{\left[\frac{1}{r'} - \frac{1}{r_o}\right]} + \frac{3qLD_a}{4\pi r_o \rho\Delta v}\right\}$$

$$C_s' = \frac{\left\{\Delta D_e C_s + \frac{3qLD_a}{4\pi r_o \rho v}\left[\frac{1}{r'} - \frac{1}{r_o}\right]\right\}}{\left\{\Delta D_e + Kr_o^2 D_a \left[\frac{1}{r'} - \frac{1}{r_o}\right]\right\}}$$

Substitution into (3) yields:

$$\frac{dq}{dt} = 4\pi D_e \frac{\left[Kr_o^2 D_a C_s - \frac{3qLD_a}{4\pi r_o \rho v}\right]}{\left\{\Delta D_e + Kr_o^2 D_a\left[\frac{1}{r'} - \frac{1}{r_o}\right]\right\}} = \frac{D_a D_e\left[4\pi K r_o^2 C_s - \frac{3qL}{r_o \rho v}\right]}{\Delta D_e + Kr_o^2 D_a\left[\frac{1}{r'} - \frac{1}{r_o}\right]} \tag{10}$$

In accordance with the shrinking-core model, the rate of release per capsule is

$$\frac{dq}{dt} = -4\pi A r'^2 \frac{dr'}{dt} \tag{11}$$

where

A = loading of drug in capsules, g/cm$^3$.

Under steady state conditions (10) and (11) are equal.

$$-4\pi A r'^2 \frac{dr'}{dt} = \frac{4\pi D_a D_e\left[Kr_o^2 C_s - \frac{LA}{\rho r v_o}(r_o^3 - r'^3)\right]}{\Delta D_e + Kr_o^2 D_a\left[\frac{1}{r'} - \frac{1}{r_o}\right]} \tag{12}$$

$$\frac{dr'}{dt}\left[Ar'^2 (Kr_o D_a - \Delta D_e) - Kr_o^2 \Delta D_a r'\right] - D_a D_e r_o^2 \left(KC_s - \frac{LA}{\rho v}\right) - \frac{D_a D_e L A r'^3}{\rho v r_o} = 0 \tag{13}$$

$$\frac{dt}{dr'} = \frac{\left[Ar'^2\left(Kr_o - \Delta\frac{D_e}{D_a}\right) - Kr_o^2 A r'\right]}{D_e r_o^2\left(KC_s - \frac{LA}{\rho v}\right) + \frac{D_e LA}{\rho v r_o} r'^3} \tag{14}$$

$$t = \int_{r_o}^{r'} \frac{[Ar'^2(Kr_o - \frac{\Delta D_e}{D_a}) - Kr_o^2 Ar']dr'}{[D_e r_o^2(KC_s - \frac{La}{\rho v}) + \frac{D_e LA}{vr_o} r'^3]}$$

$$= \frac{\rho v r_o}{D_e L} \int_{r_o}^{r'} \frac{[(Kr_o - \frac{\Delta D_e}{D_a}) r'^2 - Kr_o^2 r']dr'}{[r'^3 - r_o^3 + \frac{KC_s r_o^3 \rho v}{LA}]} \tag{15}$$

$$t = \frac{\rho v r_o}{3D_e L}[Kr_o - \frac{\Delta D_e}{D_a}]\ln[\frac{LA}{KC_s \rho v}(\frac{r'^3}{r_o^3} - 1) + 1]$$

$$+ \frac{\rho v r_o^3 K}{\sqrt{3} D_e L y}[\tan^{-1}\frac{(2r_o + y)}{\sqrt{3}y} - \tan^{-1}\frac{(2r' + y)}{\sqrt{3}y}] \tag{16}$$

$$+ \frac{\rho v r_o^3 K}{6D_e L y}\ln[\frac{(r'^2 + r'y + y^2)(r_o - y)^2}{(r_o^2 + r_o y + y^2)(r' - y)^2}]$$

where

$$y^3 = r_o^3 \frac{(1 - KC_s \rho v)}{LA}$$

Substituting (14) in (11)

$$\frac{dq}{dt} = \frac{4\pi Ar'^2[D_e r_o^2(KC_s - \frac{LA}{\rho v}) + \frac{D_e LA}{\rho v r_o} r'^3]}{[Ar'^2(Kr_o - \frac{\Delta D_e}{D_a}) - Kr_o^2 Ar']} \tag{17}$$

For the matrix-diffusion controlled case

$$\frac{\Delta D_e}{D_a} \approx 0$$

$$\frac{dq}{dt} = \frac{4\pi r'}{Kr_o(r_o - r')}[KC_s D_e r_o^2 + \frac{D_e LA}{\rho v r_o}(r_o^3 - r'^3)] \tag{18}$$

## III. MODEL PARAMETERS

The model parameters were determined experimentally using standard procedures.

### A. *Solubility of Progesterone in Plasma*

The solubility of progesterone in plasma was determined using radiolabeled drug to be 0.037 $mg/cm^3$.

### B. *Partition Coefficient*

The partition coefficient of progesterone was determined by equilibrating a known weight of polymer in various known concentrations of radiolabeled progesterone in plasma. The ratio of the liquid scintillation counts for equal weights of polymer and drug gave the value of the partition coefficient to be 14.7.

### C. *Density*

The density of the polymer was determined to be 12 $g/cm^3$.

### D. *Diffusion Coefficient*

The diffusion coefficient was determined at 37°C using a Teflon diffusion cell and a lag-time procedure. The membrane was replaced by membranes containing various loadings of unlabeled progesterone. The concentration differences in the two chambers of the cell were measured by using radiolabeled progesterone. An average of the diffusion coefficients was calculated as $7 \times 10^{-12}$ $cm^2/sec$.

## IV. MODEL PREDICTIONS

The predictions of the present model and Higuchi's model are compared with the experimental data in Figures 2 through 11. Differences in loading and size distributions are shown in the figures.

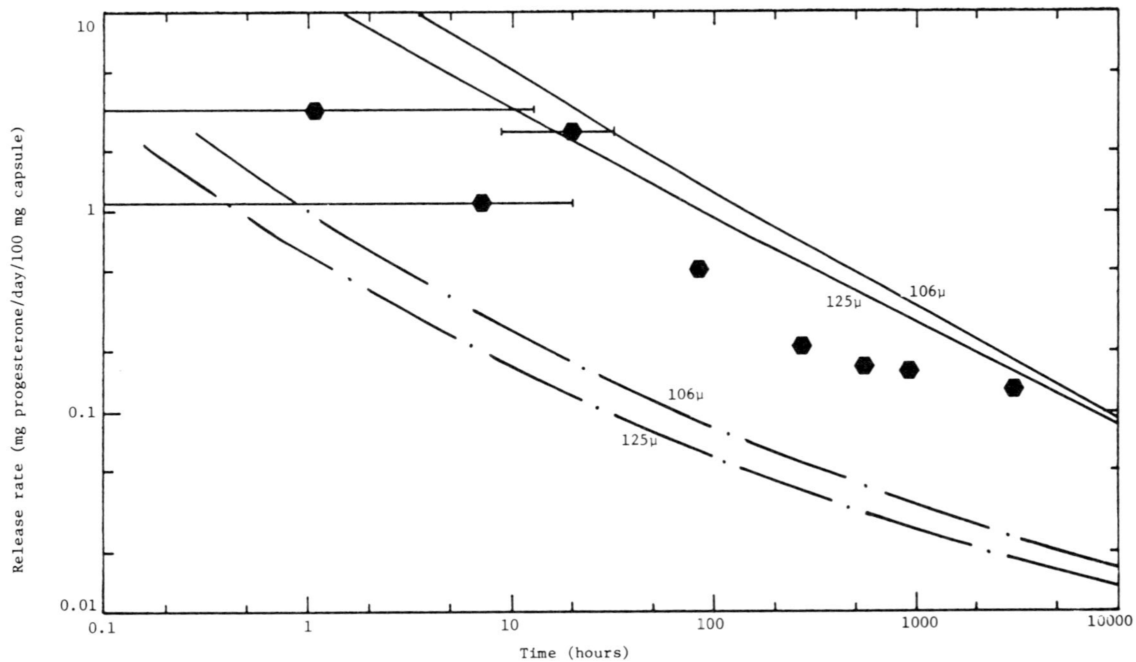


*FIGURE 2. Progesterone release rates in plasma from 106-125 μ microcapsules (12 percent loading), ⬢ experimental data; — this model; —·— Higuchi's model.*

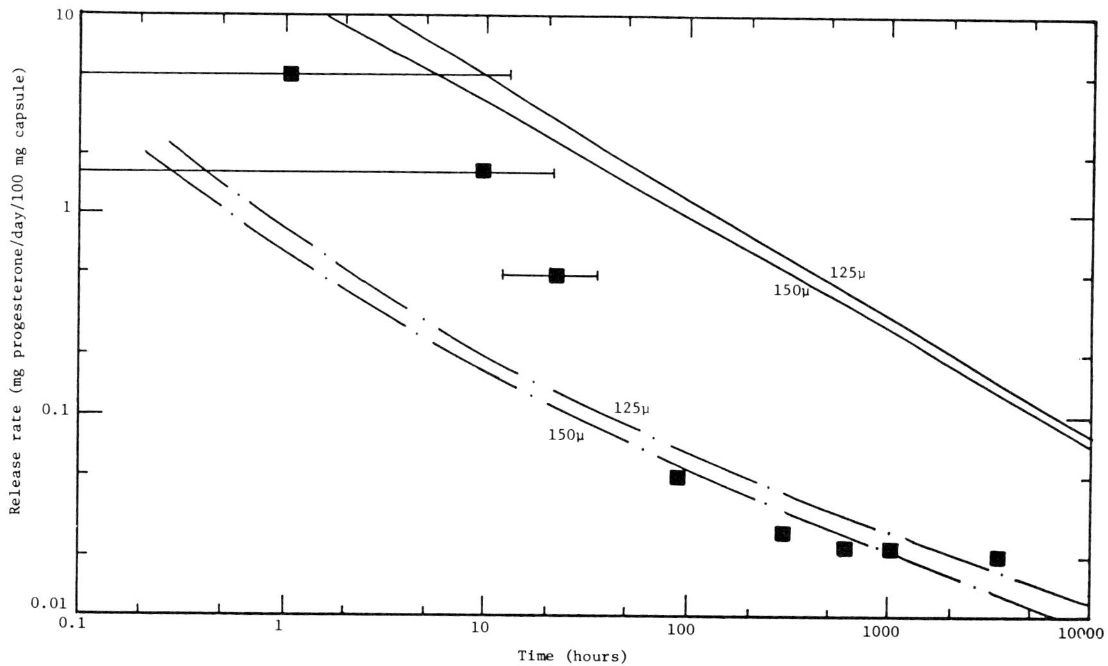


*FIGURE 3. Progesterone release rates in plasma from 125-150 μ microcapsules (12 percent loading), ■ experimental data; —— this model; —·— Higuchi's model.*

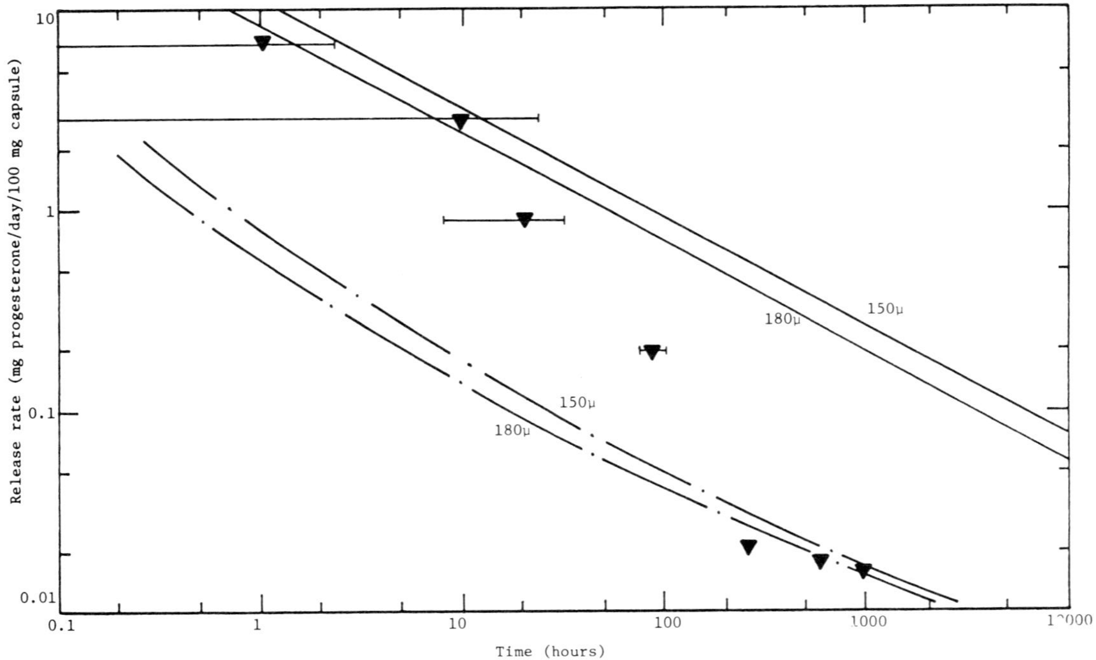


*FIGURE 4. Progesterone release rates in plasma from 150-180 μ microcapsules (12 percent loading), ▼ experimental data; —— this model; —·— Higuchi's model.*

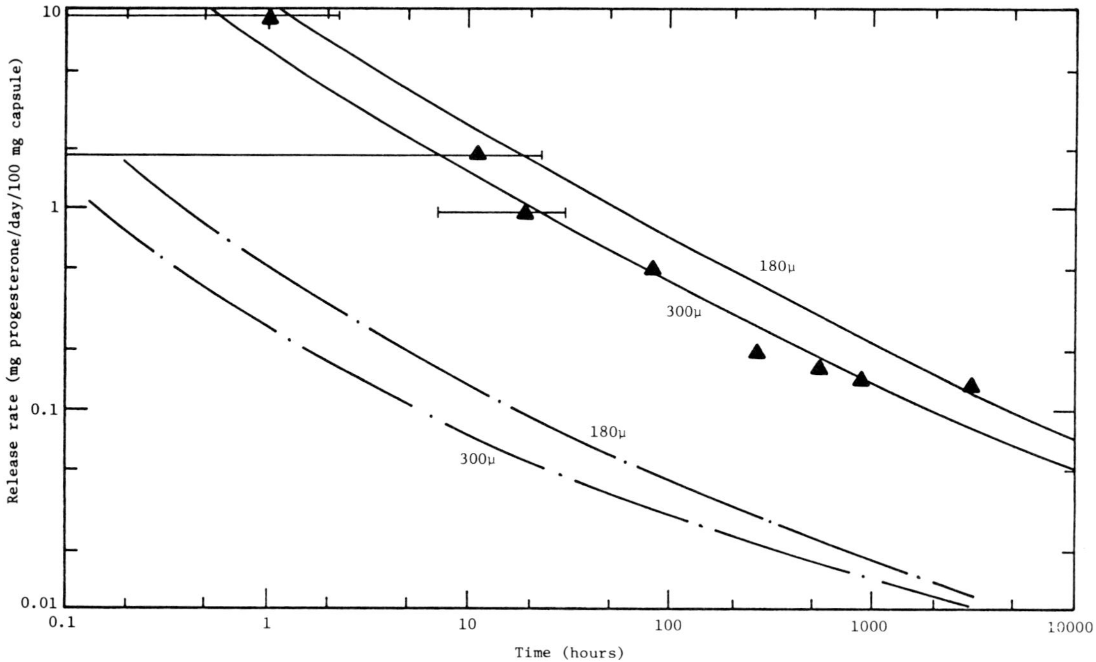


*FIGURE 5. Progesterone release rates in plasma from 180-300 μ microcapsules (12 percent loading), ▲ experimental data; — this model; —·— Higuchi's model.*

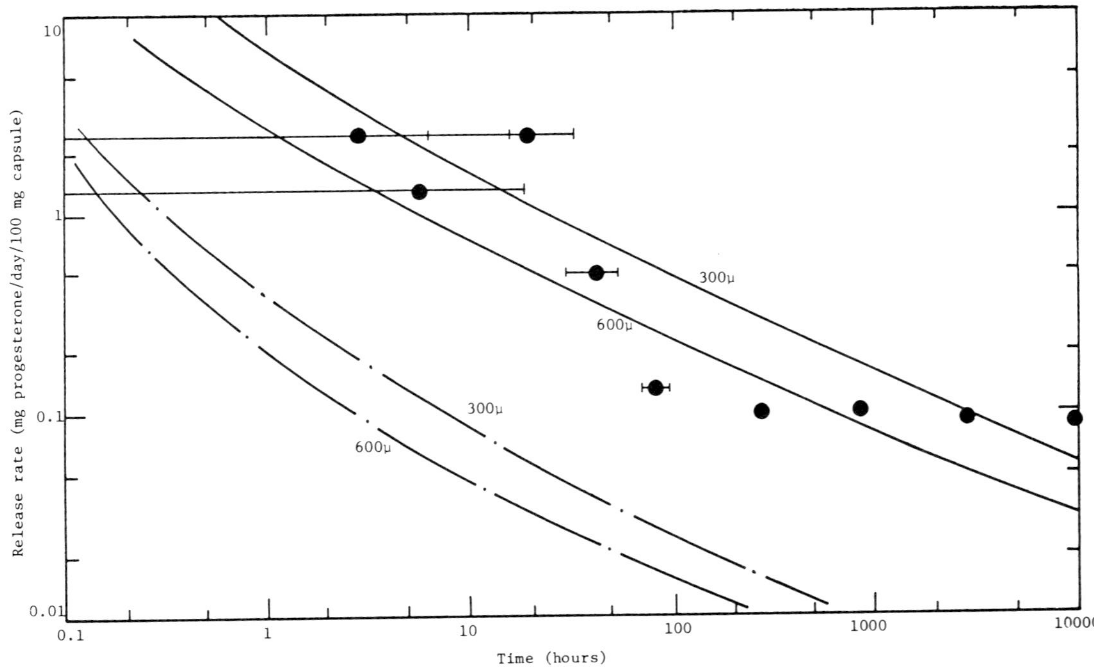


*FIGURE 6. Progesterone release rates in plasma from 300-600 μ microcapsules (12 percent loading), ● experimental data; — this model; —·— Higuchi's model.*

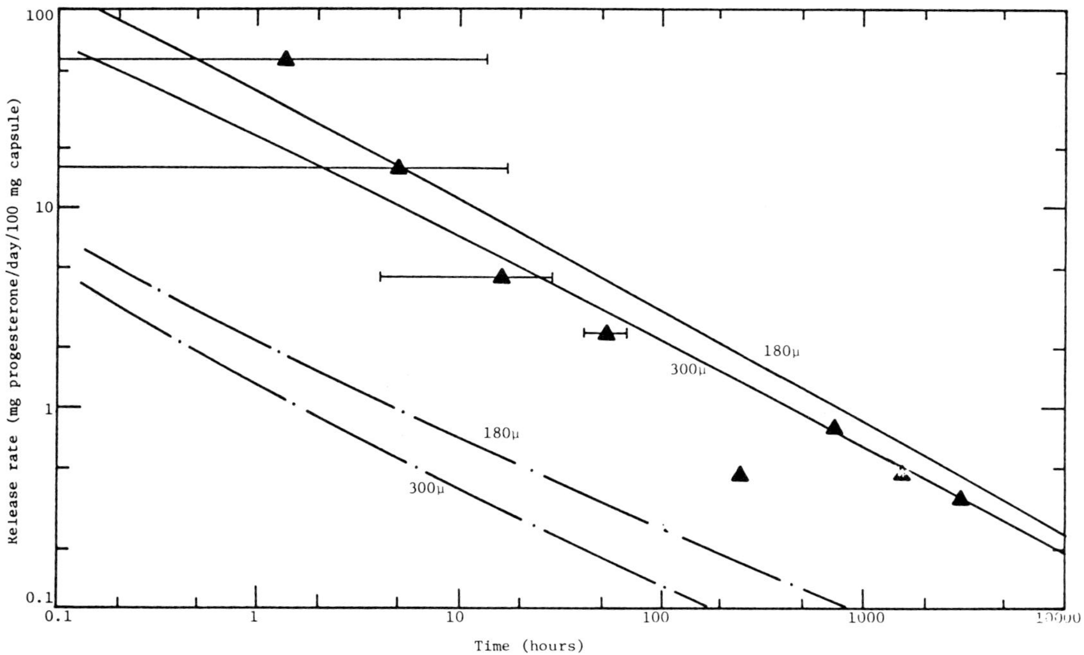


FIGURE 7. Progesterone release rates in plasma from 180-300 μ microcapsules (25 percent loading), ▲ experimental data; — this model; —·— Higuchi's model.

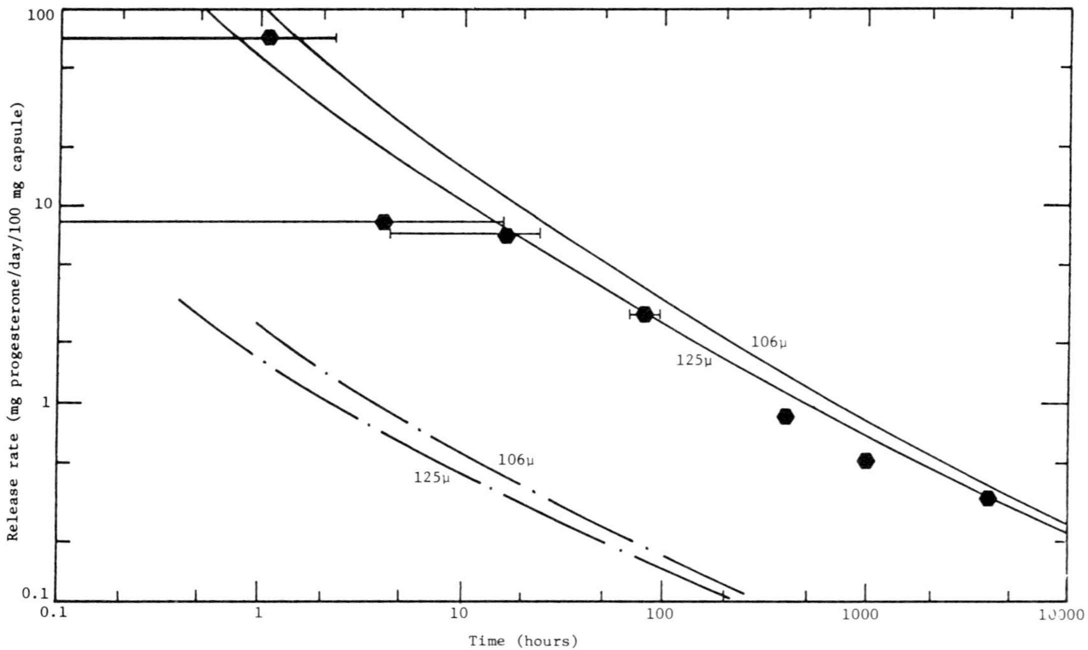


*FIGURE 8.* *Progesterone release rates in plasma from 106-125 μ microcapsules (54 percent loading),* ⬢ *experimental data; — this model; —·— Higuchi's model.*

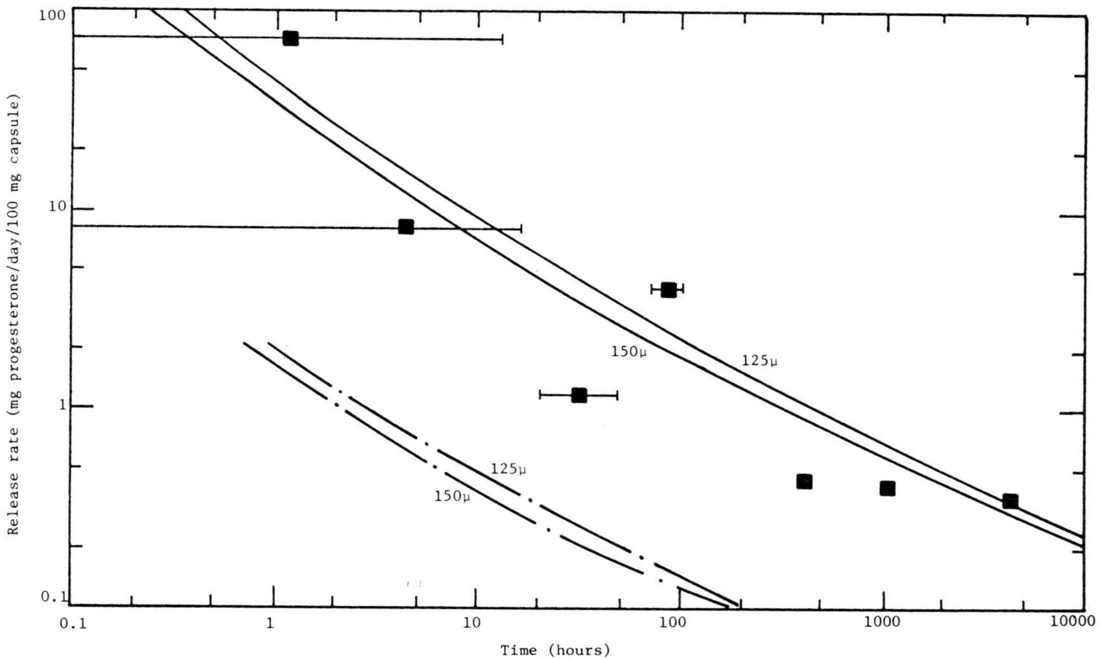


*FIGURE 9. Progesterone release rates in plasma from 125-150 μ microcapsules (54 percent loading), ■ experimental data; — this model; —·— Higuchi's model.*

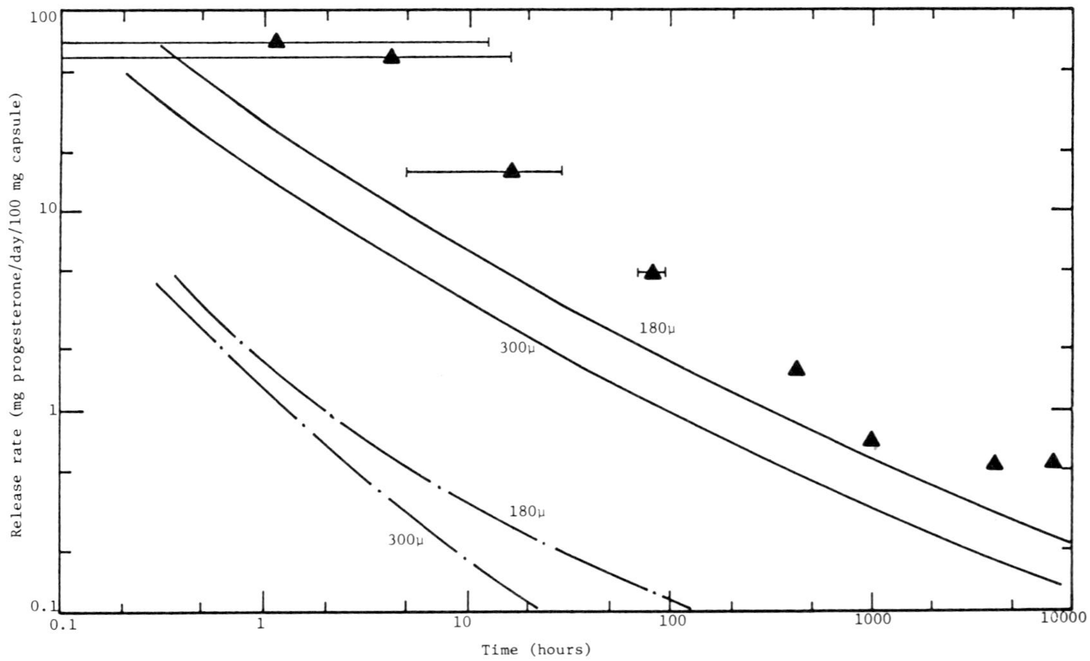


*FIGURE 10. Progesterone release rates in plasma from 180-300 μ microcapsules (54 percent loading), ▲ experimental data; — this model; —·— Higuchi's model.*

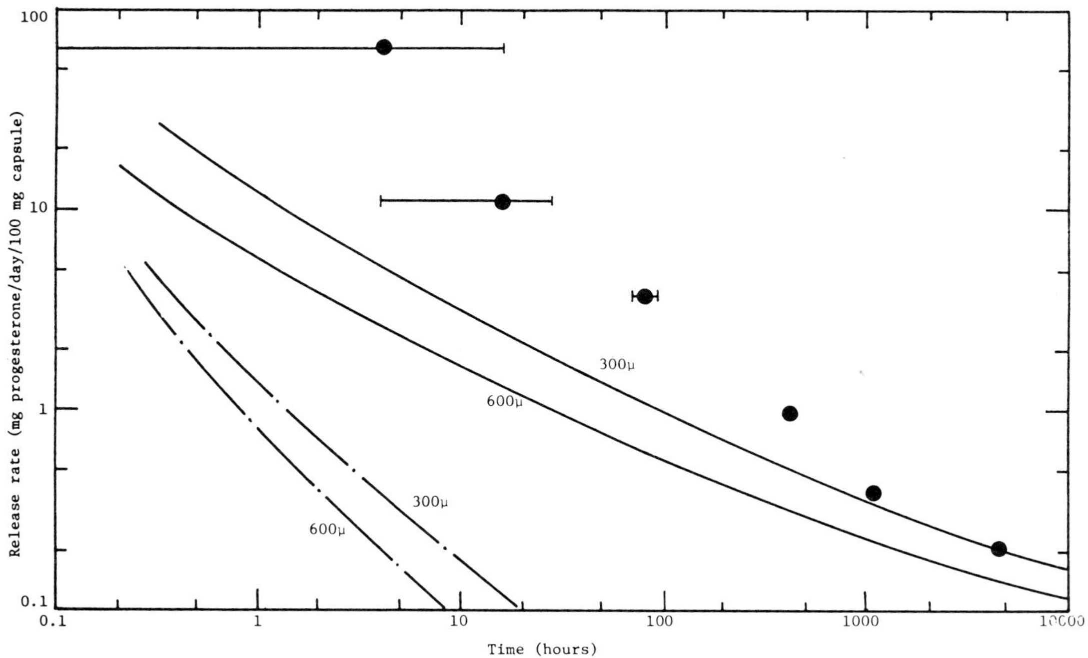


*FIGURE 11. Progesterone release rates in plasma from 300-600 μ microcapsules (54 percent loading), ● experimental data; — this model; —·— Higuchi's model.*

### A. *Sensitivity of Model*

1. *Diffusion Coefficient.* The sensitivity of the present model to the value of the effective diffusion coefficient can be seen in Figure 12. The effects on the model prediction of increasing the diffusion coefficient from the experimentally determined value of $7 \times 10^{-12}$ to $1 \times 10^{-11}$ $cm^2/sec$ and decreasing to $1 \times 10^{-15}$ $cm^2/sec$ are shown.

2. *Effect of Bulk Solution Concentration.* Figure 13 shows the effect of various constant bulk solution concentrations on release rates. The case of $x = 0$ corresponds to Higuchi's model. The dotted line is the curve predicted for the present model where the external resistance is varying. Figure 14 shows the bulk solution concentration (as a fraction of $C_s$) as predicted by the present model and as determined experimentally.

## V. DISCUSSION: MODEL APPLICABILITY TO EXPERIMENTAL RESULTS

The applicability of the present model to the prediction of experimental results depends on the validity of the assumptions made in deriving the model. These assumptions are discussed below.

It was assumed that the microcapsule matrix was homogeneous, with the drug evenly distributed throughout the polymer. Since the microcapsules were made by dissolving the drug and polymer in a single solvent, it is reasonable to assume a homogeneous region inside the capsules. Because of the size of the capsules (<600 μ) and due to practical difficulties, cross-sections could not be photographed. However, the photomicrographs of the capsules display progesterone crystals on the surface of the microcapsules which this model does not take into account. It is possible to eliminate these crystals by an ethanol wash, reducing the burst effect (4). Release rates, rather than the total quantity released, are predicted so that a valid comparison can be made to the experimental results after the first small time interval when the particles correspond more closely to the model.

The assumption of pseudo-steady state is good if there is negligible accumulation of the drug in the matrix once it dissolves. The progesterone loading, A, was always greater than solubility in the matrix. The minimum loading was 144 $mg/cm^3$ compared to $C_s = 0.544$ $mg/cm^3$.

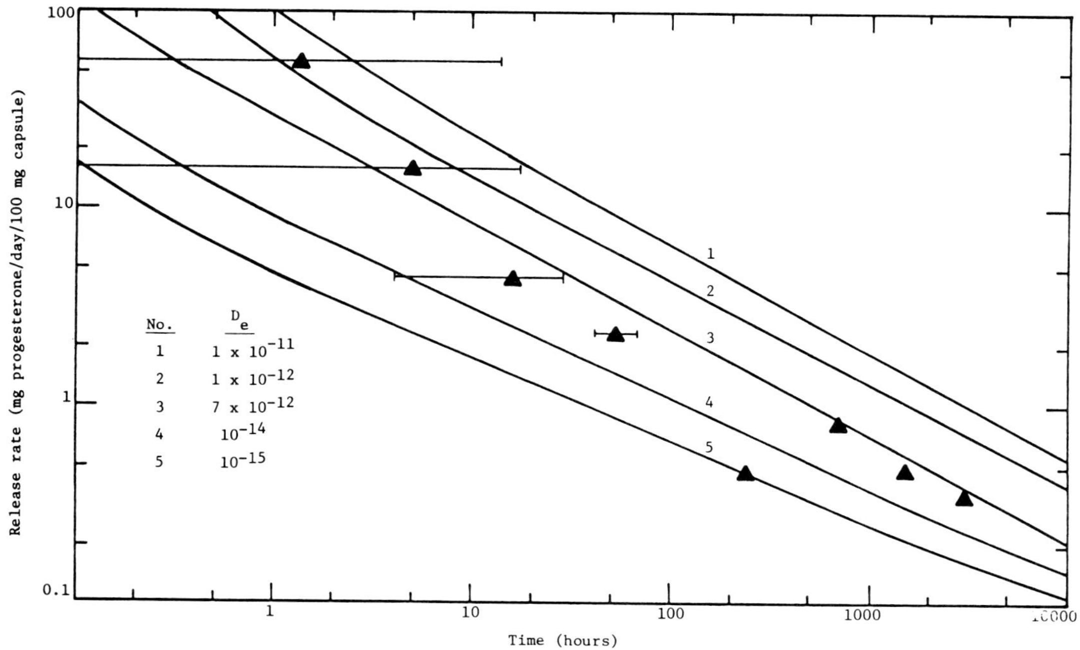


*FIGURE 12. Effect of diffusion coefficient on release rates as predicted by the model for 240 μ microcapsules (25 percent loading).*

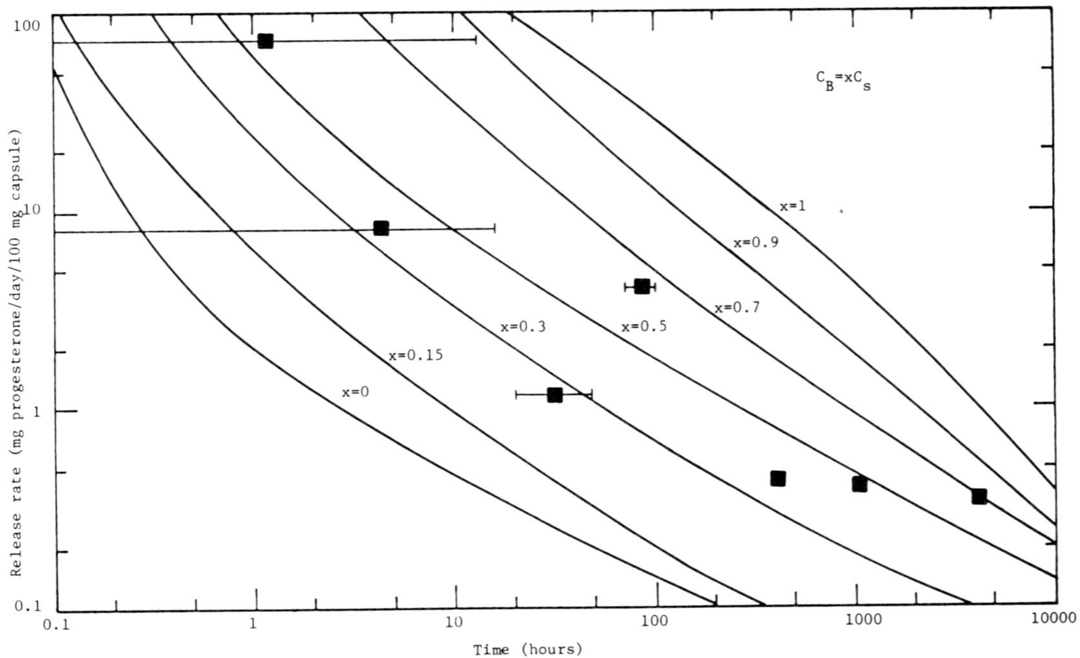


*FIGURE 13. Effect of concentration in bulk solution ($C_B$) as a fraction of solubility ($C_S$) on release rates as predicted by the model for 125-150 μ microcapsules (54 percent loading).*

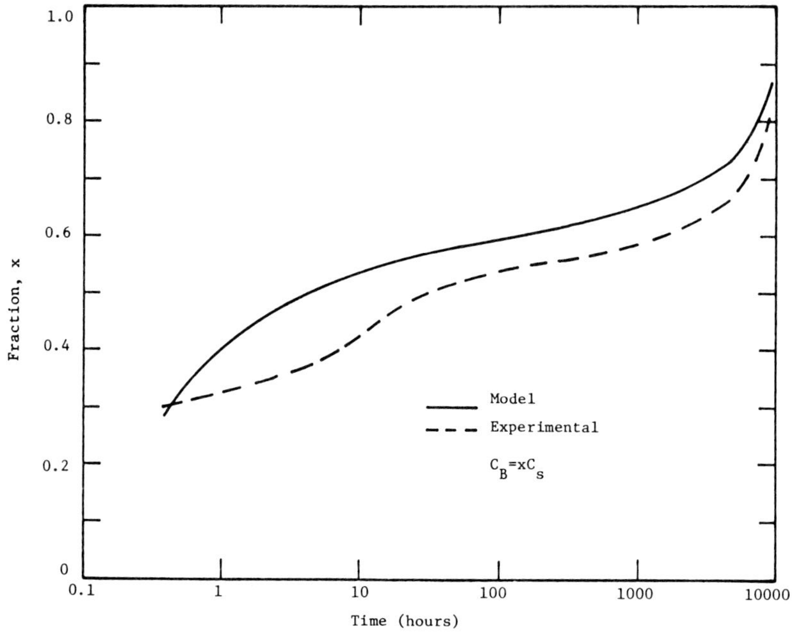


*FIGURE 14. Concentration in plasma as a fraction of solubility coefficient predicted by the present model compared with experimental results for 125-150 μ microcapsules (54 percent loading).*

The assumption of shrinking core without actual change in the overall dimensions of the capsule was checked. No difference in size distribution was observed even after the capsules were in plasma for 15 months. There is a case in literature, similar to ours, in which direct observation would have justified the use of a shrinking-core model. In that study (6), medroxyprogesterone acetate was released from a cylindrical transparent silicone matrix. Initially, the transparent material was rendered opaque by the presence of microscopic crystals of medroxyprogesterone acetate. As the drug dissolved and diffused from the matrix, a clear annular zone appeared. For this case, a shrinking-core model would be appropriate.

The diffusion coefficients calculated experimentally from flat films containing different amounts of progesterone distributed throughout the film do not change drastically compared to values calculated from film containing no progesterone. From this it may be assumed that the diffusion coefficients in the matrix do not change as the drug is released.

In the *in vitro* case, the boundary layer would be very thin because of mixing. In an *in vivo* situation, where there can be clusters of microcapsules in the tissue and there is little "flow," this assumption would be questionable; the whole model rather than the matrix-diffusion controlled case would have to be considered.

## VI. CONCLUSIONS

Even though this model does not predict the results exactly, it is superior to Higuchi's model. Considering the variables which can affect the release data and which are not considered in the model, i.e., the variables in the microencapsulation procedure, the predictions are quite good. Both models predict the same results at short times. The theoretical release rates do not fall as straight lines on a log-log plot because of the mathematical nature of the equations. Because the present model predicts the behavior better in cases where the external resistance exists, it may be assumed that it would do a better job than Higuchi's in *in vivo* situations.

## VII. NOMENCLATURE

$r$ = radius, cm
$r'$ = radius of shrinking core, cm
$r_o$ = initial radius of capsule, cm
$\Delta$ = thickness of diffusion boundary layer, cm
$A$ = loading of drug in capsules, $g/cm^3$
$C_s$ = solubility of drug in the matrix, $g/cm^3$
$C_s'$ = concentration in the matrix at capsular boundary, $g/cm^3$
$C_a'$ = concentration at liquid boundary, $g/cm^3$
$C_B$ = concentration in bulk liquid, $g/cm^3$
$q$ = amount of drug released per capsule, g
$C_a$ = solubility of drug in liquid, $g/cm^3$
$D_e$ = effective diffusion coefficient, $cm^2/sec$
$K$ = partition coefficient
$C$ = concentration, $g/cm^3$
$D_a$ = diffusion coefficient in the boundary layer, $cm^2/sec$
$Q$ = total amount of drug released by the total dose of capsules, g
$v$ = volume of liquid in release bottle, $cm^3$
$\rho$ = density of capsules, $g/cm^3$
$L$ = weight of capsule used in release experiments, g

## ACKNOWLEDGMENT

This research was supported by the program for applied research on fertility regulation under contract PARFR-80N.

## REFERENCES

1. N. Cohen, Fertility and Sterility, 3, 3, 201 (1952).
2. K. Moghissi, Fertility and Sterility, 17, 8, 663 (1966).
3. K. Moghissi and C. Marks, Fertility and Sterility, 22, 7, 424 (1971).
4. D. V. S. Gupta, R. E. Sparks, N. S. Mason, and D. W. Keller, Microencapsulation of Progesterone for Contraception by Intracervical Injection, Presented at IV International Symposium on Microencapsulation, Key West, Florida, March 1979.
5. T. Higuchi, J. Pharm. Sci., 52, 12, 1145 (1963).
6. T. J. Roseman and W. I. Higuchi, J. Pharm Sci., 59, 3, 353 (1970).

# PROSPECTS FOR CONTROLLING FOREST LEPIDOPTERA WITH CONTROLLED RELEASE PHEROMONE FORMULATIONS

*Gary E. Daterman*
*Charles Sartwell*
*Lonne L. Sower*

Pacific Northwest Forest and
Range Experiment Station
Forestry Sciences Laboratory
Corvallis, Oregon

## I. INTRODUCTION

Because of increasing environmental concerns, restrictions on use of all agents for forest pest control have dramatically increased in recent years. For future control programs, development of techniques incorporating pheromones and other behavior-modifying chemicals may represent one of our best investments. Behavior-modifying chemicals, and pheromones in particular, are generally nontoxic and highly specific for target pests. For management applications, the most progress has been with sex pheromones, and recent experiments suggest that these materials can successfully replace toxic chemicals to control some forest pests.

Various strategies have been considered for the use of pheromones to suppress insect populations (1-3). These include: the trap-out method, which reduces numbers by attracting one or both sexes to a trap where they are killed; baiting techniques in which pests are attracted to a source of poison or disease inoculum that is then spread to other members of the population; the use of marking or repellent materials to prevent feeding, oviposition, or aggregation; and mating disruption, which relies on the broadcast application of synthetic sex pheromones to inactivate communication between the sexes, thus preventing reproduction.




ISBN 0-12-074450-3

For operational pest control, mating disruption has shown the most potential. For successful reproduction, many insects depend on a sex-attractant pheromone to bring males and females together for mating. In the presence of relatively greater quantities of synthetic sex pheromone, male moths become incapable of sensing the lesser amounts of natural pheromone emitted by females. The mechanisms that cause this disruption of communication are generally unknown, although Shorey (4) indicated that sensory adaptation, habituation of the central nervous system, and confusion may all be involved. The important point for our purposes is that sexual communication can be blocked and successful suppression of populations obtained. Disruption is becoming operational for some agricultural pests, most notably the pink bollworm, *Pectinophora gossypiella* (Saunders), a pest of cotton. The ConRel Company (Needham Heights, Mass.) has secured registration from the Environmental Protection Agency for a commercial controlled-release pheromone formulation for control of the pink bollworm.

## II. STATUS OF MATING DISRUPTION TO CONTROL FOREST LEPIDOPTERA

### A. *Western Pine Shoot Borer*

The western pine shoot borer, *Eucosma sonomana* Kearfott, is vulnerable to control by mating disruption. The larvae feed in the pith of leader stems of young ponderosa pine, *Pinus ponderosa* Lawson, and several other western pine species. Trees 1.5 to 15 m tall are most frequently infested. Damage occurs because of suppressed height growth or growth distortions in the stem (5). A single larva boring in a terminal shoot is sufficient to cause economic loss. The important point in the biology of the insect for pheromone-based control—is that only a few hundred moths exist per hectare of pine plantation. This contrasts with defoliators, where economically damaging populations may number in the millions per hectare. It seems only logical that an insect sparsely distributed over its habitat will be more vulnerable to a control strategy designed to block long-distance communication between the sexes.

In 1977, the sex-attractant pheromone of western pine shoot borer was identified as a 4:1 ratio of Z- and E-9-dodecenyl acetates by Sower *et al.* (6). Small-scale field tests in 1977 showed that males could be prevented from responding to pheromone-baited traps by gridding an area

(25- x 25-m plots) with releasers of synthetic pheromone formulated in solid plastic polyvinyl chloride (PVC). In 1978 studies were conducted on 40 hectares of ponderosa pine plantations in southern Oregon using two controlled-release formulations. The first was the ConRel hollow fiber dispensed from an aircraft at 15g/ha pheromone. The second was a ground application of pheromone formulated in solid strips of PVC (7). Treated areas were gridded with 5-cm strips of PVC containing 10% pheromone by weight, to obtain two dosages: 3.5 g/ha from 100 releasers/ha spaced at 10 m, and 14 g/ha from 400 releasers/ha spaced at 5 m. The results of both 1978 tests were highly encouraging. Male-to-female orientation behavior was almost entirely disrupted in all treated plots, as measured by numbers of male moths responding to baited traps (Table I). Of greatest significance, however, was the decrease in damage to pine shoots (Table 2). Damage to terminal shoots ranged from 68 to 84% lower than on untreated controls (8,9). Considering that plot sizes were still relatively small and some immigration of mated females from surrounding untreated areas could be expected, we considered this degree of damage reduction to be excellent.

The success of the 1978 tests prompted larger scale tests in 1978 to evaluate the technique on a more operational scale. Two controlled-release formulations of the pheromone were applied for mating disruption tests in 1979. Eight

*TABLE I. Mating disruption of western pine shoot borer as indicated by response of male moths to monitoring traps.*

| *Year* | *Place* | *Dosage (g/ha)* | *No. Moths Trapped* | *Disruption (%)* |
|---|---|---|---|---|
| *1978* | *Bly, OR* | *check* | *477* | *--* |
| | | *3.5* | *0* | *100.0* |
| | | *14.0* | *0* | *100.0* |
| | *Keno, OR* | *check* | *234* | *--* |
| | | *15.0* | *1* | *99.6* |
| *1979* | *Klamath Falls, OR* | *check* | *2,262* | *--* |
| | | *10.0* | *0* | *100* |
| | *McCloud, CA* | *check* | *265* | *--* |
| | | *0.2* | *41* | *84.0* |
| | | *2.0* | *4* | *98.0* |
| | | *20.0* | *1* | *99.6* |

TABLE II. Summary of mating disruption tests to control damage by western pine shoot borer.

| Year | Location | Application | Formulation | Dosage (g/ha) | Area (ha) | Control[a] (%) |
|---|---|---|---|---|---|---|
| 1978 | Bly, OR | ground | strips[b] | 3.5 | 10 | 83 |
| 1978 | Bly, OR | ground | strips | 14.0 | 10 | 84 |
| 1978 | Keno, OR | aerial | fibers[c] | 15.0 | 20 | 68 |
| 1979 | Klamath Falls, OR | aerial | fibers | 10.0 | 800 | 76 |
| 1979 | McCloud, CA | aerial | flakes[d] | 20.0 | 110 | 88 |

[a]Percent control is based on reduction in numbers of infested terminal shoots in treated vs. untreated plots.
[b]Solid strips of polyvinyl chloride plastic (PVC) containing 10% pheromone by weight.
[c]ConRel Co., Needham Heights, MA.
[d]Hercon Div., Health Chem. Corp., New York, NY.

hundred hectares of ponderosa pine plantation were treated in southern Oregon with ConRel's hollow-fiber system at the rate of 10 g/ha pheromone. In northern California, 360 ha were treated with Hercon laminated flakes (Health Chemicals Corp., New York, NY) at dosages of 0.2, 2, and 20 g/ha. Behavioral data based on trap captures in treated and untreated plots indicated that both the ConRel and Hercon formulations were effective (Table I). Also, both formulations appeared to have good release-rate characteristics, because the treatments remained effective throughout the flight season. At 10-20 g/ha, damage to terminal shoots was again effectively reduced (76-88%) by both formulations, compared to the untreated controls. The full report of the 1979 disruption tests will be published elsewhere, but the preliminary results are compared in Table II with control levels in 1978.

Control of this pest by mating disruption apparently has excellent potential for operational use. Although some refinements in pheromone dosage and application may still be desirable, the respective companies are currently moving to obtain registrations from the Environmental Protection Agency for commercial use of their formulations to control western pine shoot borer by mating disruption.

### *B. Gypsy Moth*

The gypsy moth, *Lymantria dispar* (L.), may have received more effort for development of its sex pheromone as a control agent than all other forest Lepidoptera combined (10-14). An introduced pest from Europe, this defoliator of deciduous hardwoods is one of our most important pests of shade trees in hardwood forests in the northeastern United States. Field testing of mating disruption to control gypsy moth has produced mixed results, although the findings have been generally promising, and the potential is good for developing the pheromone as an effective control agent. The mixed results with this insect occurred largely because, until recently, the effects of (+) and (-) enantiomers of its pheromone were not considered and male gypsy moths respond only to the (+) enantiomer (14). In 1978 Schwalbe *et al.* (15) tested several different controlled-release formulations with good success. The most successful 1978 formulation was a plastic-coated gelatin microcapsule from National Cash Register Company. Mating was reduced by 83% in Massachusetts plots and by 97% in Maryland plots where insect density was much lower. The authors pointed out the apparent significance of differences in population density on the efficacy of disruption. The development of mating disruption as a management tool for gypsy moth is continuing; it appears to be an effective tool for suppression if applied in a timely way against appropriate populations.

### *C. Douglas-fir Tussock Moth*

The Douglas-fir tussock moth, *Orgyia pseudotsugata* (McDunnough), is a defoliating pest of Douglas-fir and true fir forests of Western North America. This insect normally exists at very low densities, but populations can periodically erupt to outbreak levels and cause considerable damage within a short period. Use of its sex-attractant pheromone to disrupt communication between the sexes was first demonstrated in a small-scale field test in 1976 (16). In two subsequent tests, ConRel hollow-fiber releasers were applied by aircraft against a high-density population (evident defoliation) in New Mexico and a low-density population (no apparent defoliation) in southern Oregon. As in the 1978 gypsy moth tests (15), the highest degree of disruption occurred in the lower density plots. The reduction in reproductive rates, as measured by numbers of tussock moth females producing fertile eggs, was greater than 99% in the low-density plots and 77% in the high-density locations (17). This apparent effect of population density is important, because the 23% mating that was not controlled in the high-density area might produce sufficient progeny to sustain the outbreak into the following year. Thus, population density of the pest is apparently an important factor to consider in developing mating disruption as a control strategy.

### *D. Spruce Budworms*

The eastern spruce budworm, *Choristoneura fumiferana* (Clemens), and the western spruce budworm, *C. occidentalis* Freeman, represent two of the most destructive forest insects in North America. Severe defoliation of coniferous trees (mostly Douglas-fir and true firs in the West, and balsam fir and spruces in the East) resulting from outbreaks of spruce budworm causes growth loss, top-kill, and tree mortality. Sanders (18) first demonstrated mating disruption of *C. fumiferana* by dispensing synthetic pheromone by both ground and aerial methods to small plots in Ontario. The treatments were successful in disrupting pheromone communication as indicated by reduced catches in female-baited traps. Schmidt and Seabrook (19) compared the effectiveness of mating disruption treatments against different population densities of caged *C. fumiferana;* their treatments were most effective against the lower density populations. In 1979 an aerial application of pheromone formulated in laminated plastic flakes (Hercon product) was made against the western spruce budworm in northern Oregon (Daterman *et al.* unpubl.). About 350 ha were treated at dosages of 2 and 20 g/ha. Male orientation to female-baited traps was successfully disrupted at a rate of 87% in the 20 g/ha treatment. Egg samples were taken in two of the larger plots (150 ha) and untreated check plots, but no significant differences were detected. This could have been because of a degree of successful mating within the treated plots and/or dispersal of mated females into the treated plots. As pointed out by Sanders (18), to determine the effect of disruption on production of fertile eggs by *Choristoneura,* a larger or more remote area must be treated to avoid invasion by gravid females from untreated areas.

## III. STATUS OF PHEROMONE APPLICATION TECHNOLOGY

For operationl use of pheromones in mating disruption or other pheromone strategies, a controlled-release formulation is essential. Several proprietary formulations of controlled-release pheromones are now available. These include capillary-sized, plastic hollow fibers, layered or laminated flakes of solid plastic, small-diameter microcapsules, and various types of granular formulations. Systems for aerial application exist for all of them, a factor essential in forestry where large areas of rugged, remote terrain may require control treatments.

Formulations suitable for ground applications are also available and include modifications of the hollow fibers and laminated plastic systems plus other materials such as solid polyvinyl chloride plastic (7,18,19), rubber materials impregnated with pheromone (17), or simply open reservoirs of pheromone. Such ground application systems can be useful for treating small experimental plots or small, accessible areas requiring treatments for control.

### A. *The Question of Coverage*

Although systems for controlled-release of pheromones have generally become more available and effective in recent years, several aspects could still be improved considerably. One variable that is not fully understood concerns the question of spacing or coverage of mating-disruption releasers.

This point was clearly emphasized in the 1978 western pine shoot borer tests where both aerially applied fibers and ground-applied grids of PVC releasers of pheromone provided excellent control. The hollow-fiber application used some 30,000 to 50,000 individual fiber releasers to dispense 15 g of pheromone, but the PVC strips on a 10-m grid used only 100 releasers to dispense 3.5 g/ha. For additional perspective, consider that the hollow-fiber treatment averaged 300 releasers/100 $m^2$, but the equally effective PVC application provided only 1 releaser/100 $m^2$.

To investigate further the effect of releaser spacing in 1979, we gridded 4-ha plots on a large ponderosa pine plantation in northern California with strips of PVC pheromone formulations spaced at various intervals. Lengths of PVC releasers were varied to maintain a constant total of 500 cm of formulation per hectare for a standard dosage of 3.5 g/ha of pheromone per season. Spacings between releasers ranged from 10 to 100 m, and the number of releasers/ha ranged from 1 to 100. The effectiveness of disruption of pheromone communication was assessed strictly by comparative captures of male moths in traps baited with 3 mm x 5 mm PVC pellets containing 0.5% synthetic pheromone. Four traps were placed in new positions within these plots every 14 days.

Trapping results are summarized in Table III. No significant difference in trap catch occurred among the treated

*TABLE III. Effect of mating disruption with widely spaced releasers on capture of Eucosma sonomana male moths in attractant-baited traps, McCloud, CA, 1979.*

| *Spacing (m)* | *Treatment[a] Releasers No./ha* | *Size (cm)* | *Moths Caught (avg. no./trap)[b]* | *Catch Reduction (%)* |
|---|---|---|---|---|
| *Check* | *0* | *--* | *28.9 a* | *--* |
| *100* | *1* | *500.0* | *3.2 b* | *89* |
| *50* | *4* | *125.0* | *1.8 b* | *94* |
| *33* | *9* | *56.6* | *1.9 b* | *93* |
| *20* | *25* | *20.0* | *2.4 b* | *92* |
| *10* | *100* | *5.0* | *1.6 b* | *95* |

[a]*Nominal dosage on all treated plots was 3.5 g/ha.*
[b]*Means followed by same letter do not differ (Tukey's w, $p < 0.05$).*

plots, regardless of spacing intervals between pheromone releasers. Disruption was always about 90% compared to the untreated control plots. Further, most moth captures were by traps located near the periphery of the plots, which suggests that if larger areas were treated disruption would be even greater.

These data demonstrate great flexibility in the spacing of pheromone releasers to accomplish mating disruption. The concept of wide spacing of high-strength pheromone releasers is not new; Farkas *et al.* (21) demonstrated good disruption success for cabbage looper moths with pheromone releasers spaced at up to 400 m. The effectiveness of widely spaced, higher strength releasers is significant, because it opens the way for different approaches to aerial application of pheromones, plus realistic opportunities to suppress some pest problems by hand applications. Ground applications of mating disruption may be more efficient for control of some forest pest problems associated with seed orchards, nurseries, or small plantations.

## IV. PROSPECTS FOR OPERATIONAL CONTROL

### A. *Low-Density Forest Pests*

We foresee an optimistic future for mating disruption as an operational control procedure for at least some forest insect pests. Certainly the strategy is sound for western pine shoot borer and probably other insects that can be classified as pests at low densities. We are arbitrarily categorizing low-density pests as those capable of causing damage at populations of 10,000 or fewer adult insects per hectare. The western pine shoot borer, which can cause damage at densities of only a few hundred moths per hectare, clearly fits into this low-density classification. In this respect, the western pine shoot borer probably represents a model species in terms of the biological characteristics that promote successful control by pheromone disruption. Its most important characteristics in this context are the low number of adults per hectare and the fact that individuals are scattered throughout the area and not aggregated on only a few trees. Table IV lists some examples of other low-density pests that we consider candidates for control by mating disruption or some other pheromone-based technique. Some of these species have considerably higher densities than western pine shoot

TABLE IV. *Examples of low-density forest insect pests vulnerable to control by mating disruption with synthetic pheromones.*

| Common Name | Scientific Name | Resource Damaged |
|---|---|---|
| Western pine shoot borer | *Eucosma sonomana* | Pine plantations |
| Eastern pine shoot borer | *Eucosma gloriola* | Pine plantations |
| Metallic pitch nodule moth | *Petrova metallica* | Shelterbelts |
| European pine shoot moth | *Rhyacionia buoliana* | Pine plantations |
| New Mexico pine tip moth | *Rhyacionia neomexicana* | Pine plantations |
| Nantucket pine tip moth | *Rhyacionia frustrana* | Pine plantations |
| Seedworms | *Laspeyresia* spp. | Cone and seed |
| Douglas-fir cone moth | *Barbara colfaxiana* | Cone and seed |
| Coneworms | *Dioryctria* spp. | Cone and seed |
| Sequoia pitch moth | *Synanthedon sequoiae* | Ornamental trees |
| Variegated cutworm | *Peridroma saucia* | Nursery seedlings |

borer; although they might be considered borderline for our classification system, we still consider them good candidates for suppression by mating disruption.

### *B. High-Density Forest Pests*

Those species that cause measureable damage at densities of greater than 10,000 adult insects per hectare are arbitrarily called high-density forest pests. Examples are defoliators such as, the spruce budworms, Douglas-fir tussock moth, gypsy moth, blackheaded budworms, *Acleris* sp., and others. Adult densities can become very high in outbreak conditions, sometimes exceeding a million or more moths per hectare. Under such conditions it is questionable if pheromone treatments alone could suppress a population. Schwalbe *et al.* (15) and Sower *et al.* (17) have already demonstrated that effectiveness of mating disruption of gypsy moth and Douglas-fir tussock moth decreases at high densities. Those and other studies, however, still showed definite positive effects of mating disruption on suppression of reproductive behavior at the high populations. Additional work is required to clarify whether or not pheromone disruption can play an effective role in controlling outbreak populations of defoliators or other high-density pests.

Currently, what is needed are disruption applications on large areas or against isolated defoliator populations to avoid invasion by females from untreated areas. The treatment effects on reproduction and damage can then be monitored into following generations. Possibly complete disruption of mating activity is not necessary because natural enemies of the pest are not hindered by the pheromone application and will still affect the remaining population. Even if mating disruption cannot adequately suppress high-density populations, the method may still have an important role in an integrated management plan. For example, chronic outbreak areas could be treated with a combination of a microbial pesticide plus the pheromone. The pesticide would be focused on the higher density centers of the outbreak, and the mating disruption method would be added to "mop up" by preventing reproduction in the outbreak centers and in peripheral areas. Because pheromones are nontoxic, this approach might be especially helpful in areas where environmental concerns are predominant.

Because high densities probably limit the effectiveness of control by disruption, another option might be to apply the technique as a preventive strategy. This would entail careful monitoring of the population to time the control application before large populations develop. With accurate prediction of a building outbreak, preventive measures should be much more effective than current practices, which treat defoliators only after much damage has occurred.

## V. CONCLUSIONS

Prospects appear excellent for establishing pheromone-based methods to control some forest pests. The mating disruption strategy for western pine shoot borer has been successful in controlling damage in large-scale field tests, and development has reached operational scale. For other pest species, more research is needed to resolve questions on when and how much pheromone to apply, effective means to disperse the pheromone, effects of pest density on efficacy of treatment, and degree of disruption needed to achieve satisfactory control of damage. High-density pests, such as forest defoliators, present a special problem for determining the possible role of pheromones as control agents. Because of their high numbers in outbreaks, pheromone applications may be restricted to a preventive strategy for treating before an outbreak or to strategies combining the pheromones with more conventional control agents.

Developing methods for dispersing pheromones for control by mating disruption also deserves expanded effort. The experience with western pine shoot borer—which demonstrated the equal effects of widely spaced, high-strength disruptant releasers with more numerous, aerially applied releasers—underlines the options that exist for application of pheromone for mating disruption. This principal likely extends to agricultural pests as well, and it could result in significant changes in the design and structure of controlled-release formulations and related dispensing equipment.

Assuming a continuing interest by private industry, realistic requirements by regulatory agencies for commercial registration, and continued support for related research needs, we believe that an increasing incidence of mating-disruption and other pheromone-related strategies will occur in insect pest management.

## VI. FOOTNOTE

This paper reports the results of research only. Mention of a pheromone does not constitute a recommendation for its use as a pesticide by the USDA nor does it imply registration under FIFRA as amended. Also, mention of a commercial or proprietary product does not constitute recommendation or endorsement by the USDA. This chapter was prepared by U.S. Government employees on official time, and it is therefore in the public domain.

## REFERENCES


1. M. Beroza (Ed.), Pest Management with Insect Sex Attractants, ACS Symp. Series No. 23, American Chemical Society, Washington, D.C. 1976, p. 192.
2. H. H. Shorey and J. J. McKelvy (Eds.), Chemical Control of Insect Behavior: Theory and Application, John Wiley and Sons, New York, 1977, p. 414.
3. W. L. Roelofs in Biochemistry of Insects, M. Rockstein (Ed.), Academic Press, New York, 1978, p. 649.
4. H. H. Shorey in Chemical Control of Insect Behavior: Theory and Application, H. H. Shorey and J. J. McKelvey, (Eds.), John Wiley and Sons, New York, 1977, p. 414.
5. K. J. Stoszek, J. Forest, 71, 701-705 (1973).
6. L. L. Sower, G. E. Daterman, C. Sartwell, and H. T. Cory, Environ. Entomol., 8, 265-267 (1979).
7. G. E. Daterman, USDA For. Serv. Res. Pap. PNW-180, 12 p., 1974.
8. D. L. Overhulser, G. E. Daterman, L. L. Sower, C. Sartwell, and T. W. Koerber, Mating disruption with synthetic sex attractants controls damage by *Fucosma sonomana* (Lepidoptera: Tortricidae, Olethreutinae) in *Pinus ponderosa* plantations. II. Aerially applied hollow fiber formulations, Can. Entomol., 112, 163-165 (1980).
9. C. Sartwell, G. E. Daterman, L. L. Sower, D. L. Overhulser, and T. W. Koerber, Mating disruption with synthetic sex attractants controls damage by *Euscosma sonomana* (Lepidoptera: Tortricidae, Olethreutinae) in *Pinus ponderosa* plantations. I. Manually applied polyvinyl chloride formulation, Can. Entomol., 112, 159-162 (1980).
10. M. Beroza and E. F. Knipling, Science, 177, 19-27 (1972).

11. E. A. Cameron, C. P. Schwalbe, M. Beroza, and E. F. Knipling, Science, 183, 972-973 (1974).
12. C. P. Schwalbe, E. A. Cameron, D. J. Hall, J. V. Richerson, M. Beroza, and L. J. Stevens, Environ. Entomol., 3, 589-592 (1974).
13. J. R. Plimmer, B. A. Bierl, R. E. Webb, and C. P. Schwalbe, ACS Symp. Series No. 53, American Chemical Society, Washington, D.C., 1977, p. 205.
14. J. R. Plimmer, C. P. Schwalbe, E. C. Paszek, B. A. Bierl, R. E. Webb, S. Marumo, and S. Iwaki, Environ. Entomol., 6, 518-522 (1977).
15. C. P. Schwalbe, E. C. Pascek, R. E. Webb, B. A. Bierl-Leonhardt, J. R. Plimmer, C. W. McComb, and C. W. Dull, J. Econ. Entomol., 72, 322-326 (1979).
16. L. L. Sower and G. E. Daterman, Environ. Entomol., 6, 889-892 (1977).
17. L. L. Sower, G. E. Daterman, R. D. Orchard, and C. Sartwell, J. Econ. Entomol., 72, 739-742 (1979).
18. C. J. Sanders, Environ. Entomol., 5, 868-872 (1976).
19. J. O. Schmidt and W. D. Seabrook, J. Econ. Entomol., 72, 509-511 (1979).
20. D. O. Hathaway, L. M. McDonough, D. A. George, and H. R. Moffitt, Environ. Entomol., 8, 318-321 (1979).
21. S. R. Farkas, H. H. Shorey, and L. K. Gaston, Environ. Entomol., 3, 876-877 (1974).

# EXPERIENCE IN USING A HOLLOW FIBER CONTROLLED RELEASE FORMULATION IN PHEROMONE MEDIATED SUPPRESSION OF *PECTINOPHORA GOSSYPIELLA* UNDER HUMID TROPICAL CONDITIONS

*T. W. Brooks*
*C. C. Doane*
*D. G. Osborn*
*J. K. Haworth*

Conrel
An Albany International Company
Needham Heights, Massachusetts

## I. INTRODUCTION

Hollow fiber formulations of the sex pheromone of *Pectinophora gossypiella* have proven useful for suppressing this cotton pest, in Arizona and Southern California, by the mating communication disruption method. The climate in these desert cotton growing areas is extremely warm and dry and all agriculture is conducted with extensive irrigation. Hollow fiber formulated sex pheromone has now been applied under extremely humid tropical conditions. During the 1978-79 cotton growing season in Santa Cruz, Bolivia, a demonstration project was conducted on the use of mating disruption for suppressing infestations of *P. gossypiella*. Cotton production in Bolivia is confined to a region around the city of Santa Cruz where the climate is tropical and extremely humid. Daily temperatures range from 15$^{o}$ to 35$^{o}$C (60$^{o}$-95$^{o}$F) during the growing season with seasonal rainfall measuring from 80-160 cm (32-64 inches). Rainfall tends to occur over extended periods during which time fields are virtually flooded and inaccessible. Cotton cultural practices among growers vary considerably but are generally modern. Cotton yields for the region are reported to be around 400-470 kg of lint per hectare (350-400 lbs/acre) and have been declining in recent




ISBN 0-12-074450-3

years (1). Major components of the local cotton insect pest complex are cotton leafworm (*Alabama argillacae*), pink bollworm (*Pectinophora gossypiella*), and *Heliothis spp.* Of the latter, *H. virescens* is the more serious and more difficult to control due to pesticide resistance.

*P. gossypiella* infestations which vary in severity from year to year tend to occur during the latter half of the season when visible crop damage appears and mounts rapidly. Despite the variation, considerable boll damage and yield losses can be attributed to *P. gossypiella* and the more intensive growers will attempt to suppress infestations with insecticides. Biological control approaches have only recently been considered by Bolivian cotton growers.

## II. BRIEF HISTORY OF *P. GOSSYPIELLA* SUPPRESSION WITH PHEROMONE UNDER IRRIGATED DESERT GROWING CONDITIONS

Suppression of *P. gossypiella* by the pheromone communication disruption method has been under investigation in the U.S. desert Southwest since 1972 when McLaughlin *et al.*, reported success using the pheromone mimic, hexalure (Z-7-hexadecenyl acetate), in small plot tests (2). In the intervening years Shorey and Gaston, with their colleagues, demonstrated *P. gossypiella* suppression on cotton using either hexalure or gossyplure (a 1:1 mixture of Z,Z-and Z,E-7-11-hexadecadienyl acetate) on both small and large experimental plots (3-6). The controlled release media employed by these workers included metal planchets, impregnated cotton string, small diameter Teflon tubing, gelatin microcapsules and Celcon hollow fibers sealed at one end. Commercial development of pheromone mating communication disruption by the atmospheric permeation method began when experiments demonstrated that small segments of hollow fiber sealed at one end were a useful controlled release medium for gossyplure and other insect pheromones (7-9). This work produced a commercial formulation known as Nomate PBW (formerly called Gossyplure H. F.) now registered by the U.S. Environmental Protection Agency for use on cotton, and gaining acceptance by cotton producers in the U.S. desert Southwest. Nomate PBW is essentially a hollow fiber controlled release formulation of gossyplure designed for broadcast dissemination onto growing cotton.

Prior to the Bolivian project described in this paper, demonstrations of the efficacy of pheromone communication disruption for suppressing *P. gossypiella* had been confined to growing areas which are characterized by extremely dry desert climates, namely Arizona and Southern California. Growing cotton in this part of the U.S. requires a strict schedule of irrigation, and the climatic conditions are thought to be particularly favorable for development of *P. gossypiella* populations (10). The basic technology and strategies for using Nomate PBW were developed for the growing practices in this area since this is the only part of the U.S. where *P. gossypiella* is economically significant.

Claims for the efficacy of Nomate PBW in suppressing *P. gossypiella* populations on cotton are based upon pheromone trap orientation disruption results, differences in levels of larval infestation on treated versus untreated control fields, and mating table experiments in which mating success of female moths in a pheromone permeated atmosphere were observed visually. These data have been reported elsewhere together with summaries of purely commercial experience in Arizona and Southern California (9).

## III. DEMONSTRATION OF NOMATE PBW IN BOLIVIA—RESULTS AND DISCUSSION

The Bolivian test farm selected for demonstrating Nomate PBW under humid tropical conditions is located some 45 km east of Santa Cruz. The farm was comprised of eight cotton fields ranging in size from eight ha to 87 ha and totaling 320 ha. It was completely isolated from neighboring agriculture by dense jungle on all sides. Only one other crop, about 80 ha of sugar cane, was being cultivated within the isolation zone. All fields on the farm, both sugar cane and cotton were contiguous. The farm had a history of cotton insect pests typical of the Santa Cruz region. Major pests were *A. argillicae, P. gossypiella* and the *H. zea-H. virescens* complex. Other secondary insect pests were present but were minor or insignificant during the 1978-79 test season. The proprietor of the test farm was a graduate agronomist well informed in entomology and the fundamentals of biological pest control. The test farm had abundant populations of predacious insect species including a natural population of parasitic wasps. A 30 ha check field was selected some 4 km from the test farm and connected with the test farm by a winding road through dense jungle. The check farm employed conventional insecticide materials for pest control.

The test and check fields were monitored with pheromone baited Delta traps (11) at densities of approximately one trap per 16 ha on test fields. Traps were checked four times weekly except when flooding rains made fields inaccessible.

Cotton was planted in the months of October and November and harvest completed by June 20, 1979. Applications of Nomate PBW on test fields commenced during late October through early January as plants reached the pre-square stage of growth. Subsequent applications of pheromone were made at approximately 20-day intervals. A total of five pheromone applications were made, with the final applications occurring from February 7 through 10. All applications of Nomate PBW were by air at a rate of 50 g/ha for a total dose of 250 g/ha/season, corresponding to an active ingredient dose of 3.8 g/ha/application or 19 g/ha/season. The design and function of aerial application equipment have been described by Funkhouser (12).

Insecticide applications were made on an as-needed basis on both test and check fields throughout the season. Insecticide applications were generally directed at *A. argillacae* or *Heliothis spp.* No insecticide was applied specifically for *P. gossypiella* on either the check or test farms. Early season control of *A. argillacae* on the test farm was accomplished with *Bacillus thuringiensis*. Other insecticides used included organochlorine, organophosphate, and carbamate types of materials and a synthetic pyrethroid. The test field grower favored a synthetic pyrethroid while the check field grower relied almost exclusively on organophosphates and carbamates.

In Figure 1 male moth captures through the season are shown for both test fields and check fields. Pheromone treatments of the test field produced trap orientation disruption results similar to those which have been observed in previous work (9). Figure 2 shows boll infestation data for the test and check fields obtained from twice weekly sampling and inspection of susceptible bolls using a published sampling protocol (13). Both trap readings and boll infestation data on the check field indicate a relatively low population of *P. gossypiella* during the first 10-12 weeks of the season in spite of a strong flight of moths at the outset on both check and test fields. This may be attributable in part to heavy rains in this period which often kept the fields flooded. Flooding in irrigated cotton growing areas such as the Sudan is reported to produce high larval mortality (14). By the end of January, however, boll infestations in the check field were beginning to rise and continued to go up at an exponential rate through the end of the season. By April 19, or about halfway into their harvest, the difference in boll infestation on the check versus the pheromone treated fields was a spectacular 98% versus 7.5%.

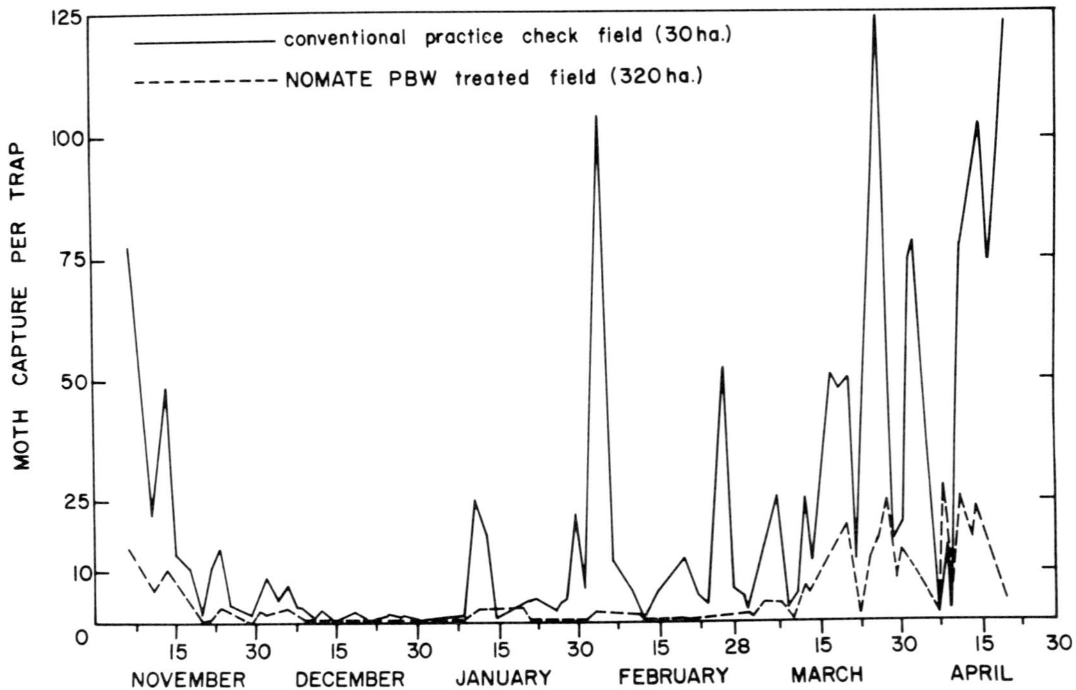


*FIGURE 1.* *Captures of male P. gossypiella in gossyplure baited traps in a conventional practice check field and eight gossyplure permeated test fields in 1978-79 near Santa Cruz, Bolivia.*

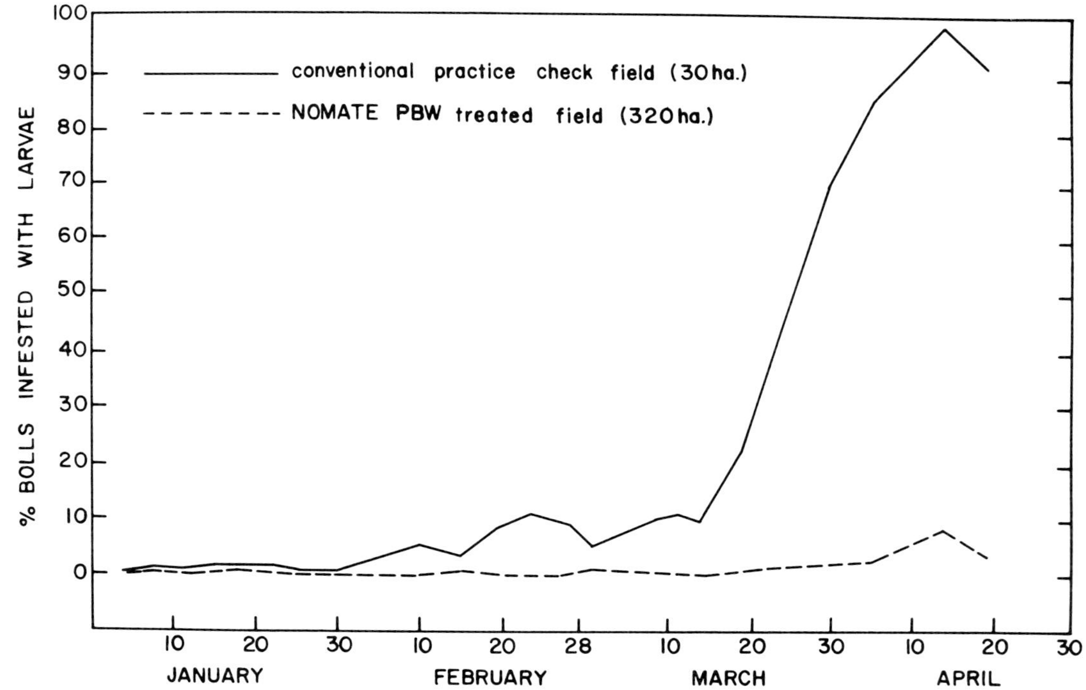


*FIGURE 2. P. gossypiella larval infestations in cotton bolls from eight fields permeated with gossyplure and a conventional practice check field near Santa Cruz, Bolivia in 1978-79.*

Insecticide applications, mainly for *Heliothis spp.* and *A. argillacae* numbered 11 on the check field versus an average of 5.7 on the pheromone treated fields. The check field grower, relying heavily on organophosphate and carbamate insecticides, experienced severe insecticide burning on his cotton plants and sustained considerable defoliation by *A. argillacae*. On pheromone treated fields, where insecticide use was limited mostly to the synthetic pyrethroid, plant foliage remained healthy and vigorous throughout the season. Insect control treatment regimes and costs for the test and check fields are summarized in Table I. Pheromone treatment resulted in a 67% reduction in insecticide usage relative to the conventional practice check field, and contributed to better overall insect control. Costs ran considerably higher for insect control materials on the test field than on the check field. This is attributable partly to pheromone treatment costs and partly to the test field growers preference for the more expensive new pyrethroid material (fenvalerate) over cheaper organophosphate and carbamate materials favored by the conventional practice grower. Since boll damage by

*TABLE I. Comparison of Nomate PBW Versus Conventional Practice Insect Control Materials Usage and Costs—Santa Cruz, Bolivia 1978-79 Season*

| | Conventional practice field | Nomate PBW treated field |
|---|---|---|
| Avg. no. insecticide applns. | 11 | 5.7 |
| Avg. no. B.T. applns.[a] | 0 | 1.5 |
| Avg. no. pheromone applns.[b] | 0 | 5 |
| Total amt. of insecticide active ingredient | 7.55 kg/ha | 2.51 kg/ha |
| Total amts. of gossyplure active ingredient | 0 | 19 g/ha |
| % reduction in insecticide use | -- | 67% |
| Avg. insect control materials cost per ha | $82/ha | $174/ha |
| Total hectares analyzed | 30 ha | 320 ha |

[a] B.T. = *Bacillus thuringienis*
[b] Nomate PBW

*P. gossypiella* was dramatically suppressed on the Nomate BPW fields, the conventional practice grower essentially neglected control of *P. gossypiella*, apparently concentrating on control of *A. argillacae and Heliothis spp.* While the conventional practice grower may have trimmed expenses for insect control, savings were achieved at considerable expense in yield and fiber quality as revealed in the harvest data.

In Table II are shown harvest results from the pheromone treated and conventional practice fields. Average harvest results for the entire Santa Cruz area are also shown for comparison. Both fiber quantity and quality were superior from the pheromone demonstration fields affording the demonstration grower a net yield advantage over the conventional practice check grower worth about $390 per hectare, more than offsetting the higher insect control expenses. The demonstration grower also out-performed area averages significantly. The value of yield and fiber quality for the pheromone

*TABLE II. Comparison of Harvest Results from Nomate PBW Treated, Conventional Practice Fields, and Santa Cruz Area Averages—Santa Cruz, Bolivia 1978-79 Season*[a]

| | Nomate PBW treated | | Check grower | | Santa Cruz area average | |
|---|---|---|---|---|---|---|
| *Fiber grades* | *%* | *value* | *%* | *value* | *%* | *value* |
| 1-3 | 65.4 | $535 | 6.0 | $ 29 | 48.2 | $259 |
| 4-6 | 33.5 | 254 | 45.0 | 199 | 37.8 | 187 |
| 7-9 | 1.1 | 8 | 49.0 | 178 | 14.0 | 58 |
| Yield per ha | 1441 lbs | | 839 lbs | | 947 lbs | |
| Value per ha | $796 | | $406 | | $504 | |
| Av. value per lb | $0.552 | | $0.484 | | $0.532 | |
| Ha analyzed | 320 | | 250 | | 32,000 | |

[a]*All figures are on a per hectare basis. Values are expressed in U.S. dollars and were calculated from anticipated prices for 1978-79 Bolivian cotton. The price used for the premium SM grade was $0.58 per lb with lower grades valued according to the standard price de-escalation schedule used internationally.*

fields exceeded area averages by 58% or $292 per hectare. These remarkable differences are attributable to the combined effect of yield and fiber quality on economic performances. The economic benefit of higher fiber quality is reflected in the calculated average values of harvest lint which turn out to be $0.552, $0.484, and $0.532 per pound for the test, check, and area average results respectively. Dominant though not necessarily exclusive factors influencing fiber quality are harvest technique and pest damage. Control of pest damage is thought by local experts to be the most significant factor and the results of this demonstration project would tend to support such a belief.

## IV. CONCLUSIONS

The hollow fiber controlled release formulation of gossyplure found to be effective in suppressing *P. gossypiella* under irrigated desert cotton growing conditions appears to be equally effective in a humid tropical climate. An insect control regime based on pheromone suppression of *P. gossypiella* coupled with *Bacillus thuringiensis*, and synthetic pyrethroid for control of other major cotton insect pests, affords considerable economic advantages over conventional chemical insect control measures in Bolivian cotton culture.

## ACKNOWLEDGMENTS

We gratefully acknowledge the splendid cooperation provided us by the proprietor of the demonstration farm, Ing. Carlos Malatesta. His knowledge of the local insect complex and biological control approaches to crop protection were an important contributing factor in the success of the project. We are also indebted to Asociacion de Productores de Algodon for background information and assistance in conducting the project.

## REFERENCES

1. Anon., Cotton-World Statistics, Quarterly Bulletin of the International Advisory Committee, Secretariat of the Intl. Committee, So. Agr. Bldg., Washington, D.C., October 1977.

2. J. R. McLaughlin, H. H. Shorey, L. K. Gaston, R. S. Kaae, and F. C. Stewart, Environ. Entomol., 1, 654 (1972).
3. H. H. Shorey, R. S. Kaae, and L. K. Gaston, J. Econ. Entomol., 67, 347 (1974).
4. H. H. Shorey, L. K. Gaston, and R. S. Kaae, in Pest Management with Insect Sex Attractants, M. Beroza, (Ed.), ACS Symposium Series 23, American Chemical Society, Washington, D.C., pp. 67-74 (1976).
5. L. K. Gaston, R. S. Kaae, H. H. Shorey, and D. Sellers, Science, 196, 904 (1977).
6. H. E. Hummel, L. K. Gaston, H. H. Shorey, R. S. Kaae, K. J. Byrne, and R. M. Silverstein, ibid, 181, 873 (1973); B. A. Bierl, M. Beroza, R. T. Staten, P. E. Sonnet, V. E. Adler, J. Econ. Entomol., 67, 211 (1974).
7. T. W. Brooks, E. Ashare, and D. W. Swenson, in Textile and Paper Chemistry and Technology, J. C. Arthur, (Ed.), ACS Symposium Series 49, American Chemical Society, Washington, D.C., pp. 111-126 (1977).
8. E. Ashare, T. W. Brooks, and D. W. Swenson, in Controlled Release Polymeric Formulations, D. R. Paul and F. W. Harris, (Eds.), ACS Symposium Series 33, American Chemical Society, Washington, D.C., pp. 273-282 (1976).
9. T. W. Brooks, C. C. Doane, and R. T. Staten, in Chemical Ecology: Odour Communication in Animals, F. J. Ritter, (Ed.), Elsevier/North-Holland Biomedical Press, Amsterdam, pp. 375-388 (1979).
10. L. W. Noble, Fifty Years of Research on the Pink Bollworm in the United States, Agr. Handbook No. 357, Agr. Res. Svc., U.S. Department of Agriculture, Washington, D.C. 1969.
11. R. N. Foster, R. T. Staten, and E. Miller, J. Econ. Entomol., 70, 289 (1977).
12. W. Funkhouser, Proceedings of the 1979 Beltwide Cotton Production Research Conference, National Cotton Council of America, Memphis, Tennessee (in press).
13. R. A. Van Steenwyck, G. R. Ballmer, and H. T. Reynolds, J. Econ. Entomol., 69, 579 (1976).
14. J. S. A. Kisha, Paper Presented at the Symposium on Crop Pest Management in the Sudan, February 1978.

# IMPLANTABLE SYSTEMS FOR DELIVERY OF INSECT[1] GROWTH REGULATORS TO LIVESTOCK

*H. Jaffe*
*J. A. Miller*
*P. A. Giang*
*D. K. Hayes*

Livestock Insects Laboratory
Agricultural Research, Science
And Education Administration, USDA
Beltsville, Maryland
and
US Livestock Insects Laboratory
Agricultural Research, Science
And Education Administration, USDA
Kerrville, Texas

## I. INTRODUCTION

Systems for the controlled release of narcotic antagonists, steroids and other drugs have been the subject of intensive development (1). For the past 3 years our laboratories have been investigating the application of this technology to the control of arthropod pests of livestock. As a result, we have developed methods whereby a systemic pesticide or insect growth regulator is released into the circulatory system of the host animal via implanted controlled release pellets or microcapsules to affect feeding parasitic insects or ticks. We first have reported the preparation and

[1]*Parts of this paper have been presented at the 5th International Symposium on Controlled Release of Bioactive Material, August 14-16, Gaithersburg, Maryland (1978).*




ISBN 0-12-074450-3

bioevaluation of injectable microcapsules of the systemic pesticide famphur (O-[p-dimethylsulfamoy-l)phenyl] O,O-dimethyl phosphorothioate) in biodegradable polymers (2). Subcutaneous injections of a suspension of these microcapsules in guinea pigs were shown to kill feeding ticks over an extended period by continuously releasing active ingredient into the host's circulatory system. However, large injections were required because of the relatively low activity of famphur and the volume of sesame oil used to suspend the microcapsules. The use of this treatment for cattle was thus impractical. We therefore sought to avoid these problems by developing controlled release implants of systemic insect growth regulators which are active at much lower dosages.

In this paper we describe the preparation of controlled release microcapsules and pellets of the insect growth regulators methoprene and AI3-36206 in biodegradable polymers and the results of their implantation into ears of cattle against cattle grubs. In addition, methods are described for analysis of the active ingredient in these systems and the measurement of its release rate *in vitro* and *in vivo*. Our hope is to develop a formulation that will control cattle grubs for an entire season with one treatment.

Cattle grubs, larval stage of the heel fly, are among the most destructive pests that attack cattle (3). The grubs are commonly found in cysts on the backs of cattle. Both larvae and adults are destructive. The grubs cause irritations, allow secondary bacterial infections, and create holes in the hides. Losses at slaughter include damaged hides and reduced value of the carcasses due to trimming of the grub infested area. In addition, the adults in attempting to lay their eggs on the hairs of the cattle, cause the animals to run wildly (gadding). Cattle do not graze properly, are difficult to handle, and occasionally injure themselves. Total losses in the U.S. due to these insects were estimated to be about $300 million annually (4).

Methoprene (isopropyl (E,E)-11-methoxy-3,7,11-trimethyl-2,4-dodecadienoate) is a non-persistent insect growth regulator of minimal mammalian toxicity. The compound has been shown to have systemic activity (5) and to be effective at very low levels against the horn fly, *Haemotobia irritans* (L.), and the common cattle grub, *Hypoderma lineatum* (de Villers) (6), in drinking water (7-9) or in boluses permitting controlled release (4). AI3-36206 (2-ethoxy-9(p-isopropylphenyl)-2,6-dimethylnonane) is an experimental systemic (10) insect growth regulator developed at our institute (11). In all cases these insect growth regulators act on the larval stage of insects to prevent development of the adult. Thus in the case of cattle grubs, the effect of the insect growth regulators is not to prevent damage done by the present generation of grubs, but to limit future infestations by reduction of the adult population. In both pellet and microcapsule formulation the insect growth regulator acts as a plasticizer for the polymer matrix from which it is believed to be released into the animal's circulatory system via a diffusion/hydrolysis mechanism.

## II. MATERIALS AND METHODS

Poly(d,l-lactic acid) and lactic-glycolic copolymers were synthesized for us by Dynatech Corp., Cambridge, Massachusetts, and purified by us by precipitation in water from dioxane followed by washing with methanol. The methoprene used was Altosid[R] technical material (95.1% active ingredient AI) obtained from Zoecon Corp., Palo Alto, California. The Synovex Implanter[R] was obtained from Syntex Laboratories, Inc., Animal Health Division, Palo Alto, California. Empty gelatin capsules, size 5, P-D Stock 511-000000 were obtained from Parke-Davis Company, Industrial Products Division, Detroit, Michigan.

### A. *Preparation of Pellets*

A mixture of 1.4 g methoprene and 2.6 g poly(d,l-lactic acid) in a 300 ml round bottom flask was completely dissolved in 150 ml methylene chloride. Complete removal of solvent on the rotary evaporater was followed by evacuation at high vacuum to form 4.0 g of a clear plastic film. Pieces of film weighing 0.10 g were placed into the center of 15 cm lengths of 3 mm ID Teflon[R] tubing. The center of the tubing was immersed in a water bath at 54-55$^{o}$ for 5 seconds to soften the film, which was then compressed by application of pressure applied by glass rods (3 mm dia) inserted into both ends of the tubing to form a pellet. To ensure good pellet formation, the softening and compression cycle was repeated. The resulting pellets (3 x 17 mm) were stored in the tubing from which they could easily be removed before use. Pellets with lower levels of AI were similarly prepared using both poly(d,l-lactic acid) and poly(L(+)lactic acid) (Table I).

*TABLE I. Microcapsules and Pellets of Methoprene in Biodegradable Polymers*

| *Type*[a] | *Initial % AI* | *Polymer*[b] |
|---|---|---|
| *P* | *35.0* | *d,l-PLA* |
| *P* | *25.0* | *L(+)-PLA* |
| *P* | *15.0* | *d,l-PLA* |
| *M* | *40.0* | *d,l-PLA* |
| *M* | *33.3* | *d,l-PLA* |
| *M* | *26.5* | *d,l-PLA* |
| *M* | *33.3* | *90/10 L(+)-PLA-PGA* |
| *M* | *26.5* | *90/10 L(+)-PLA-PGA* |

[a]*P=pellet, M=microcapsules.*
[b]*d,l-PLA=poly(d,l-lactic acid), L(+)-PLA=poly(L(+)-lactic acid), 90/10 L(+)-PLA-PGA=copolymer of 90% L(+)-PLA and 10% poly (glycolic acid).*

## B. *Preparation of Microcapsules*

Microcapsules of methoprene were prepared by precipitation techniques reported previously (2). A 600 ml beaker charged with 0.9 g poly(d,l-lactic acid), 0.6 g methoprene and 25 ml $CH_2Cl_2$ was carefully agitated to effect complete solution. To the clear solution was added 250 ml water and 0.11 g sodium dodecyl sulfate. The encapsulating mixture was stirred and after 5 min chilled in an ice-water bath. The stirring was continued for a total of 2 hr. Our yield was 0.94 g of a coarse powder. Differential interference contrast microscopy showed the particules to be large aggregates of microcapsules. Microcapsules which contained lower levels of AI in both poly(d,l-lactic acid) and a lactic-glycolic copolymer were similarly prepared (Table I).

Pellets and microcapsules containing AI3-36206 were similarly prepared (Table II). Both formulations were analyzed for active ingredient by ultraviolet and infrared spectroscopic techniques.

*TABLE II. Microcapsules and Pellets of AI3-36206 in d,l-PLA*[a]

| *Type* | *Nominal % AI*[b] | *Actual % AI* |
|---|---|---|
| *M*[c] | *33.3* | *28.2* |
| *M* | *35.0* | *33.4* |
| *P*[d] | *35.0* | *28.6* |
| *P* | *45.0* | *35.5* |
| *P* | *60.0* | *43.2* |

[a]*d,l-PLA = poly(d,l-lactic acid)*
[b]*per cent AI3-36206 used to prepare formulation*
[c]*M = microcapsules*
[d]*P = pellet*

## C. *In Vitro Release Rate of AI3-36206*

The release of AI3-36206 from its microcapsules and pellets into pH 7.40 buffer at 37°C was measured *in vitro* by use of a flow system previously described (2) but since modified as illustrated in Figure 1. In the apparatus, phosphate buffer at 37°C was passed over a sample of microcapsules or a pellet. The active ingredient was released into the aqueous buffer in which it has very low solubility. The released active ingredient was continuously removed from the buffer into carbon tetrachloride by means of a liquid-liquid extractor built into the system. The carbon tetrachloride was periodically drained from the liquid-liquid extractor for analysis of released AI3-36206. The sample was contained in 1/4 inch glass-tubing bent as shown in Figure 1. An area of the tubing was enlarged to accommodate the sample which was contained in glass wool in the case of the pellet or mixed with sand in the case of the microcapsules. All connections (indicated by a dot) were made with Tygon® tubing (type R-3603). Connections were clamped just above the level of the water bath to prevent swelling and leakage. The flow rate of buffer was in the range of 7-8 ml min. The opening on top of the liquid-liquid extractor was sealed with Parafilm® to prevent loss of buffer by evaporation.

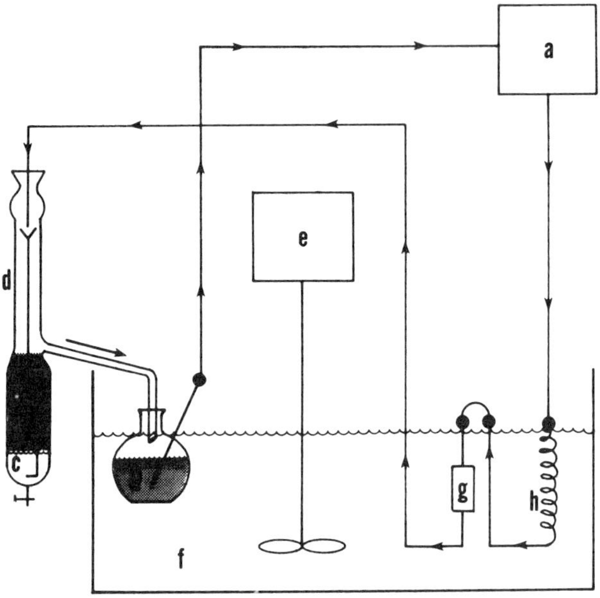


*FIGURE 1. Apparatus for the determination of the in vitro release rate from microcapsules and pellets. (a-peristaltic pump, b-pH 7.40 phosphate buffer, c-carbon tetrachloride, d-liquid-liquid extractor with stopcock drain, e-heater-stirrer, f-water bath at 37°C, g-microcapsules in sand column or pellet in glass wool, h-temperature equilibration coil.)*

Cumulative percent AI released and average percent of total AI released were plotted versus time for both microcapsules and pellets (Figure 2-4).

*D. In Vivo Release (Beltsville)*

The release of AI3-36206 from pellets *in vivo* was examined by implanting one 0.1 g pellet in each ear of four Black Angus steers (ca. 200 kg) with the Synovex Implanter. Seven of the pellets contained 28.6% AI. In addition one control pellet containing pure poly(*d,l*-lactic acid) was implanted. Pellets were removed periodically, washed free

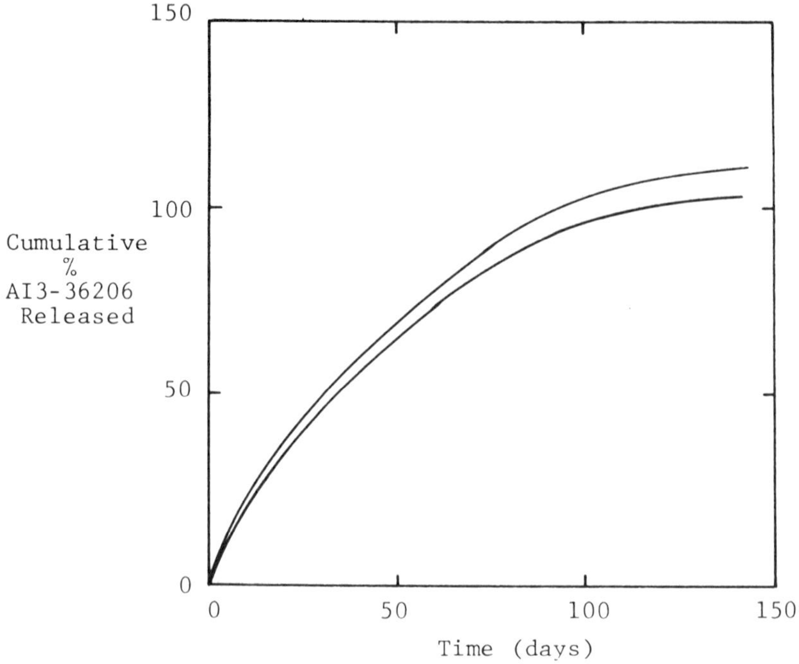


*FIGURE 2. Cumulative % AI3-36206 released in vitro. Upper curve, microcapsules; lower curve, pellet.*

of tissue, air dried and analyzed for percent AI. The percent of starting AI remaining in the implant was calculated only for those implants that we were confident had been recovered in entirety.

## *E. Efficacy Experiments (Kerrville)*

Experiments were run designed to evaluate the practicality of our implant systems and their efficacy against cattle grubs. Untreated controls were also run. In the first series of experiments done in 1977, Hereford calves (ca. 200 kg) infested with *H. lineatum* were implanted with 0.1 g methoprene containing pellets with the Synovex Implanter with standard needle when the grubs were visible on the back.

In the first experiment of this series, three Hereford calves received four 0.1 g implants via two injections at the base of each ear for a total of eight implants per animal.

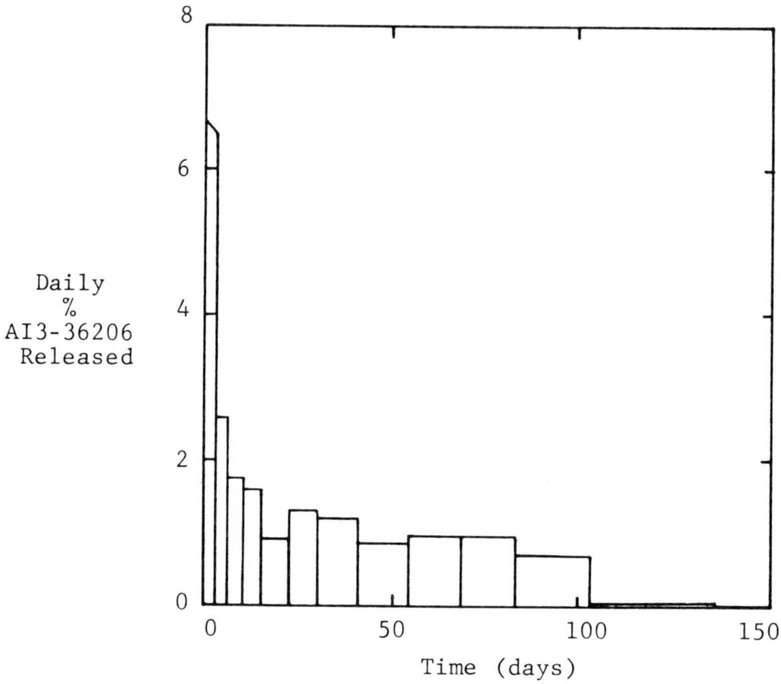


*FIGURE 3. Daily % AI3-36206 released in vitro from microcapsules.*

Each animal received pellets containing the same AI but a different AI was used for each animal. *H. lineatum* larvae were collected as they emerged from the animals backs and allowed to pupate, and the subsequent emergence of the adult heel flies was recorded. Abbott's formula was used to calculate the percent inhibition of emergence of flies from pupae from treated animals as compared to untreated controls.

In the second experiment, three calves were implanted as before but the 35 percent methoprene implant was used. Each animal was treated with eight implants. In contrast to the first experiment, *H. lineatum* grubs were already emerging from the backs of the cattle at the time of implantation.

In a second series of experiments done in 1978, Hereford calves (ca. 200 kg) infested with *H. lineatum* were implanted with 0.1 g AI3-36206 pellets and microcapsules. In the first experiment of this series the first animal received four 35.5% (AI) implants via two injections at the base of each ear, a total of eight implants per animal. A second infested

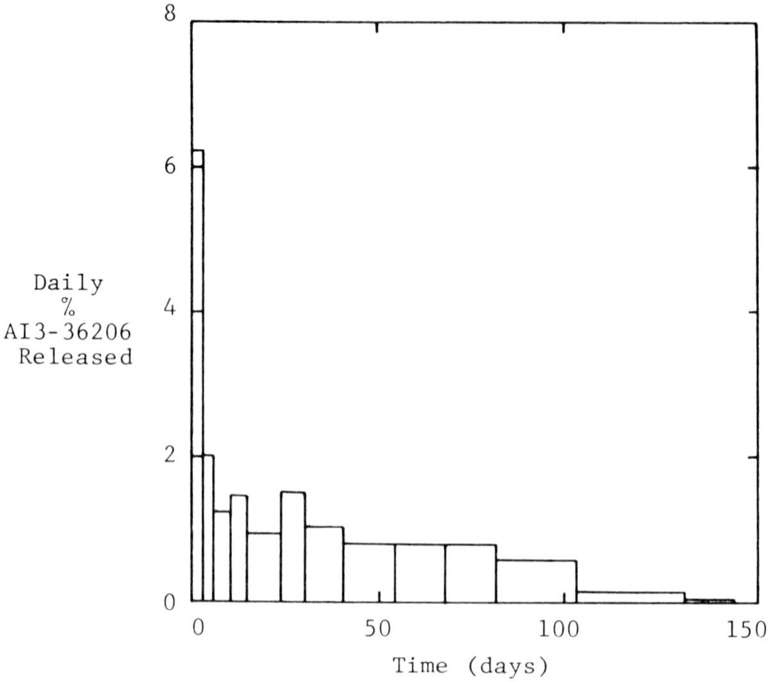


*FIGURE 4. Daily % AI3-36206 released in vitro from pellet.*

animal received eight 28.6% implants. A third animal received one 35.5% AI implant. Pellets were recovered after 79 days. In this first experiment, the percent of AI remaining in the implant could not be determined for those animals that received multiple implants because we were uncertain that each implant had been completely recovered. However, the percent of AI remaining in the implant was determined for the animal that received a single implant.

In the second experiment of this series, one animal received 12 gelatin capsules filled with 0.05 g of microcapsules containing 33.4% AI. The gelatin capsules were implanted with a Synovex Implanter modified with a slightly enlarged needle so as to accommodate the smallest commercially available gelatin capsule (no. 5). Because of binding in the needle, we were unable to administer more than one capsule per injection for a total of six capsules per ear. In the future, with greater enlargement of the needle, we anticipate that up to four gelatin capsules could be implanted per injection. As expected, the gelatin capsules were quickly absorbed releasing the microcapsules. No attempt was made to recover any material in this series.

In both experiments, *H. lineatum* larvae were collected as they emerged from the animal's backs and allowed to pupate, and the subsequent emergence of the adult heel flies was recorded. Abbott's formula was used to calculate percent inhibition of emergence of flies from pupae from treated animals as compared to untreated controls.

### *F. Analysis of Methoprene and AI3-36206 Formulations*

Both pellets and microcapsules of methoprene and AI3-36206 were analyzed for percent AI by UV or IR spectroscopic methods. When stored at room temperature, all methoprene formulations seemed to exhibit a rapid loss of free methoprene as determined by our analysis. Only a few percent methoprene remained after 1 month (Figure 5). Refrigerated samples show only slight losses during the same period.

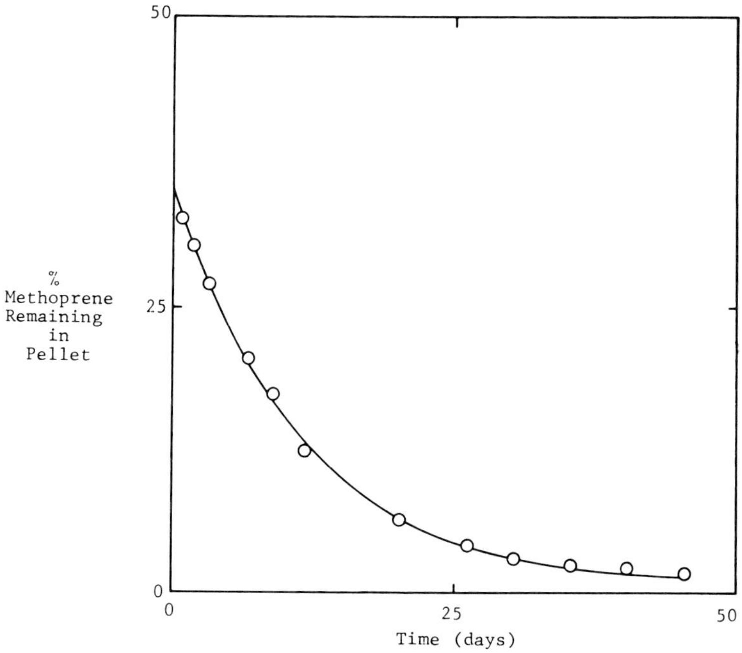


*FIGURE 5. Loss of methoprene from pellets stored at room temperature.*

Thus the nominal percentage of AI methoprene used refers to the starting percentage of AI. Formulations of AI3-36206 were stable.

## III. RESULTS AND DISCUSSION

The AI of the microcapsules and pellets of methoprene and AI3-36206 are shown in Table I and II. Because of the lability of methoprene and its rapid loss from its formulations, *in vitro* and *in vivo* release rates could only be measured for AI3-36206 formulations. As shown in Figure 2, the *in vitro* cumulative percent AI3-36206 released from both microcapsules and pellets was non-linear overall. After a rapid release during the first 10 days (burst effect), a near linear release was maintained throughout a period of about 3 months. This near linearity can be seen clearly in Figures 3 and 4 where the data is plotted as an average daily percent release of the total available active ingredient. After the initial burst effect an average daily release of 0.65 - 1.5% was maintained until the formulation was exhausted of AI.

As seen in Table III, AI3-36206 was released from its pellets at a much slower rate *in vivo* than *in vitro*. This is not unexpected because a large volume of buffer is pumped

*TABLE III. In Vivo Release of AI3-36206 From Pellets.*

| *Starting AI (%)* | *Location*[a] | *Time (days)* | *% AI Remaining*[b] |
|---|---|---|---|
| 28.6 | B | 6 | -[c] |
| 28.6 | B | 10 | 81 |
| 28.6 | B | 17 | 77 |
| 28.6 | B | 42 | -[c] |
| 28.6 | B | 73 | 87 |
| 28.6 | B | 119 | 80 |
| 35.5 | K | 79 | 67 |

[a]*B = Beltsville, MD; K = Kerrville, TX*

[b]$\%\ AI\ remaining = \frac{Recovered\ wt \times \%\ AI}{Starting\ wt \times \%\ AI} \times 100$

[c]*Entire pellet may not have been recovered*

past the pellet *in vitro*. The situation *in vivo* is quite different because the observed low level of tissue reaction and the onset of tissue encapsulation probably combine to cause a decreased flow of biological fluids past the implant. Poor correlation between *in vivo* and *in vitro* results has been widely reported in the literature (12).

In the first efficacy experiment of our methoprene series there was 89 percent inhibition of adult heel fly eclosion with the pellets containing 35 percent AI. This fell off as concentration of active ingredient decreased (Table IV). No trend of increased emergence over the 60 day collection period was observed.

In the second experiment of this series a decline in emergence of adults from pupae was observed in the treated cattle. Examination of the grubs collected 1 week post-treatment and beyond indicated that 100 percent control was achieved in all three animals.

In contrast to our results with methoprene in the first series of experiments, pellets and microcapsules containing AI3-36206 in the second series of efficacy experiments were not effective against cattle grubs, presumably because of their low rate of release of AI3-36206.

We believe that these initial experiments demonstrate the practicality of our implant systems and the feasibility of the use of subcutaneous implants in cattle to control livestock pests throughout an entire season. Currently we are investigating polymer matrices that show greater release rates so as to achieve efficacy against cattle grubs with our AI3-36206 formulation. During future cattle grub seasons, we intend to extend our research against the northern cattle grub, *Hypoderma bovis* (L.).

*TABLE IV. Methoprene Implants for Control of Cattle Grubs.*[a]

| *Initial % AI* | *Pupae collected* | *Adults emerged* | *% Inhibition of emergence (corrected)* |
|---|---|---|---|
| *35*[b] | *12* | *1* | *89* |
| *25*[b] | *28* | *14* | *30* |
| *15*[b] | *3* | *2* | *6* |

[a]*Hypoderma lineatum*
[b]*35 and 15%, d,l-PLA: 25%, l(+)-PLA*

We believe that an ester interchange reaction between the methoprene and the polymer may account for the observed loss of free methoprene from the pellets and microcapsules. This reaction could be auto-catalyzed by the carboxyl end group of polymer and could take place at the alcohol end of the chain or at an internal ester linkage with concurrent scission of the polymer chain. It is thus possible that the observed activity of these implants could result from the release of some other active derivative of methoprene such as a lactyl ester from the polymer along with release of the remaining free methoprene. In addition the formation of low molecular weight molecules as a result of these reactions may serve to further plasticize the polymer matrix and provide for a more rapid release of AI from the formulation.

This paper reports the results of research only. Mention of a pesticide in this paper does not constitute a recommendation for use by the USDA nor does it imply registration under FIFRA as amended. Also, mention of a commercial or proprietary product in this paper does not constitute an endorsement by USDA.

## ACKNOWLEDGMENTS

The authors are grateful to Dr. M. Schwarz, Biologically Active Natural Products Laboratory, AR, SEA, USDA, for providing us with samples of purified AI3-36206.

## REFERENCES

1. See, for example, S. Yolles in Polymers in Medicine and Surgery, R. L. Kronenthal, Z. Oser, and E. Martin, (Eds.) Polymer Sci. Technol., 8, 245 (1975) and references contained therein.
2. H. Jaffe, D. K. Hayes, P. Giang, R. O. Drummond, and T. M. Whetstone, Proc. Controlled Release Pesticide Symposium, August 22-24, 1977, Corvallis, Oregon, p. 272.
3. R. E. Pfadt and T. R. Robb, Control Livestock Pests, Bull, 327R, Agricultural Experiment Station, Univ. of Wyoming, Laramie, Wyoming, 1966, pp. 16-20.
4. J. A. Miller, M. L. Beadles, and W. L. Gladney, Proc. Controlled Release Pesticide Symposium, August 22-24, 1977, Corvallis, Oregon, p. 253.

5. R. L. Harris, E. D. Frazar, and R. L. Younger, J. Econ. Entomol., 66, 10-9 (1973).
6. R. L. Harris, W. F. Chamberlain, and E. D. Frazar, J. Econ. Entomol., 67, 384 (1974).
7. M. L. Beadles, J. A. Miller, W. F. Chamberlain, J. L. Eschles, and R. L. Harris, J. Econ. Entomol., 68, 781 (1975).
8. J. A. Miller, W. F. Chamberlain, M. L. Beadles, M. O. Pickens, and A. R. Gingrich, J. Econ. Entomol., 69, 330 (1976).
9. J. L. Eschle, J. A. Miller, and C. D. Schmidt, Nature, 265, 325 (1977).
10. J. A. Miller, unpublished results.
11. M. Schwarz, R. W. Miller, J. E. Wright, W. F. Chamberlain, and D. E. Hopkins, J. Econ. Entomol., 67, 598 (1974).
12. See, for example, J. H. R. Woodland, S. Yolles, D. A. Blake, M. Helrich, and F. J. Meyer, J. Med. Chem., 16, 897 (1973) and references contained therein.

# CONTROLLED RELEASE FORMULATIONS AND CONTROL OF THE IMPORTED FIRE ANT: WHAT ARE THE POSSIBILITIES

*Robert K. Vander Meer*
*Clifford S. Lofgren*

Insects Affecting Man and
Animals Research Laboratory
Agricultural Research
Science and Education Administration
USDA
Gainesville, Florida

*Danny H. Lewis*
*William E. Meyers*

Southern Research Institute
Birmingham, Alabama

The imported fire ant, *Solenopsis invicta,* infests large areas of 9 southeastern and southern states. The importance of the ant as a pest and its unique foraging habits offer a tremendous challenge to controlled release technology. The present paper presents background information, introduces requirements for control of the fire ant, and discusses recent results and future prospects for several types of controlled release systems.

Two species of fire ant, *S. invicta* and *S. richteri,* were accidently imported into the United States at Mobile, Alabama, about 40 and 60 years ago, respectively. The more aggressive *S. invicta* has proved to be the most important of the two, and has spread rapidly over most of the South (Figure 1). *S. richteri* occupies small areas in northern Mississippi and Alabama. The major factors that now limit further migration of the ants are the cold, northern winters and the desert areas of central Texas.



ISBN 0-12-074450-3

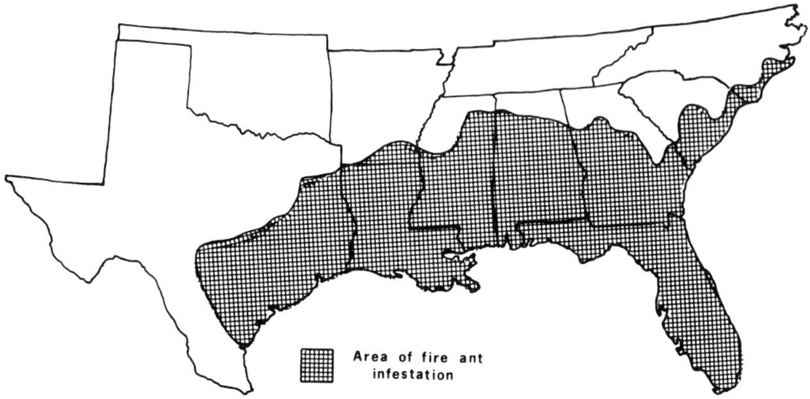


*FIGURE 1. Infestation of Solenopsis invicta in the United States.*

Initial efforts at developing control methods were begun in the early 1950's (Table I). However, the rapid spread of *S. invicta* into grazing and croplands, roadsides, parks and lawns, along with their mound-building, aggressive behavior and potent sting caused such great concern among the general public and farmers that a cooperative federal-state program was instituted by the U.S. Congress in 1957.

Based on earlier work in Alabama (1, 2), granular formulations of heptachlor and dieldrin were used to obtain long-term residual control of fire ants. However, in the late 1950's and early 1960's, the attention of the public and the scientific community was focused on the detrimental effects of these residual chlorinated pesticides, and their use was greatly restricted (3). Consequently, in 1962 these chemicals were replaced with a newly-developed bait formulation consisting of mirex dissolved in soybean oil and applied to a corncob grit carrier (4, 5). Although mirex was another chlorinated hydrocarbon, it was considerably less toxic to non-target organisms than dieldrin and heptachlor (6). It was very effective. As a result, over 140 million acres of land were treated with mirex from 1962 to 1978. Research during the same period resulted in reductions in the amount of the toxicant required for control from 4.2 to 0.46 g/acre. However, in 1978 the Environmental Protection Agency canceled the registration for mirex because residues of the chemical were being found in the environment (7) and in human tissues (8). Also, laboratory studies suggested it might be a

*TABLE I. Important Historical Events in Control of the Imported Fire Ant (Solenopsis invicta).*

| Year | Event |
|---|---|
| 1940 | Accidental introduction of Solenopsis invicta into the Mobile, Alabama, area. |
| 1949-1953 | Surveys of nurseries showed rapid dissemination of S. invicta in southeast U.S. due to shipment of infested plants from Mobile, Alabama, area. |
| 1957 | Federal-state control program initiated. |
| 1952-1962 | Heptachlor and dieldrin used for long-term residual control. |
| 1962 | Mirex replaced heptachlor and dieldrin. |
| 1962-1978 | Over 140 million acres of land were treated with mirex. |
| 1978 | Mirex registration cancelled by the EPA eliminating the only chemical registered for control of fire ants on farmland. |
| 1979 | Experimental use permit issued by EPA for field testing American Cyanamide 217,300, a new bait toxicant. |

carcinogen (9,10). The ban on mirex as a control agent for fire ants, particularly on farm land, has left a void that we are working hard to fill.

The difficulty in finding suitable insecticides for use in baits for fire ants is directly related to the behavior and ecology of the insect. Foraging worker ants, only a small fraction of a colony's population, leave the central nest through a system of radiating tunnels (Figure 2). These tunnels can extend as far as 15-30 m with outlet holes to the foraging area surface about every 30 cm. By using this foraging scheme, the ants thoroughly search the area around a mound for food sources. Once a foraging ant locates food,

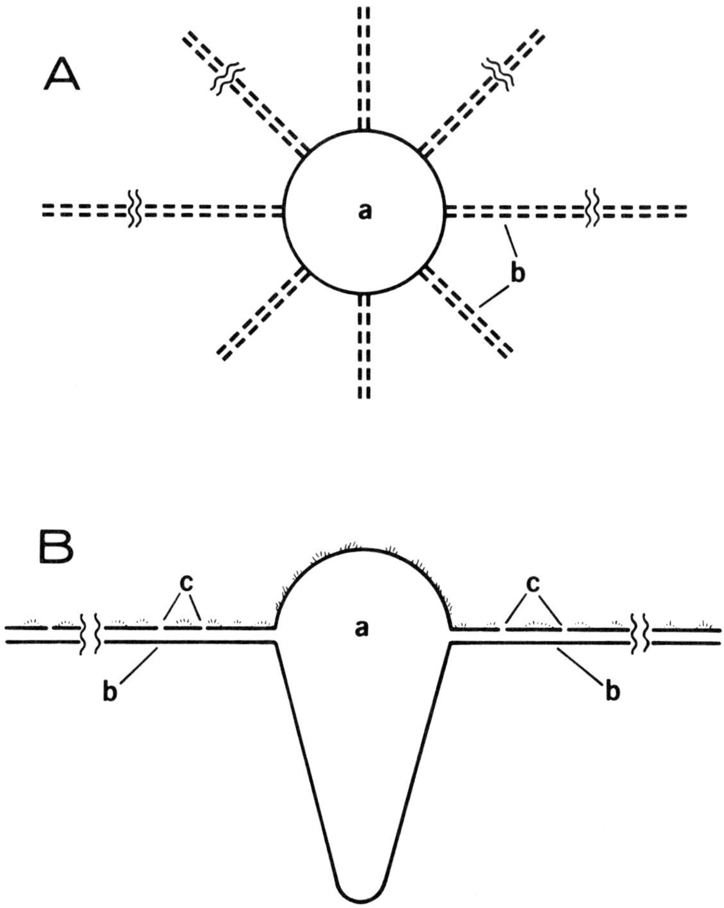


*FIGURE 2. Fire ant mound diagram. Top view (A) and cross section (B), showing mound (a), radiating foraging tunnels (b), and outlets to the foraging areas (c).*

it returns to the foraging tunnel and nest laying a trail with the trail pheromone. In this way other workers are recruited to the food source and a continuous stream of ants soon develops between the food and nest. The foraging ants store food in their crops, and through regurgitation and food exchange (trophallaxis), they quickly disperse the material to other members of the colony (11).

Two major qualifications for a toxicant are apparent. If the toxicant acts too quickly, the foraging workers will die before they can distribute the material to other members of the colony and ultimately to the queen. Therefore, delayed toxicity is required. Tests with dyed soybean oil indicated that complete colony distribution is achieved within 24-72 hr. Secondly, the process of trophallaxis greatly dilutes a toxicant (a mature colony may contain more than 250,000 workers) making it necessary to have delayed toxicity over a wide range of dosages (preferably > 100).

Currently, the Insects Affecting Man and Animals Research Laboratory in Gainesville, Florida, and the USDA Imported Fire Ant Research Laboratory in Gulfport, Mississippi, are actively conducting research to develop new control methods based on a sound knowledge of the biology, ecology and biochemistry of the fire ant. Included in this holistic approach is the search for pathogens of *S. invicta* in its native South American habitat. Also, we have bioassayed over 5000 potential toxicants and are actively looking for additional candidates. Among these are traditional toxicants, insect growth regulators, chitin inhibitors, and sterilants.

For much of this work we depend on the following procedure developed for the rapid laboratory screening of candidate toxicants: The toxicant is dissolved to the desired concentration in either soybean oil or honey-water (1:1) depending on its solubility. Test groups of 20 worker ants are allowed to feed for 24 hr on cotton swabs saturated with the formulation. After 48 hr, the ants are fed unadulterated soybean oil. Mortality counts are made at 1, 2, 3, 6, 8, 10 and 14 days after the initial exposure. Each material is tested at 3 concentrations: 1, 0.1 and 0.01% (12).

All chemicals tested are classified according to the scheme shown in Table II. Class II compounds are good toxicants but they do not have the required delayed toxicity, whereas class III compounds have delayed action, but their range of activity is too narrow. The type of activity we are looking for in a toxicant is exemplified by a class IV or V response, i.e., it exhibits delayed toxicity over a wide range of concentrations.

As expected most of the 5000 compounds screened (Table III) fall into the non-toxic class I category. Most of the 20 class IV compounds are organophosphorous insecticides that did not work well in field tests or are derivatives of mirex, including ACD-1189 (known as Kepone). As expected the only class V compound is mirex.

Of primary interest to this symposium is our effort to utilize the information gained through our basic physiological studies and intensive screening program to support

TABLE II. Classification System for Imported Fire Ant Bait Toxicants

| Class | Definition |
|---|---|
| I | Compounds that give insufficient kill at the preliminary concentrations (less than 90% at the end of the test period). |
| II | Compounds that kill too quickly at the higher concentrations but give insufficient kill at the lower concentrations, that is, higher concentrations give 15% or more kill after 24 hr and 90-100% at the end of the test period, but lower concentrations give less than 90% kill at the end of the test period. |
| III | Compounds that show no greater than a 9-fold difference between the minimum and maximum concentrations that exhibit delayed toxicity.[a] |
| IV | Compounds that showed at least a 10-fold but not greater than 99-fold difference between the minimum and maximum concentrations that exhibited delayed toxicity. |
| V | Compounds that show at least a 100-fold difference between the minimum and maximum concentrations that exhibit delayed toxicity. |

[a]Delayed toxicity is defined as mortality of less than 15% after 24 hr and more than 89% at the end of the test period.

research projects whose aim is to modify or formulate fast-acting toxicants into delayed-action materials. The possibilities presently considered can be categorized as in Table IV.

Matrix bound and microencapsulated toxicants (A and B) have several similarities. Both are presented to the ants as a solid formulation dispersed in soybean oil. Ideally, after ingestion and dispersal within the colony, lethal concentrations of toxicant will gradually build up via diffusion

*TABLE III. Laboratory Evaluation of Bait Toxicants Against Imported Fire Ants*

| Class | No. of chemicals tested | Percent of chemicals in class |
|---|---|---|
| I | 4269 | 87.0 |
| II | 374 | 7.6 |
| III | 255 | 5.0 |
| IV | 20 | 0.2 |
| V | 1 | 0.02 |
| Total | 4919 | |

to the surface of the matrix or through the wall of the microcapsules. Another possible release mechanism depends on the composition of the matrix or wall material, since toxic concentrations can also be released through chemical or enzymatic degradation of the polymer used.

In collaboration with USDA and private laboratories we have tested several formulations of both types. The results (Tables V and VI) have touched the two extremes—either very rapid kill similar to that of the unformulated compound or no appreciable effect. We had estimated from research with electron microscopy that particles 5 μ or less would easily pass into the ant's digestive system. This size constraint is achievable for both microcapsules and microparticle matrixes. However, we have recently discovered that worker

*TABLE IV. Four Methods of Modifying or Formulating Fast-Acting Toxicants into Delayed-Acting Materials*

A. Matrix bound toxicants

B. Microencapsulated toxicants

C. Modification of toxicant structure

D. Pendent toxicants

TABLE V. Results When Matrix Bound Toxicant Formulations Were Tested Against *S. invicta* Workers.

| Matrix bound toxicant | Matrix | Concn. (%) | Percent kill after indicated number of days | | | | | | |
|---|---|---|---|---|---|---|---|---|---|
| | | | 1 | 2 | 3 | 6 | 8 | 10 | 14 |
| Coumaphos[a] | Polystyrene | 0.01 | 0 | 2 | 2 | 2 | 2 | 3 | 13 |
| | | 0.1 | 25 | 70 | 83 | 93 | 93 | 100 | |
| | | 1.0 | 42 | 92 | 98 | 98 | 98 | 98 | 98 |
| Coumaphos[a] | Polymethylmethacrylate | 0.175 | 88 | 98 | 98 | 100 | | | |
| Coumaphos[a] | Polymethyl-acrylate | 0.19 | 2 | 2 | 2 | 3 | 3 | 5 | 12 |
| | | 0.38 | 83 | 87 | 87 | 90 | 90 | 95 | 95 |
| | | 0.75 | 100 | | | | | | |
| Coumaphos[b] | Natural polymer | 1.0 | 98 | 100 | | | | | |
| | | 2.0 | 98 | 100 | | | | | |
| Trichlorfon[c] | | 0.1 | 0 | 0 | 0 | 0 | 2 | 2 | 3 |
| | | 0.5 | 5 | 12 | 12 | 13 | 13 | 15 | 20 |

[a]Formulations supplied by Dr. J.P. Kochansky, USDA, Beltsville, MD.
[b]Formulation supplied by Griffith Laboratories, Chicago, IL.
[c]Formulation supplied by Dr. E. Gabbay, Chemistry Department, University of Florida.

TABLE VI. Results of Microencapsulated Toxicant Formulations against *S. invicta* Workers.

| Encapsulated toxicant[a] | Concn. (%) | Percent kill after indicated number of days: 1 | 2 | 3 | 6 | 8 | 10 | 14 |
|---|---|---|---|---|---|---|---|---|
| Ethyl pirimiphos[b] (3.5 μ) | 0.01 | 78 | 85 | 100 | | | | |
| | 0.1 | 100 | | | | | | |
| | 1.0 | 100 | | | | | | |
| Methyl pirimiphos[b] (4.2 μ) | 0.01 | 7 | 12 | 18 | 22 | 23 | 25 | 35 |
| | 0.1 | 60 | 100 | | | | | |
| | 1.0 | 100 | | | | | | |
| Permethrin[b] (4.9 μ) | 2.0 | 8 | 20 | 22 | 22 | 22 | 22 | 23 |
| Diazinon in honey-water[c] (4 μ) | 0.01 | 0 | 0 | 2 | 12 | 13 | 22 | 40 |
| | 0.1 | 100 | | | | | | |
| | 1.0 | 100 | | | | | | |
| Diazinon in soybean oil[c] (4 μ) | 0.01 | 2 | 3 | 10 | 15 | 22 | 27 | 37 |
| | 0.1 | 13 | 33 | 67 | 98 | 98 | 98 | 100 |
| | 1.0 | 92 | 100 | | | | | |

[a]Numbers in parentheses = mean particle size.
[b]Formulations supplied by ICI Americas.
[c]Numbers in parentheses = mean particle size.

ants have a very sophisticated mechanism for preventing solid particles from passing into the pharynx and esophagus. Food material is compressed in the infrabuccal pocket with liquids forced through a series of hairs or setae; this traps particles before they have a chance to enter the pharnyx (Figure 3). The liquid then moves on to the crop and/or digestive tract. When the infrabuccal pocket is filled with solids the compressed mass is expelled in the form of a pellet.

Studies with fluorescent latex microspheres have shown that particles as small as 0.88 μ (Table VII) in diameter are trapped by the filtration mechanism. Since this size approaches the limit of light microscopy, we do not know how small the particles must be to readily pass through the filter. A similar filtration mechanism is known for another ant, *Camponotus americanus* (13). However, in this case, particles under 100 μ readily passed through to the crop. Therefore it is reasonable that fast toxic action with the matrix and encapsulated formulations is caused by the insecticide contaminating the surface or by its rapid diffusion into the soybean oil or honey-water solution. Lack of toxicity with the formulation is simply due to the particles being filtered. This ultra-fine filtration mechanism

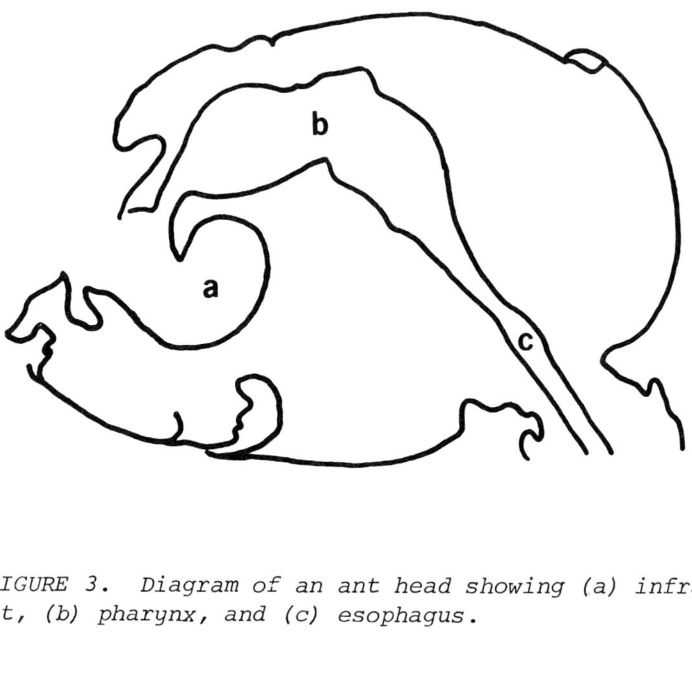


*FIGURE 3. Diagram of an ant head showing (a) infrabuccal pocket, (b) pharynx, and (c) esophagus.*

TABLE VII. Microfiltration by Fire Ant Workers. Analysis of Ants Fed Honey-Water Containing Fluorescent Microspheres

| | | Avg. no. particles/microscopic field | |
|---|---|---|---|
| Microsphere size (μ) | No. ants examined | Crop[a] | Infrabuccal pocket[b] |
| 0.88 | 5 | 3.16 | 77.28[c] |
| 1.8 | 8 | 0.002 | 44 |
| 4.6 | 5 | 0.007 | 26.0 |

[a]Total sample area examined.
[b]Five microscopic fields per ant examined.
[c]Buccal pellet ejected in 3 cases. One pellet count = 1760 particles.

of the ants appears to make utilization of the matrix and microencapsulation approaches impractical. However, we are still investigating the possibility that the type of polymer or wall material may influence acceptance and ingestion.

Method C, the modification of a toxicant structure (Table IV), is defined as the chemical alteration of a known fire ant toxicant (preferably a class II compound) to a relatively nontoxic material that is so structured that the parent toxicant can be regenerated by an enzymatic or hydrolytic mechanism present in the ants. This pro-agent approach has been well developed in the pharmaceutical industry (14) and has also led to remarkable selective activity in some herbicides and insecticides (15).

The commercially available insecticides are conspicuous for their lack of functionality required for simple modification. Only trichlorfon has a functional group capable of forming an ester linkage (Figure 4). Therefore, in collaboration with the USDA Insect Physiology Laboratory in Beltsville, Maryland (16), we attempted to design delayed action toxicants using trichlorfon and sodium fluoroacetate as precursors (Figure 4). (Although sodium fluoroacetate is too highly toxic to mammals for practical use, it is a good model for studying the feasibility of this approach).

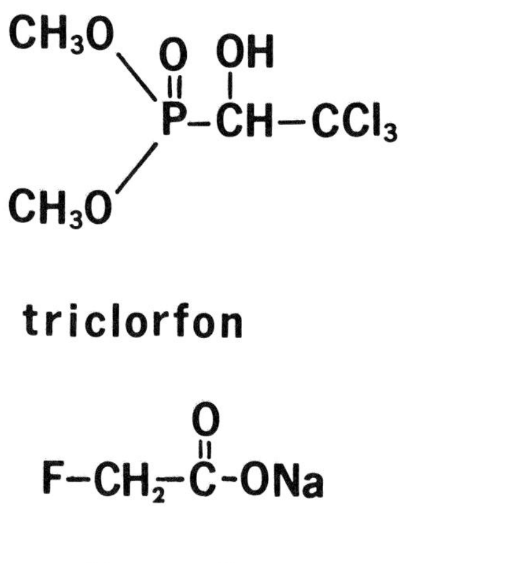


*FIGURE 4. Structures of trichlorfon and sodium fluoroacetate.*

Trichlorfon esters were prepared by reaction in ether with the appropriate acid chloride and pyridine. Esters of fluoroacetic acid were prepared by reaction with phosphoryl chloride followed by the addition of the appropriate alcohol and pyridine. The results (Tables VIII and IX) show that delayed toxicity was achieved with the caprate ester of trichlorofon and the two sterol esters of fluoroacetic acid. Unfortunately the range of toxicity is too narrow; therefore, in effect, we had changed class II compounds into class III compounds. Although these experiments did not yield the perfect delayed toxicant, they illustrated the feasibility and potential of the approach.

The pendent toxicant method for slow release has been successfully utilized in the area of controlled release herbicides (17,18). This approach is similar to that of pro-toxicants since the active agent is slowly released as the covalent linkages are hydrolyzed in the digestive system of the ant. Our laboratory has been interested in pendent toxicants for some time, and in September 1978, researchers at the Southern Research Institute, Birmingham, Alabama, were given a USDA Grant that allowed them to embark on a pendent toxicant feasibility study. Preliminary results look encouraging, and we will present a full description of the work next year.

In summary, matrix-bound and microencapsulated toxicants probably cannot be used for controlled release of toxicants against fire ants because of the remarkable filtration mechanism found in the ants. Structure modification of existing insecticides to form pro-toxicants has shown promise in achieving a degree of delayed toxicity. Although this approach is greatly limited by the lack of insecticides with suitable functional groups, we hope work in this area will continue.

---

This paper reports the results of research only. Mention of a pesticide in this paper does not constitute a recommendation for use by the U.S. Department of Agriculture nor does it imply registration under FIFRA as amended.

## REFERENCES

1. W. G. Eden and F. E. Garrett, Control of soil insect pests in Gulf Coast Irish potato fields. Ala. Polytech. Inst., Agric. Exp. Stn. Prog. Rep. Ser. No. 60, 1955.

TABLE VIII. Toxicity of Trichlorfon and its Esters When Fed in Soybean Oil to *S. invicta* Workers.[a]

| Compound | Concen. | Percent kill after indicated number of days | | | | | | |
|---|---|---|---|---|---|---|---|---|
| | | 1 | 2 | 3 | 6 | 8 | 10 | 14 |
| Trichlorfon | 0.01 | 7 | 12 | 15 | 23 | 23 | 23 | 25 |
| | 0.1 | 95 | 98 | 98 | 98 | 98 | 98 | 98 |
| | 1.0 | 100 | | | | | | |
| Caproate | 0.1 | 0 | 0 | 1 | 6 | 10 | 11 | 14 |
| | 1.0 | 12 | 28 | 44 | 57 | 61 | 66 | 67 |
| Caprate | 0.1 | 1 | 1 | 2 | 4 | 4 | 6 | 10 |
| | 1.0 | 5 | 30 | 53 | 79 | 82 | 83 | 86 |
| | 4.0 | 26 | 39 | 39 | 45 | 46 | 49 | 49 |
| Myristate | 0.1 | 0 | 0 | 0 | 2 | 3 | 3 | 8 |
| | 1.0 | 5 | 22 | 42 | 67 | 68 | 72 | 75 |
| | 4.0 | 15 | 25 | 37 | 40 | 40 | 40 | 40 |
| Stearate | 0.01 | 0 | 0 | 1 | 2 | 5 | 6 | 10 |
| | 0.1 | 2 | 3 | 9 | 24 | 26 | 29 | 35 |
| | 1.0 | 46 | 79 | 95 | 96 | 98 | 99 | 99 |

[a]Kochansky et al. (16).

TABLE IX. Toxicity of Fluoroacetyl Derivatives when Fed in Soybean Oil to *S. invicta* Worker Ants[a]

| Compound | Concen. (%) | Percent kill after indicated number of days | | | | | | |
|---|---|---|---|---|---|---|---|---|
| | | 1 | 2 | 3 | 6 | 8 | 10 | 14 |
| Sodium fluoroacetate[b] | 0.01 | 3 | 8 | 15 | 22 | 22 | 23 | 25 |
| | 0.1 | 88 | 100 | | | | | |
| | 1.0 | 100 | | | | | | |
| Cholesteryl fluoroacetate | 0.01 | 2 | 3 | 3 | 4 | 4 | 6 | 10 |
| | 0.1 | 0 | 0 | 0 | 2 | 2 | 3 | 6 |
| | 1.0 | 3 | 15 | 38 | 84 | 91 | 95 | 100 |
| | 4.0 | 88 | 98 | 98 | 100 | | | |
| Sitosteryl fluoroacetate | 0.01 | 2 | 2 | 5 | 9 | 11 | 13 | 22 |
| | 0.1 | 1 | 2 | 2 | 9 | 13 | 19 | 37 |
| | 1.0 | 7 | 21 | 42 | 66 | 79 | 87 | 96 |
| | 4.0 | 23 | 37 | 58 | 78 | 87 | 90 | 93 |
| Octadecyl fluoroacetate | 0.01 | 0 | 1 | 1 | 6 | 10 | 11 | 15 |
| | 0.1 | 1 | 1 | 2 | 17 | 33 | 44 | 62 |
| | 1.0 | 98 | 99 | 99 | 99 | 99 | 99 | 99 |

[a]Kochansky et al. (16).
[b]Fed to the ants in a honey-water solution.

2. G. H. Blake Jr., Imported fire ant on the march in Alabama. Ala. Polytech. Inst., Agric. Expt. Sta. Highlights of Agric. Res. 9(3) 1956.
3. D. W. Coon and R. R. Fleet, Environment, 12(10), 28-38 (1970).
4. C. S. Lofgren, F. J. Bartlett, and C. E. Stringer, J. Econ. Entomol., 54(6), 1096-1100 (1961).
5. C. S. Lofgren, F. J. Bartlett, and C. E. Stringer, J. Econ. Entomol., 56(1), 62-66 (1963).
6. F. Bellinger, R. E. Dyer, R. King, and R. B. Platt, the imported fire ant, a report by the Fire Ant Committee of the Georgia Academy of Science, Inc., July 1964.
7. D. P. Wojcik, W. A. Banks, W. B. Wheeler, D. P. Jouvenaz, C. H. Van Middelem, and C. S. Lofgren, Pestic. Monit. J., 9, 124-133 (1975).
8. F. W. Kutz, A. R. Yobs, W. G. Johnson, and G. G. Wiersma, Environ. Entomol., 3(5), 882-884 (1974).
9. J. R. Innes, M. B. Ulland, M. G. Valerio, L. Petrucelli, L. Fishbein, E. R. Hart, A. J. Pallotta, R. R. Bates, H. L. Falk, J. J. Gart, M. Klein, I. Mitchell, and J. Peters, J. Natl. Cancer Inst., 42(6), 1101-1114 (1969).
10. B. M. Ulland, N. P. Page, R. A. Squire, E. K. Weisburger, and R. L. Cypher, J. Natl. Cancer Inst., 58(1), 133-140 (1977).
11. C. S. Lofgren, W. A. Banks, and B. M. Glancey, Ann. Rev. Entomol., 20, 1-30 (1975).
12. W. A. Banks, C. S. Lofgren, C. E. Stringer, and R. Levy, Laboratory and field evaluation of several organochlorine and organophosphorous compounds for control of imported fire ants. ARS-S-169, p. 13 (1977).
13. T. Eisner and G. M. Happ, Psyche, 69, 107-116 (1962).
14. V. Stella in Pro-drugs as Novel Drug Delivery Systems, T. Higuchi and V. Stella (Eds.), ACS Symp. Series 14, American Chemical Society, Washington, D.C. 1975, pp. 1-115.
15. A. Albert in Selective Toxicity and Related Topics, Methuen and Co., Ltd., London 1968, p. 531.
16. J. P. Kochansky, W. E. Robbins, C. S. Lofgren, and D. F. Williams, J. Econ. Entomol., (in press).
17. F. W. Harris, M. R. Dykes, J. A. Baker, and A. E. Aulabough, in Controlled Release Pesticides, H. B. Scher, (Ed.), pp. 102-111, ACS Symp. Series 53, American Chemical Society, Washington, D.C. 1977.
18. C. L. McCormick and M. Fooladi, in Controlled Release Pesticides, H. B. Scher, (Ed.), pp. 112-115, ACS Symp. Series 53, American Chemical Society, Washington, D.C. 1977.

# ENCAPSULATION OF DIMILIN AND ITS EFFECT ON INSECTICIDAL ACTIVITY AGAINST *AEDES AEGYPTI* (L.) AND *MUSCA AUTUMNALIS* DEGEER

*F. W. Knapp*
*Chad Nontapan*

Department of Entomology
University of Kentucky
Lexington, Kentucky

## I. INTRODUCTION

In recent years man has become conscious of environmental pollution by pesticides that persist for long periods of time and that may eventually harm plants, animals, beneficial insects, other non-target species or even himself. Pesticides with effective control of certain pest populations at a particular time and place, are considered to be inimical to man's welfare (1). The use of pesticide-polymer combinations as a means of obtaining controlled release of biodegradable and/or short life pesticides in correct amounts over a period of time may provide effective control of insect pests and at the same time alleviate the problem of environmental pollution (2).

Biologically active agents, including pesticides, may suffer a number of disadvantages when applied in conventional ways. The persistence of pesticides is shortened by mechanisms such as biodegradation, chemical degradation, evaporation, and radiation, so that multiple applications may be needed to maintain an effective concentration over the time span required. Extending the effective residual life of highly toxic pesticides with a short persistence, even for a few days, could vastly improve the efficacy of the materials and reduce the hazards caused by the necessity of frequent applications (3). Controlled release formulations of pesticides offer the advantages of a longer period of control over




ISBN 0-12-074450-3

the target species and lesser toxicity to higher forms of life. In a controlled release formulation the toxicant is incorporated into a polymer which protects it from natural degradation and makes it more stable. The continuous release of a toxicant at a controlled level may mean that a lower dosage is needed to maintain pest populations below economic thresholds (4). The lower dosage thus minimizes the amount of pesticides released into the environment.

The insect growth regulator Dimilin, [(1-(2,6-difluorobenzoyl)-3-(4-chlorophenyl) urea], is a selective insecticide with a wide range of biological activity (5). This compound proved to be a powerful stomach toxicant for the larvae of several insect species, yet it was quite safe to mammals (6-9). It is one of the more promising new compounds for the control of mosquito and fly larvae in breeding areas (10-12). It is relatively safe to non-target species associated with mosquito breeding habitats (13) including the mosquito fish, *Gambusia affinis* Baird and Girard, an important biological control agent (14). Dimilin rapidly degrades in water and water-hydrosoil systems, especially above pH 9 (15-17). As a feed through to livestock for control of insects breeding in the feces, it has a half life of 4 to 5 days (18) but residues have been found in eggs, milk, and animal tissues (19-22).

Dimilin applied in a conventional manner is a short lived pesticide and its rapid breakdown eliminates it as a good residual mosquito larvicide. Because of the residues found in animal products, Dimilin is also eliminated as a potential feed through pesticide for livestock. However, because of its wide range of biological activity, its rapid degradation and its safety to mammals, Dimilin would be an ideal compound if ways could be found to increase its residual effectiveness against target species and to reduce its residue in animal products and the environment. Therefore, this study was initiated with the objective of determining if Dimilin could be encapsulated into formulations that would release the toxicant slowly so as to give effective control of target species over a longer period of time than with conventional formulations.

## A. *Controlled Release Insecticides*

The concept of controlled release insecticides has been in existence for many years but recently there has been a renewed interest among researchers in this method of application. Mosquito control workers might be considered as the pioneers in the field of insect control utilizing the concept of controlled release insecticides. Raley and Davis (23) used toxic briquettes, a mixture of 200 g casting plaster, 1 cup sawdust, 50 g DDT (50 wt%), 20 g lindane (25 wt%), and 200 ml water, to control mosquito larvae in ponds and streams in the San Joaquin Valley of California. They found that these briquettes gave very effective mosquito control for three months.

Bruce-Chwatt and Fitz-John (24) used Gammexane pellets [consisting of 1 part Gammexane powder (BHC wetable powder containing 6.5% Gamma isomer), 3 parts of cement, 7 parts of sand and a sufficient quantity of water to produce a coarse dough] to control mosquitoes breeding in crab-holes. They found that application of Gammexane pellets prevented the breeding of *Aedes irritans* Theo. for five months. Elliott (25) reported that 10 g of sand cement pellets containing 16% dieldrin in a 2 gallon pot filled with water gave complete control of *A. aegypti* for up to one year. Similarly, Evans and Fink (26) observed that all *A. aegypti* larvae in fire barrels were dead 24 hours following treatment with briquettes containing 16% dieldrin at the rate of 5 g/gallon of water, and no *A. aegypti* were recovered from the treated barrels during the 150 day inspection period. In an extensive field test, dieldrin briquettes were found to retain larvicidal properties after almost five years. Chromatographic analysis of a sample of these briquettes showed that 40 to 50% of the initial available dieldrin remained in the briquettes after five years of use.

The earlier concept of utilizing insecticidal briquettes in mosquito control was closely related with the use of chlorinated hydrocarbon insecticides which have long residual properties. Recently, investigators began to consider using briquettes impregnated with the non-persistant organophosphorus insecticides which are rapidly hydrolysed in water and have relatively low toxicity to non-target species. Barnes *et al.* (27) reported that briquettes containing 2% Abate at the rate of 45 mg active ingredient (AI)/ℓ of water and briquettes containing 5% malathion at the rate of 117 mg AI/ℓ gave 100% mortality of *C. p. pipiens* larvae throughout a 35 day laboratory test. In a six month field test, Barnes and Webb (28) obtained excellent control of *A. canadensis* Theo., *A. cantator* Coq., *A. vexans* Meigan, and *Culex tarsalis* by using briquettes containing 30% Abate at 2.1 mg AI/ℓ of water.

Insecticides in briquette form were not easily handled and because of the porosity of the briquettes, the insecticides diffused much too quickly and sporadically (29). To overcome these problems, researchers began to look for other substrates such as organic polymers.

Mathis *et al.* (30) formulated a montan wax hydrogenated cottonseed oil - DDVP formulation containing 25% dichlorvos, and tested this against adult *Anopheles quadrimaculatus* Say. in Peet-Grady chambers and in simulated primitive huts. The formulation gave 100% control for 16 weeks. Miles *et al.* (31) developed a montan wax-dibutyl phthalate - DDVP formulation containing 25% dichlorvos and tested it against *A. quadrimaculatus* in primitive huts in Africa. The results were superior to those obtained by Mathis *et al.* (30) against the same species of mosquito.

Whitelaw and Evans (32) incorporated Abate, Dursban, Dibrom, and malathion at selected concentrations into plastic pellets composed of polyvinyl chloride, polyurethane foam, and polyamide and bioassayed their formulations with *C. p. quinquefasciatus*. Single pellets of the formulations were used repeatedly and larval mortality was recorded at the end of 24 hours. All nonfoam formulations except Dibrom and malathion at 0.5 and 1.0% concentrations gave greater than 90% mortality throughout the 37 day test period. Polyurethane foam pellets gave 90 to 100% mortality for only 2 days at the 20% rate for each insecticide.

At almost the same time that the plastic formulations were being investigated, the use of rubber as a carrier for mosquito larvicides was developed by the B.F. Goodrich Aero-Space and Defense Products Division of the B.F. Goodrich Company, Akron, Ohio. The toxic components, generally compounds of tributyltin, were incorporated in a rubber base. The toxicity of such compounds to mosquito larvae was first shown by Cardarelli and Mercier (33) and later by Cardarelli *et al.* (34) who tested tributyltin oxide impregnated neoprene rubber and found that it gave good kill of mosquito larvae for more than four months. Schultz and Webb (35) bioassayed *C. p. quadrimaculatus* with neoprene pellets containing 9.6% tributyltin oxide and 5.7% Abate in EPDM (floating rubber) at rates of 78 and 45 mg of pellets in 800 ml of water, respectively. The results showed that 100% mortality of mosquito larvae was obtained for about 12 days for the tributyltin oxide impregnated neoprene rubber pellets, and 90 to 100% mortality throughout the 105 day test with the formulation of Abate in EPDM. Tributyltin oxide impregnated rubber formulation was also found to be highly toxic to *A. taeniorhynehus* and *Culex nigripalpus* larvae but not to the eggs of both species (36).

Parker (37) reported that a single application of a 9.9% Dursban in polyethylene pellet formulation at 2.5 ppm reduced *A. canadensis* Theo. larvae for 13 weeks in woodland pools, and was highly effective against naturally occurring populations of *C. p. quinquefasciatus* and *C. restuans* for more than 21 weeks. A single application of Dursban emulsifiable concentrate at 0.7 lb AI/acre gave 3 weeks of effective mosquito control (38). A single application of a charcoal formulation containing 1% Dursban at the rates of 0.5 and 2.5 ppm gave effective control of mosquito larvae for 20 and 26 weeks, respectively (39).

The effectiveness of chlorpyrifos in various polymer carriers has been investigated by several researchers. Polyethylene and polyvinyl chloride formulations containing 9.9 and 10% chlorpyrifos, respectively at the rate of 5 ppm provided effective control (94 to 100%) of *C. p. quinquefasciatus* larvae for at least 18 months (covering two breeding seasons) without producing excessive residues in the treated water (40,41). However, this rate seems to be very high. Nelson, *et al.* (42) showed that 0.23 ppb of chlorpyrifos concentration released from polyethylene pellets gave 99.7% control of laboratory populations of *C. p. quinquefasciatus*.

In subsequent studies, Roberts *et al.* (43) found that controlled release formulations of chlorpyrifos appeared to reduce populations of non-target species, *Gerris* spp., *Chironomus* spp., *Chaoborus punctipennis* Say, and *Laccophilus fasciatus* Aube, in treated pools. In order to develop the guidelines for recommending the use of controlled release polyethylene formulation of chlorpyrifos, field tests were conducted by Evans *et al.* (44) to evaluate the effectiveness of this formulation as a larvicide. Sufficient chlorpyrifos residues were found to effectively control mosquitoes in pools treated at dosage rate of 1 ppm throughout a 22-week study. *Psorophora confinnis* larvae in rice field habitats were found to be susceptible for more than 11 weeks to a polyethylene formulation of chlorpyrifos at the rate of 2 ppm (45).

Some types of formulations fail to release the amount of insecticide necessary for efficient mosquito kill. Roberts and Miller (38) indicated that a single application of a 10% Abate polyvinyl chloride (PVC) formulation at 10 ppm gave effective control of *C. p. quinquefasciatus* and *C. restuans* populations for only 2 weeks, whereas a single application of Abate emulsifiable concentrate at the same rate gave 3 weeks of control.

Nelson *et al.* (29) observed that Dursban was slowly released from PVC formulations under static laboratory conditions. Varying the PVC resin or plastizer had little effect on the release rate of Dursban maintained in water. The differences in percentage of Dursban in PVC formulations (9.6 to 37.8%) had little effect on the levels of Dursban maintained in water (46). Dursban in PVC formulation applied at 2.5 ppm gave effective control of mosquitoes for 8 weeks, while Fenthion in PVC at 2.5 ppm provided effective control for 20 weeks (39). A single application of chlorpyrifos in PVC at 5 ppm provided effective control of *C. p. quinquefasciatus* for 24 weeks (40). However, the large amount of polymer formulations required at the 5 ppm dosage might make them impractical for larger-scale field use (41).

The release rates of toxicants from polymer formulations studied by Nelson *et al.* (47) indicated that formulations with the same dosages significantly increased in average release rates with an increase in surface to volume ratio. This means that the greater the surface to volume ratio the greater release rate of toxicant. Similar results were obtained by Lawson *et al.* (48) working with different surface to volume ratios of chlorpyrifos-polyethylene pellets. Their results demonstrated that the formulations with smaller particle size showed a higher release rate.

Recently, encapsulated formulations of 75% malathion (National Cash Register Company, Dayton, Ohio), 2% malathion and 1.1% Abate (3M Company, St. Paul, Minnesota) were evaluated both under laboratory and simulated field conditions (49). Technical information concerning the chemical nature of the capsule walls of these formulations was not made available in this report. The laboratory results showed that 5 ppm malathion residues were maintained in water for 8 weeks by the 2% encapsulated formulation and 22 ppm for the 75% encapsulated formulation. In the simulated field tests, none of the encapsulated formulations were effective in providing extended control of *C. restuans* larvae, although the 1.1% Abate formulation gave about 7 weeks of greater than 90% control at dosage rates of 0.5 and 1 ppm. However, this does not actually present significantly extended control, since it was possible to achieve 3 weeks of control by treatment with an EC formulation of Abate alone (38). Also from the standpoint of the amount required/acre, the 1.1% Abate encapsulated formulation might not be practical for use at the dosages evaluated in this study (41).

Methoprene (ZR-515 an insect growth regulator produced by Zoecon Corp., Palo Alto, California), was the first juvenile hormone (JH) mimic formulated in a controlled release polymer and used as a mosquito larvicide. Dunn and Strong (50) applied polyurethane foam formulations containing 0, 1, and 3% methoprene (W/W) into catch basins to implement treatment. They found that a single application of a polyurethane foam formulation containing 3% methoprene resulted in a high degree of kill for almost 2 months, while a single application of Technical methoprene at 1.0 ppm was effective for only 2-3 days. These preliminary studies demonstrated that the controlled release polymers of methoprene could effectively overcome the short residual problem. Schaefer and Wilder

(51) reported that the application of a 10% flowable liquid formulation (controlled release) of methoprene at 0.1 lb AI/acre gave 100% mortality of second, third, and fourth instars of *Aedes nigromaculis* throughout a 48 hour test period; whereas the application of 4.0 lb/gal EC formulation of methoprene at the same rate provided 85, 30, and 10% mortality of the third instar of this species at 0, 24, and 48 hours after treatment, respectively.

In livestock insect control numerous attempts have been made to develop a means of passing various insecticides through the digestive tract of animals with a resulting minimal loss of insecticide so as to maintain the insecticide in the feces to control fly larvae breeding there. Lloyd and Matthysee (52) incorporated technical diazinon, dimethoate, famphur, dimetilan, and dichlorvos into three types of selected polymer protectants. These were ureaformaldehyde resin, polyester, and a polyvinylchloride resin. Each polymer was made up to 25% with each insecticide. Each polymer-insecticide was administered to steers at the rate of 40 mg/kg of body weight per day, and the feces of the steers were analyzed for residue 24 hours after feeding. These results showed that diazinon, dichlorvos, and dimethoate eluted much more rapidly from the cow than *in vitro* aqueous. Dimethoate demonstrated a rapid elution from the cow within the first 24 hours, and a slow elution rate thereafter. Of the polymers tested, polyvinylchloride appeared most ideal for livestock feed additives. It showed the resiliency of retaining its shape after feeding. These authors suggested that incorporation of the insecticides in polyvinylchloride could insure the presence in feces of larvicides that normally would be degraded if fed unprotected.

Investigations on the effects of ureaformaldehyde formulations of three systemic insecticides were conducted by Medley and Drummond (53) to evaluate the systemic activity of insecticides against the screwworm, *Collitroga hominivorax* Coquerel, and stable flies. Administration of the materials to Angora goats infested by these insects indicated that ureaformaldehyde resin containing 25% Dipterex applied at dosage rates of 50, 75, and 100 mg/kg body weight showed little increase in duration of effectiveness over technical Dipterex. Ureaformaldehyde resin containing 25% dimethoate applied at dosage rates of 50, 75, and 100 mg/kg body weight afforded a longer systemic activity than the technical material. Ureaformaldehyde resin containing 50% dimethoate reduced the duration of the effectiveness below that of the technical grade material. Ureaformaldehyde resins containing 25% and 10% dimethoate were found to increase the duration by factors of approximately 1 1/2 and 2, respectively. With

Bayer 37342 [O, O-dimethyl O-(3, 5-dimethyl-4-methylthiophenyl) phosphorothioate], there was variation among individual animals, but at dosages of 100 mg/kg body weight, resins containing 25% and 10% AI increased the duration of effectiveness over that of the technical material by a factor of approximately 1 1/2 and 1 3/4, respectively.

Breeden, *et al*. (54) recently investigated an encapsulated 7% dry premix controlled-release formulation of the insect growth regulator methoprene as a feed additive for control of house flies breeding in chicken feces. The results showed that the technical formulation of methoprene at the rate of 50 ppm AI in feed gave 98% inhibition of adult house fly emergence. The encapsulated formulation of methoprene at 5 ppm AI in feed also achieved the same level of control. Neither formulation showed an effect on house fly emergence at 2 days post-treatment. Herald (55) reported much lower concentrations of technical methoprene were required in the feed of hens to completely inhibit the development of house flies than reported by Breeden *et al*. (54). Technical methoprene at 5 and 15 ppm AI in feed gave, respectively, 92 and 100% inhibition of house fly emergence at the sixth day of treatment. He also reported that the technical formulation of an insect juvenile hormone analog RO 20-3600 ((E)-4-[(6,7-epoxy-3,7-dimethyl-2-nonenyl)oxy]-1,2-(methylenedioxy) benzene), and its encapsulated formulation containing 10% AI fed to cattle at a rate of 2.5 mg/kg of body weight per day prevented, respectively, 94 and 97% emergence of house fly adults in cattle feces.

Some controlled release insecticides can be used as poison baits applied around dairy barns, poultry farms, pig pens, and similar problem sites for house fly control. Kilpatrick *et al*. (56) applied an ureaformaldehyde formulation containing 36% DDVP, and a DDVP-wax formulation containing 25% DDVP as poison baits in bait dispensers. They found that at a chicken ranch treated with the DDVP-ureaformaldehyde formulation excellent fly control was obtained for the entire 10-week observation period. At a dairy barn treated with the DDVP-wax bait, excellent to satisfactory control was achieved during the entire 10-week test period.

Dichlorvos (20%) impregnated resin strips (3/16 x 2 1/2 x 10 in.) installed at the rate of 1/400 $ft^3$ of space in one milking barn has been found to reduce house fly populations (57). Knapp (58) reported that water forced through 20% dichlorvos resin pellets and offered to house flies was effective in killing house flies for more than 90 days. Knapp and Ellis (59) found that the amount of dichlorvos released from dichlorvos (20%) resin pellets subjected to water leaching was about 850 ppm of dichlorvos during the first 10 minutes of operation, and the amount of dichlorvos contained in the leachate during subsequent flushing decreased sharply until less than 100 ppm were found on the 4th day. Dichlorvos leachate then declined slowly until about 1-2 ppm were found on the 85th day. They also found that the dichlorvos released from a constant amount of 20% dichlorvos-resin pellets was dependent upon the water volume. As the water volume increased the amount of dichlorvos increased proportionally and reached a plateau between 12 and 16 hours. They suggested that this probably was due to a concentration-equilibrium effect between the dichlorvos within the resin pellets and that within the water.

Controlled release formulations of insecticides have also been used for control of insects on pets. Plastic cat collars containing 3 and 5% dichlorvos were found to be effective against adult cat fleas, *Ctenocephalides felis* Bouche on cats for at least 3 months (60).

Price (61) reviewed the use of ureaformaldehyde resins containing dieldrin, DDT, malathion, and diazinon against German cockroaches, *Blatella germanica* L., bedbugs, and pharaoh and Argentine ants, in England. Kydonieus *et al.* (62) reported that the use of Hercon-Roach Tape$^R$ strips containing diazinon and Baygon$^R$ were effective in controlling the German cockroach for more than 52 weeks. Another product, Killmaster$^R$, containing chlorpyrifos provided effective control of German cockroaches for more than 105 days (63).

## II. MATERIALS AND METHODS

### A. *Test Compound*

Technical Dimlin [1-(2,6-diflurobenzoyl)-3-(4-chlorophenyl)urea] as a 99.4% air milled powder, was obtained from Thompson-Hayward Chemical Company, Kansas City, Kansas.

### B. *Bioassay Insects*

*1. Mosquitoes (Aedes aegypti).* The mosquitoes were from a colony which had been maintained in the Veterinary and Medical Entomology Laboratory, Department of Entomology, University of Kentucky, Lexington, Kentucky, since 1972.

The rearing procedure was identical to that described by Gerberg (64), except a liver powder solution was used for feeding of larvae instead of finely ground prepared animal food. A temperature of 26±1$^\circ$C and a relative humidity of 65±5% were maintained through the experiments.

*2. Face Flies (Musca autumnalis).* A laboratory colony was maintained at a temperature of 26±1°C, and 70±5% relative humidity. The adult fly diet consisted of a mixture of whole egg solids, nonfat dry milk, and granulated sugar in a 1:1:4 ratio (V/V), and water (65).

To obtain face fly eggs, a 3.8 x 16 cm plastic lined oviposition dish containing fresh cattle feces was placed in the face fly cage for 2 hours, then removed to a seed germination chamber for egg hatching and larval development. The temperature and relative humidity in the chamber were 27±2°C and 80±5%, respectively.

## *C. Test Animals*

Six Holstein steer calves approximately six months old ranging from 111 to 140 kg at the initial phase of investigation and 172 to 222 kg at the last phase were used in feed-through studies. The steer calves were housed in individual 1.5 x 2 x 3 m stalls held at 25±2°C. Each morning the calves received a grain ration equal to 1% of the body weight, and chopped alfalfa hay mixed with 10% molasses equal to 2% of the body weight. The grain ration consisted of 35 kg oat grain, 57 kg cornmeal, 8 kg molasses, 500 g mineral salt, and 200 g ADE vitamin combination. This was top dressed on the molasses-chopped alfalfa hay in a half bushel rubber bucket. In addition, excess molasses-chopped alfalfa hay was also fed to each calf in the afternoon.

## *D. Preparation of Encapsulated Formulations of Dimilin*

Four types of polymers, ureaformaldehyde, polyamide, polyurethane foam, and polyvinylchloride were used in the incorporation of Dimilin to form encapsulated formulations. Each type of polymer was formulated to contain 0, 5, and 10% Dimilin (W/W). In addition, Carbopol[R] was formulated at 0, 50, and 75% formulations. The polymers containing no Dimilin were used as control formulations. Each formulation contained 20 g of Dimilin.

*1. Ureaformaldehyde Resin Formulations.* Ureaformaldehyde resin formulations containing 0, 5, and 10% Dimilin were designated as UF-0, UF-5, and UF-10, respectively.

The method of preparation was similar to that of Medley and Drummond (53), with the exception of the insecticide used and the pellet production procedure. To form encapsulated ureaformaldehyde resin formulations, glycerol, granular gelatin, 40% formaldehyde solution, and granular urea were combined at the ratio of 5:5:5:4 (W/W). The glycerol and gelatin were mixed and allowed to set for 6 hours to form a plasticizer, then the urea and desired amounts of technical Dimilin to provide 0, 5, and 10% AI Dimilin (W/W) were added. The mixture was stirred to a uniform granular consistency to form a resin, the 40% formaldehyde was added, and the ingredients again thoroughly mixed. Six to 8 drops of technical grade sulfuric acid were added drop by drop with a medicine-dropper as a catalyst while stirring with a glass rod. This mixture was allowed to harden by polymerization in a circular aluminum pan (20 x 3 cm).

2. *Polyamide Formulations*. The polyamide formulations, containing 0, 5, and 10% AI Dimilin, designated as PA-0, PA-5, and PA-10, respectively, were prepared following the procedure of Whitlaw and Evans (32), with the exception of the insecticide used. The desired amounts of Elvamide[R] Nylon grade 8061 (E.I. Du Pont De Nemour and Co., Wilmington, Delaware) were dissolved in 800 ml absolute methyl alcohol. The mixture was heated and agitated, and then the desired amounts of technical Dimilin (W/W) were added and thoroughly mixed by stirring with a glass rod for 30 minutes. The mixture was poured into a 30 x 22 x 3 cm aluminum pan to allow any excess alcohol to evaporate.

3. *Polyurethane Foam Formulations*. Polyurethane foam formulations containing 0, 5, and 10% AI Dimilin, designated as PU-0, PU-5, and PU-10, respectively, were prepared following the procedure of Whitlaw and Evans (32) with the exception of the insecticide used. The desired amounts of technical Dimilin (W/W) were mixed with desired amounts of polyurethane foam component "W" of Isofoam PE-12[R] (Witco Chemical, Isocyanate Products Division, New Castle, Delaware). The desired amounts of component "A" of Isofoam PE-12 at a ratio of 75 parts component "W" to 100 parts component "A" (W/W), were added and thoroughly mixed using a glass rod and then placed in a circular aluminum pan (20 x 3 cm) to harden.

4. *Polyvinylchloride Formulations*. Polyvinylchloride formulations containing 0, 5, and 10% AI Dimilin, designated as PVC-0, PVC-5, and PVC-10, respectively, were prepared following the method of Wilkinson *et al*. (39), with the exception of the insecticide used. The plasticsol was formed by mixing 2 parts Diamond PVC-71[R] (Diamond Shamrock Corporation, Cleveland, Ohio), and 3 parts W.W dioctylphthalate plasticizer (Eastman Kodak Co., Rochester, New York). The

desired amounts of technical Dimilin were added and thoroughly mixed. The mixtures were then transferred into circular aluminum pans (20 x 3 cm) and cured at 130°C for one hour.

5. *Carbopol Resin Formulations.* Carbopol resin formulations containing 0, 50, and 75% AI Dimilin, designated as CP-0, CP-50, and CP-75, respectively, were prepared following the procedure of the manufacturer (66). The desired amounts of Carbopol 934[R] (B.F. Goodrich Chemical Company, Cleveland, Ohio) was mixed with desired amounts of technical Dimilin (W/W). Distilled water was added drop by drop and stirred with a glass rod to form the granulation. The granules were left for two weeks to allow the water to evaporate. All formulations (2 to 3 mm thick) were cut into ribbons and blended with a heavy duty laboratory blender and sieved to 10/20 and 20/32 mesh sizes.

Throughout all of the experiments the 10/20 mesh size was used in the studies on the effects of the formulations on mosquito larvae in water. Similarly, the 20/32 mesh size was used for face fly larvae breeding in cattle feces.

## E. *Establishment of the $IC_{50}$ and $ED_{50}$ of Dimilin Formulations for A. aegypti and M. autumnalis*

1. *$IC_{50}$ for A. aegypti.* Dosage-inhibition of adult emergence tests were conducted using 5 concentrations of technical Dimilin and Dimilin-encapsulated formulations. The concentrations were prepared by placing the appropriate amount of each Dimilin encapsulated formulation into tap water to make 1 ppm Dimilin. These solutions were allowed to stand for 48 hours to release Dimilin in the water. After 48 hours the treated water was stirred with a glass rod and left to stand for about 5 minutes to partition the solid particles from the aqueous portion. An aliquote of the 1 ppm Dimilin stock solutions of each formulation was then diluted with tap water to provide the desired dosage rates. The total volume of 200 ml used for each test was placed into 500 ml insulated foam cups for the mosquito bioassay.

The 1 ppm stock solution of technical Dimilin was prepared by serial dilutions of a 1000 ppm Dimilin solution.

Encapsulated formulations containing no Dimilin were used as untreated checks for each Dimilin encapsulated formulation treatment, and a water-acetone solution containing no Dimilin was used as untreated checks for technical Dimilin.

The amounts of the encapsulated formulations and the water-acetone solution used were the same as for the highest rates used in the treatments.

Twenty 3rd instar *A. aegypti* were released in each of the 100 ml insulated styrofoam cups containing the test solution. Each test solution was replicated 4 times. All larvae were fed liver powder solution during the test period. The numbers of adults eclosed were recorded.

The inhibition of adult emergence in these tests was subjected to Abbot's formula (67) to correct for natural inhibition of emergence. The encapsulated check formulations were used for correcting the inhibition for the encapsulated treatments and the acetone check was used for correction of inhibition for the technical Dimilin treatments. Probit regression lines were prepared by Finney's method (68). The $IC_{50}$s (the concentration which produced 50% inhibition of adult emergence of the tested population) and their 95% fiducial limits were calculated with the SAS computer program for probit analysis (69). The effectiveness of formulations were compared based on the $IC_{50}$ values.

*2. $IC_{50}$ and $ED_{50}$ for M. autumnalis.* The $IC_{50}$ and $ED_{50}$ of technical grade and encapsulated formulations of Dimilin were determined in the laboratory against face fly larvae in cattle feces. Two methods, that of the incorporation of Dimilin into the feces and as a feed-through to cattle, were used.

For the direct incorporation of Dimilin into feces a 10 ppm Dimilin-cattle feces stock mixture was prepared. Technical Dimilin was dissolved in acetone to form a 1000 ppm solution. The appropriate volume of this solution was then mixed with fresh cattle feces to form 10 ppm Dimilin-feces stock mixture.

In the preparation of the 10 ppm stock mixtures of encapsulated formulations, the appropriate amounts of each formulation were weighed and directly mixed with fresh cattle feces.

The stock mixtures were further diluted to obtain desired concentrations for testing by adding the appropriate amounts of stock mixtures to untreated fresh cattle feces and blending thoroughly with an electric hand mixer.

Immediately prior to use, 4 100 g-samples were placed in souffle cups. Twenty newly emerged face fly larvae were placed on the surface of the prepared samples. The number of adults eclosed were recorded.

For the cattle feed-through method, six Holstein steer calves were fed the previously described diets during a 5 day adjustment period prior to testing. One day before the initiation of a 3 day test period, each calf was weighed and placed in individual 1.5 x 2 x 3 m stalls.

Five dosage rates, plus a control of each formulation, based on the mg of Dimilin per kg of body weight per day were fed to the calves (1 rate for one calf) for 3 consecutive days. To obtain the desired dosage rate for each calf, the appropriate amount of each encapsulated formulation was mixed with 200 g cornmeal. The desired dosage rates of technical Dimilin formulation were obtained by serial dilutions of a 1000 ppm Dimilin feed mix.

Because the same calves were used for all formulations, they remained on an untreated ration for one week between each 3 day treatment period. Preliminary bioassay of feces from a calf fed the highest Dimilin rate had shown no face fly mortality 6 and 7 days post-treatment. Therefore, a one week period between each 3 day treatment was considered sufficient to prevent any residual effects of Dimilin against face fly larvae.

Fresh fecal samples were collected from each treated calf at days 1, 2 and 3 during the 3 day treatment. Post-treatment fecal sample collections were at days 1 and 3 after the calves were returned to normal feed. Bioassays were conducted as previously described. The $ED_{50}$ (the dosage fed to cattle which inhibited 50% adult emergence from larvae in feces) values for face fly were determined from feces collected at day 3 during treatment.

### F. *Residual Activity of Dimilin Formulations Against Adult Emergence*

*1. Residual Activity Against A. aegypti.* To evaluate the residual activity of technical Dimilin and the Dimilin-encapsulated formulations against the mosquito, each formulation was made up in 3600 ml of tap water at concentrations of 0, 0.05, 0.1, and 0.5 ppm, in a 3.8 liter wide mouth glass jar. The jars were covered with muslin cloth and held at $25 \pm 2^{\circ}C$ and 65±5% relative humidity.

To compensate for water loss by evaporation on each bioassaying date, tap water was added up to the levels previously marked on the jars. The water in each jar was gently stirred with a glass rod and allowed to stand for about 5 minutes for particle settlement before drawing samples.

After samples were removed, the water levels in the jars were again marked. This procedure was repeated at each bioassay date (after 150 days, water in the jars was removed and replaced 24 hours prior to the next bioassay).

Tap water containing acetone, and non-Dimilin-encapsulated formulations served as untreated checks for the technical Dimilin and Dimilin-encapsulated formulation treatments, respectively. The amounts of acetone and non-Dimilin-encapsulated formulations used were the same as for the highest rates used in the treatments.

At days 1, 7, 15, 30, 45, 60, 75, 90, 105, 120, monthly thereafter for 360 days, four 20 ml aqueous samples from each jar were transferred into individual 20 x 150 ml test tubes and bioassayed with ten 3rd instar *A. aegypti*. The bioassay procedure was the same as previously described except one drop of 10% liver powder solution was provided to larvae in each tube on the same day of testing and two drops per day for other consecutive days until pupation. The criterion for effectiveness was based on the inhibition of adult emergence.

*2. Residual Activity Against M. autumnalis.* Two kg of 100 ppm Dimilin-feces stock mixture of each encapsulated formulation, the technical Dimilin and untreated checks were prepared as previously described, and transferred into individual 3.8 liter wide-mouth glass jars. All jars were tightly closed with jar lids to prevent water loss by evaporation. At days 1, 7, 15, 30, 45, 60, 75, 90, 105, 120, and 150, each stock mixture was mixed with a spatula and an appropriate amount of the stock sample was diluted with fresh cattle feces to form 0.05, 0.1 and 0.5 ppm of each formulation; except technical Dimilin which was diluted to form 0.01, 0.05 and 0.1 ppm because 0.05 to 0.5 ppm technical Dimilin resulted in 100% inhibition of adult emergence. These were then bioassayed with face fly larvae as previously described.

Calculation and evaluation of results were the same as previously described for the mosquito test.

## III. RESULTS AND DISCUSSION

### A. *Establishment of the $IC_{50}$ and $ED_{50}$ of Dimilin Formulations for A. aegypti and M. autumnalis*

*1. $IC_{50}$ for A. aegypti.* Laboratory bioassays of technical Dimilin against 3rd instar *A. aegypti* demonstrated that it effectively inhibited *A. aegypti* adult emergence (Figure 1A). Five dosage rates from 0.0005 to 0.005 ppm gave between 25 to 94% inhibition of adult emergence. Probit analysis showed that technical Dimilin has an $IC_{50}$ of 0.0011 ppm.

The effects of ureaformaldehyde encapsulated formulations of Dimilin at dosage rates ranging from 0.1 to .5 ppm are shown in Figure 1B. UF-5 and UF-10 and 0.1 ppm gave 16 and 3% inhibition of adult emergence, respectively. At 0.5 ppm both UF-5 and UF-10 resulted in 93% inhibition of adult emergence. An $IC_{50}$ of 0.25 ppm for UF-5 was not significantly different from that of 0.26 ppm for UF-10 at the 5% level.

PA-5 at concentrations from 0.3 to 0.8 ppm gave inhibition of adult emergence ranging from 11 to 99%. PA-10 gave no inhibition at 0.2 ppm and 90% inhibition at 1.0 ppm (Figure 1C). An $IC_{50}$ of 0.45 ppm for PA-5 was significantly different at the 5% level from that of 0.63 ppm for PA-10.

PU-5 at dosage rates from 0.2 to 1.0 ppm resulted in 1 to 98% inhibition of adult emergence, whereas PU-10 at rates from 0.1 to 0.5 ppm showed inhibition of adult emergence ranging from 10 to 97% (Figure 1D). PU-5 with an $IC_{50}$ of 0.55 ppm was significantly different at the 5% level from that of 0.23 ppm for PU-10. The PU-10 formulation was about twice as effective in the inhibition of adult emergence as the PU-5 formulation.

PVC-5 at concentrations from 0.2 to 1.0 ppm ranged from 7 to 81% inhibition of adult emergence. PVC-10 at rates from 0.2 to 1.0 ppm gave inhibition of adult emergence ranging from 18 to 97% (Figure 1E). The $IC_{50}$ of 0.43 ppm for PVC-10 was significantly different at the 5% level from that of 0.59 ppm for PVC-5.

Figure 1F shows that CP-50 at concentrations from 0.2 to 1.0 ppm gave inhibition of adult emergence from 13 to 94%. CP-75 at rates from 0.04 to 0.2 ppm resulted in inhibition of adult emergence ranging from 20 to 90%. CP-75 with an $IC_{50}$ of 0.069 ppm was 8 times more effective in the inhibition of adult emergence than CP-50 with an $IC_{50}$ of 0.55 ppm.

The results from this study indicated that technical Dimilin was highly effective in the inhibition of *A. aegypti* adult emergence with an $IC_{50}$ requiring only 0.0011 ppm. This $IC_{50}$ value is about 10 times higher than the 0.00012 ppm of technical Dimilin against *A. aegypti* reported by Hsieh and Steelman (70). However, Jakob (10) reported an $IC_{95}$ of 0.005 ppm Dimilin for *A. aegypti* and the results reported herein indicated that technical Dimilin at 0.005 ppm (Figure 1) produced 94% inhibition of *A. aegypti* adult emergence. Thus, it appears that Hsieh and Steelman were working with an extremely susceptible strain of *A. aegypti*. They also used a low larval density in the Dimilin suspension (10 larvae/ℓ) as compared to the results reported herein which was based on 20 larvae/200 ml Dimilin suspension. However, overcrowding of larvae would have the tendency to decrease the effect of insecticide owing to depletion of the amount of Dimilin available (71,72).

The inhibition of *A. aegypti* adult emergence, based on $IC_{50}$ values, showed that each of the encapsulated formulations of Dimilin had controlled release insecticidal activity, although there were differences between formulations (Table I). All encapsulated formulations of Dimilin had a higher $IC_{50}$ ratio than that for the technical formulation. A higher $IC_{50}$ value indicates that the Dimilin was tied up in the formulation and allowed less to be released during the 48 hours proceeding the $IC_{50}$ determination.

Among the same type of polymers containing different levels of Dimilin, only UF-5 and UF-10 gave similar $IC_{50}$ values. The other polymers showed considerable difference between $IC_{50}$ values. It is assumed that the $IC_{50}$ values relates to the amount of Dimilin that is released, then the higher the $IC_{50}$ value, the less Dimilin is released. For instance, the CP-75 formulation released 8 times more Dimilin than in the CP-50 formulation. Similarly, PU-10 released 2.4 times as much Dimilin as did PU-5 and PVC-10 released 1.4 times more than the PVC-5. Only one formulation differed in this trend. The lower concentration of Dimilin PA-5 had a lower $IC_{50}$ value than the PA-10 formulation. Except for the PA-5 formulation, the polymers containing higher levels of Dimilin gave the lower $IC_{50}$ values. These results were similar to the work of Stockman *et al.* (73) with the polymer neoprene containing organotin. They found that the release rate of organotin from neoprene was directly dependent upon the concentration of organotin in neoprene. Similar results were reported by Nelson *et al.* (29,47) with polyvinyl chloride formulations containing various levels of Dursban. They found that the greater the amount of Dursban in the PVC formulations, the greater the amounts of Dursban which was released into water. Besides levels of toxicants in polymers, Lawson *et al.* (48) reported the difference in the surface area per volume of the particles also makes the polymer formulations of insecticides release the toxicants at different

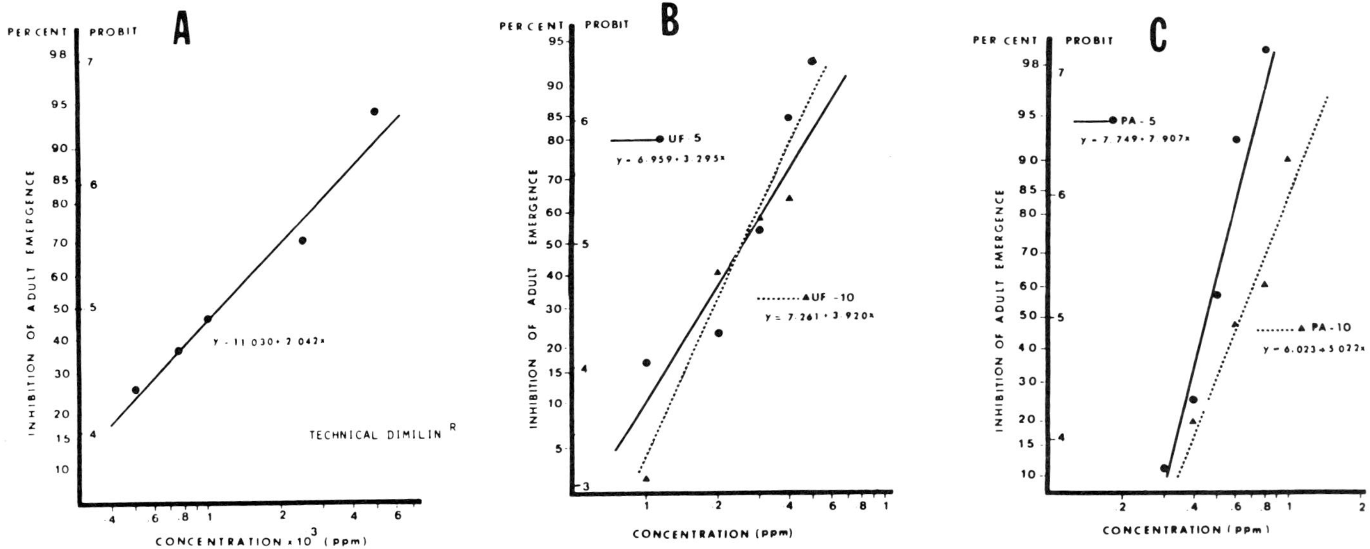


*FIGURE 1. Probit regression lines for effectiveness of technical dimilin and polymer formulations of dimilin against A. aegypti adult emergence.*

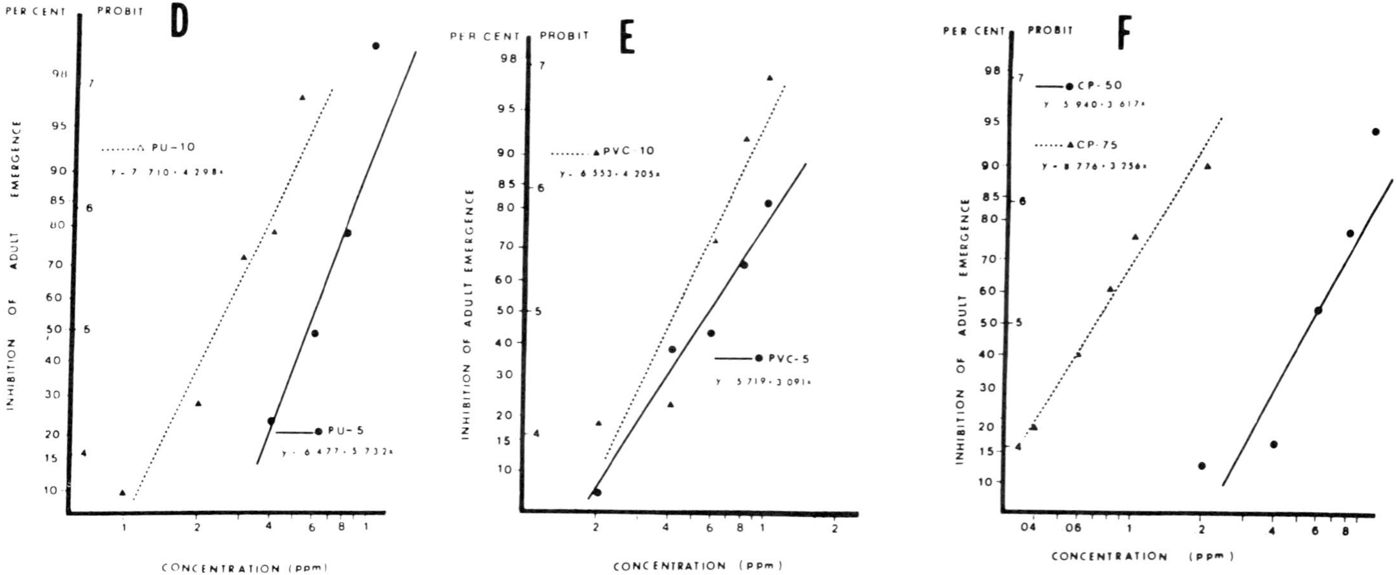


FIGURE 1d, e, and f.

*TABLE I. Effectiveness of Encapsulated Formulations of Dimilin Against A. aegypti Adult Emergence*[a]

| Formulation of dimilin[b] | ppm $IC_{50}$[c] | ppm 95% Fiducial limits | $IC_{50}$ Ratio values[d] |
|---|---|---|---|
| Technical dimilin | 0.0011 | 0.0009-0.0014 | - |
| CP-75 | 0.069 | 0.058-0.081 | 60 |
| PU-10 | 0.23 | 0.20-0.27 | 210 |
| UF-5 | 0.25 | 0.21-0.30 | 230 |
| UF-10 | 0.26 | 0.23-0.30 | 240 |
| PVC-10 | 0.43 | 0.37-0.49 | 390 |
| PA-5 | 0.45 | 0.42-0.49 | 410 |
| CP-50 | 0.55 | 0.47-0.64 | 500 |
| PU-5 | 0.55 | 0.49-0.61 | 500 |
| PVC-5 | 0.59 | 0.49-0.70 | 530 |
| PA-10 | 0.63 | 0.55-0.70 | 570 |

[a]Data summarized from SAS program for probit analysis.
[b]Freshly prepared solution bioassay for technical Dimilin; 48 hours old solution bioassay for encapsulated formulations of Dimilin; twenty 3rd instar A. aegypti larvae in each of 4 replications.
[c]Concentration that inhibited 50% adult emergence from exposed larvae.
[d]Comparative values relative to the $IC_{50}$ of technical Dimilin $\left(\frac{IC_{50} \text{ of encapsulated formulation}}{IC_{50} \text{ of technical Dimilin}}\right)$.

rates. They reported that the surface area per volume ratio is a function of particle size. An increase in the surface area per volume ratio of the particles results in an increase in the release rate. In the case of encapsulated formulations of Dimilin, all were the same particle size. At an equivalent particle size and dosage (i.e. active ingredient) the total amounts of 5% formulation used was twice that amount used for the 10% formulation. However, in this study only one formulation, the PA-5, resulted in a higher $IC_{50}$ value than the PA-10 formulation. In this situation, it is possible that relative surface area per volume ratio was more influential than relative percent composition as was noted by Nelson *et al.* (42). In the other formulations tested, the higher concentrations, except for ureaformaldehyde, did result in a smaller $IC_{50}$ value. No difference in inhibition of *A. aegypti* adult emergence resulted with the two concentrations of Dimilin-ureaformaldehyde formulations. This may be due to the fact that these two levels did not differ sufficiently in the percent composition in this type of polymer.

*2. $IC_{50}$ and $ED_{50}$ for M. autumnalis.* Laboratory bioassays of the technical Dimilin-fresh cattle feces mixtures with face fly larvae showed that technical Dimilin was highly effective in the inhibition of face fly adult emergence (Figure 2A). The inhibition of adult emergence ranged from 0 to 90% at concentrations from 0.001 to 0.01. Probit analysis of the data obtained from these tests place the $IC_{50}$ at 0.0062 ppm.

The effectiveness of ureaformaldehyde formulations of Dimilin against face fly adult emergence shown in Figure 2B indicated that the UF-5 and UF-10 formulations at dosage rates from 0.025 to 0.125 resulted in an inhibition of adult emergence ranging from 7 to 100%. An $IC_{50}$ of 0.052 ppm for UF-5 was not significantly different at the 5% level from that of 0.058 for the UF-10 formulation.

The effectiveness of polyamide formulations of Dimilin against face fly adult emergence is shown in Figure 2C. PA-5 and PA-10 at 0.5 ppm resulted in 7 and 10% inhibition of adult emergence, respectively. The 2 ppm resulted in complete inhibition of adult emergence. PA-10 with an $IC_{50}$ of 0.84 ppm was significantly different at the 5% level in the inhibition of adult emergence from that of PA-5 with an $IC_{50}$ of 1.01 ppm.

PU-5 at 0.05 ppm exhibited no effect on face fly adult emergence whereas at 0.25 ppm it showed 67% inhibition of adult emergence (Figure 2D). PU-10 with an $IC_{50}$ of 0.10 ppm was twice as effective in the inhibition of adult emergence as PU-5 with an $IC_{50}$ of 0.21 ppm.

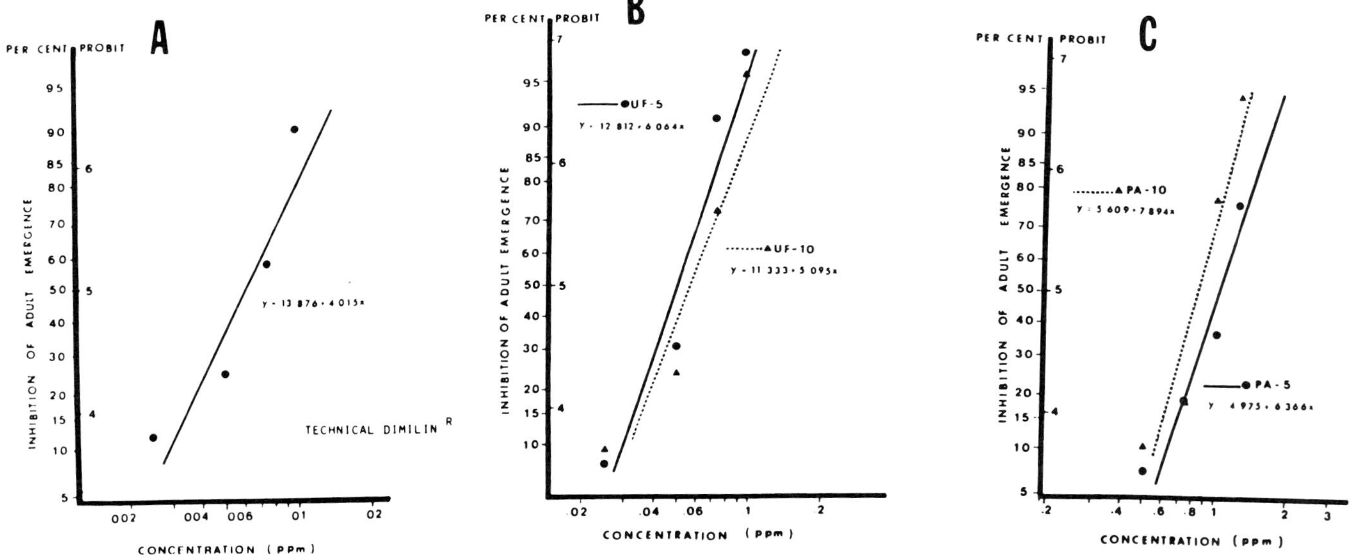


*FIGURE 2. Probit regression lines for effectiveness of technical dimilin and polymer formulations of dimilin against face fly adult emergence from cattle feces.*

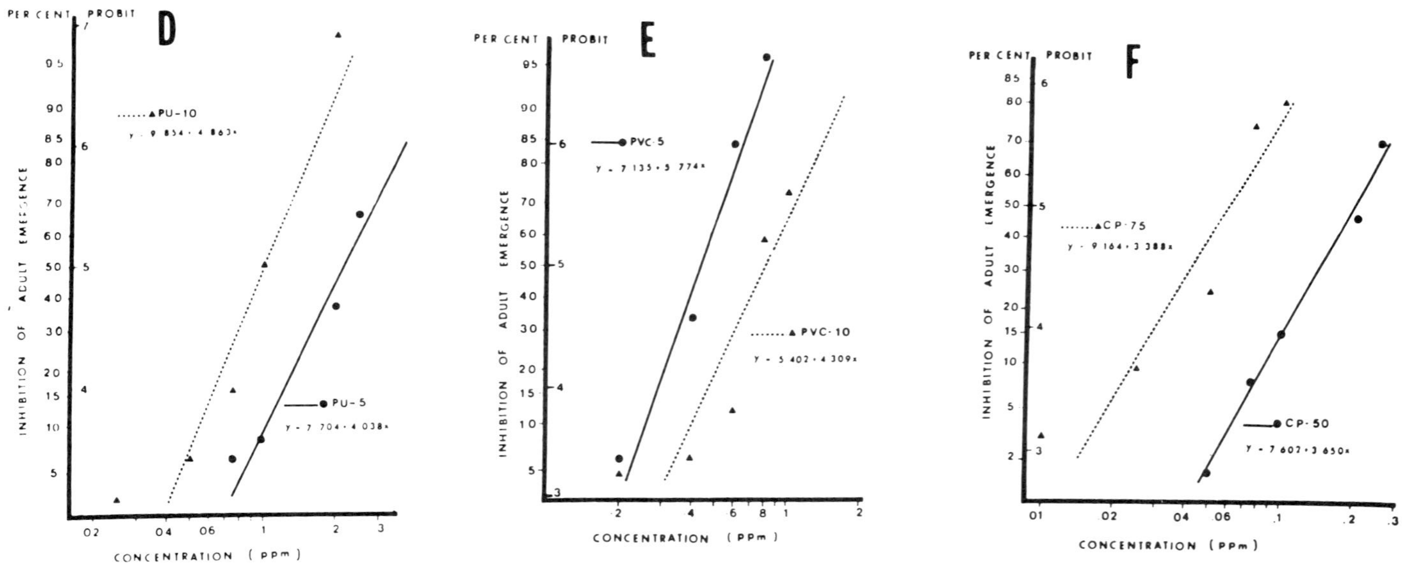


FIGURE 2d, e, and f.

The effectiveness of polyvinylchloride formulations of Dimilin on the face fly adult emergence is shown in Figure 2E. PVC-5 at 0.2 and 1 ppm in fresh cattle feces resulted in 6 and 100% inhibition of adult emergence, respectively. PVC-10 at 0.2 and 1 ppm, resulted in 4 and 73% inhibition of adult emergence, respectively. PVC-5 with an $IC_{50}$ of 0.43 ppm was about twice as effective in the inhibition of adult emergence as PVC-10 with an $IC_{50}$ of 0.81 ppm.

Figure 2F shows the effect of Carbopol formulations of Dimilin on the face fly adult emergence. CP-50 at concentrations from 0.05 to 0.25 ppm showed variations of the inhibition of adult emergence from 2 to 70%. CP-75 at 0.01 and 0.10 ppm, showed 3 and 81% inhibition of adult emergence, respectively. A probit analysis showed that CP-75 with an $IC_{50}$ of 0.059 ppm was 3 times more effective in the inhibition of face fly adult emergence than was the CP-50 with an $IC_{50}$ of 0.19 ppm.

Results from these studies indicated that technical Dimilin is effective in the inhibition of face fly adult emergence. A concentration of 0.0062 ppm was required to obtain an $IC_{50}$ for technical Dimilin. At the higher concentration of 0.01 ppm, technical Dimilin inhibited 90% of adult emergence. These results are very similar to those reported by Miller (74) who obtained 89% inhibition of face fly adult emergence with 0.01 ppm Dimilin in cattle feces.

Results from these studies based on $IC_{50}$ values (Table II) indicate that the encapsulated formulations extend the residual activity of Dimilin in feces for longer periods of time than for the technical material.

Among polymers of the same type but differing in the levels of Dimilin, UF-5 and UF-10 showed no difference in the inhibition of face fly adult emergence. Other encapsulated formulations of Dimilin showed different degrees of inhibition. The higher concentration of Dimilin in Carbopol, polyurethane foam, and polyamide formulations released more Dimilin than the same polymers having a lower concentration; indicating that Dimilin release from the polymers depended more on the Dimilin concentration than the surface area per volume ratio of the resin granules. However, the PVC-5-formulation, which contained the lower concentration of Dimilin, released more Dimilin than did the PVC-10 formulation. The fact that 2 times the amount of PVC-5 particles were used, thus increasing the surface area per volume of PVC-5 exposed to feces, may have been responsible for this difference. The release rates of Dimilin from the encapsulated formulations in cattle feces and in water did not react similarly. Considerable variation existed as to which formulation was superior in each test medium. In water,

*TABLE II. Effectiveness of Dimilin Formulations Against Face Fly Adult Emergence from Cattle Feces*

| | ppm | | |
|---|---|---|---|
| Formulation of dimilin[a] | $IC_{50}$[b] | 95% Fiducial limits | $IC_{50}$ Ratio values[c] |
| Technical Dimilin | 0.0062 | 0.0053-0.0072 | - |
| UF-5 | 0.051 | 0.043-0.060 | 8 |
| UF-10 | 0.057 | 0.047-0.068 | 9 |
| CP-75 | 0.059 | 0.050-0.071 | 10 |
| PU-10 | 0.10 | 0.088-0.116 | 16 |
| CP-50 | 0.19 | 0.16-0.24 | 31 |
| PU-5 | 0.21 | 0.18-0.27 | 35 |
| PVC-5 | 0.43 | 0.37-0.48 | 69 |
| PVC-10 | 0.81 | 0.71-0.96 | 130 |
| PA-10 | 0.84 | 0.77-0.91 | 140 |
| PA-5 | 1.01 | 0.92-1.11 | 160 |

[a]Freshly prepared Dimilin-feces mixture bioassayed with 20 face fly larvae in each of 4 replications.
[b]Concentration that inhibited 50% adult emergence from exposed larvae.
[c]Comparative values relative to the $IC_{50}$ of technical Dimilin $\left(\frac{IC_{50} \text{ of encapsulated formulation}}{IC_{50} \text{ of technical Dimilin}}\right)$.

based on $IC_{50}$ values against *A. aegypti*, the CP-75 formulation released the greatest amount of Dimilin while the PA-10 released the smallest amount of Dimilin compared to the other 8 encapsulated formulations (Table I). In cattle feces, based on $IC_{50}$ values against *M. autumnalis*, the UF-5 and PA-5, respectively, released the greatest and the smallest amount of Dimilin (Table II). However, some of the same variation obtained with the water tests with *A. aegypti* were also seen in the cattle feces tests. The UF-5, UF-10, CP-75, and PU-10 released the most Dimilin in both the cattle feces and in the water. In general the encapsulated formulations released Dimilin at greater rates in cattle feces than in water. The $IC_{50}$ ratio values ranged from 8 to 160 in the cattle feces tests, whereas the $IC_{50}$ ratio values ranged from 60 to 570 in the water tests. This may be due to the difference in the test media and the method of applications. In the water tests, only the aliquot portions were sampled while the samples from the cattle feces also contained granular particles of the encapsulated formulations which continuously released Dimilin into the feces. In cattle feces, Dimilin was also found to be tied up with organic matter as supported by Schaeffer and Dupras (17). This is also indicated in the residual test (Table III).

Face fly bioassay of feces from calves fed the Dimilin formulations varied in the inhibition of adult emergence. The effect of technical Dimilin on face fly emergence when administered to calves as a feed additive at dosage rates of 0.01, 0.02, 0.03, 0.04, and 0.05 mg/kg of body weight per day is shown in Figure 3A. The inhibition of adult emergence at dosage rates of 0.01 mg/kg of body weight of calf per day began at the second day during treatment. In feces from calves fed technical Dimilin at dosage rates ranging from 0.02 to 0.05 mg/kg of body weight per day, the inhibition of adult emergence began at day 1 during treatment. Feces from calves fed technical dimilin showed greater inhibition of adult emergence at days 2 and 3 than of day 1 during treatment. However, the inhibition of adult emergence declined sharply 1 day post-treatment. No inhibition of adult emergence was observed at day 3 post-treatment. A rate of 0.01 mg/kg of body weight per day gave 2% inhibition of adult emergence at day 3 during treatment, whereas the 0.05 mg/kg of body weight per day gave 88% inhibition. A dosage rate of 0.033 mg/kg of body weight per day was required for an $ED_{50}$ at day 3 during the treatment.

Figure 3B presents the effect of ureaformaldehyde formulations of Dimilin on face fly emergence when administered to cattle as a feed additive. UF-5 at rates from 0.001 to 0.01 mg/kg of body weight per day gave inhibition of adult emergence ranging from 9 to 97%, at day 3 during treatment. UF-10 at 0.001 mg/kg of body weight per day showed no effect on face fly emergence whereas the 0.01 mg/kg of body weight per day gave 96% inhibition of adult emergence at day 3 during the treatment. An $ED_{50}$ of 0.0046 mg/kg of body weight per day for the UF-5 was not significantly different at the 5% level from that of 0.0047 mg/kg of body weight per day for the UF-10.

The effect of polyamide formulations of Dimilin on face fly emergence when administered to cattle as a feed additive at the dose ranging from 0.04 to 0.2 mg/kg of body weight per day is shown in Figure 3C. The feces from day 1 during treatment was less effective in the inhibition of adult emergence than at days 2 or 3 during the treatment except for the 0.2 ppm formulation, which gave 100% control all 3 days. Similar effects on face fly emergence were observed at days 2 and 3 during the treatment. PA-5 at 0.04 and 0.2 mg/kg of body weight per day resulted in 6 and 100% inhibition of adult emergence, respectively, at day 3 during the treatment. PA-10 showed an effect similar to that shown by PA-5. An $ED_{50}$ of 0.0699 mg/kg of body weight per day for PA-5 was not significantly different at the 5% level from that of 0.0726 mg/kg of body weight per day for PA-10.

Polyurethane foam formulations of Dimilin at dosage rates from 0.01 to 0.1 mg/kg of body weight per day produced inhibition of face fly adult emergence ranging from 0 to 88% (Figure 3D). PU-5 and PU-10 at dosage rates of 0.02 and 0.01 mg/kg of body weight per day, respectively, had no effect on face fly adult emergence. PU-10 was more effective in the inhibition of adult emergence than the PU-5 formulation at day 3 during the treatment. PU-5 at 0.1 mg/kg of body weight per day resulted in an 80% inhibition of adult emergence versus 83% inhibition for the PU-10 at 0.08 mg/kg of body weight per day. The effectiveness of PU-5 and PU-10 against face flies declined sharply at day 1 post-treatment. No effect on face fly emergence could be observed at day 3 post-treatment. The $ED_{50}$ of 0.0569 mg/kg of body weight per day for PU-10 was significantly different at the 5% level than that of 0.0783 mg/kg of body weight per day for PU-5.

The effect of polyvinylchloride formulations of Dimilin at dosage rates from 0.04 to 0.2 mg/kg of body weight per day on face fly emergence is shown in Figure 3E. The effect of PVC-5 and PVC-10 on face fly emergence were less at day 1 of treatment than was at days 2 or 3 during the treatment. No difference was observed with these formulations at day 2 and 3 during the treatment. PVC-5 and PVC-10 at 0.04 mg/kg of body weight per day gave 16 and 21% inhibition of face fly adult emergence, respectively, at day 3 during the treatment.

TABLE III. Residual Activity of Dimilin Formulations Against *A. aegypti*

| Formulation and concentration (ppm)[a] | Corrected percent inhibition of adult emergence from treated water bioassayed at indicated days post-treatment[b] | | | | | | | | | | | | | | | |
|---|---|---|---|---|---|---|---|---|---|---|---|---|---|---|---|---|
| | 1 | 7 | 15 | 30 | 45 | 60 | 75 | 90 | 105 | 120 | 150 | 180 | 210 | 270 | 330 | 360 |
| Technical dimilin | | | | | | | | | | | | | | | | |
| 0.05 | 25 | 9 | 6 | 6 | 3 | 0 | 0 | - | - | - | - | - | - | - | - | - |
| 0.10 | 100 | 100 | 100 | 69 | 22 | 0 | 0 | - | - | - | - | - | - | - | - | - |
| 0.50 | 100 | 100 | 100 | 100 | 100 | 80 | 27 | 0 | 0 | - | - | - | - | - | - | - |
| UF-5 | | | | | | | | | | | | | | | | |
| 0.05 | 0 | 27 | 24 | 86 | 100 | 100 | 100 | 97 | 100 | 100 | 92 | 30 | - | - | - | - |
| 0.10 | 9 | 30 | 100 | 100 | 100 | 100 | 100 | 100 | 100 | 100 | 100 | 100 | 95 | 70 | 55 | 38 |
| 0.50 | 27 | 97 | 100 | 100 | 100 | 100 | 100 | 100 | 100 | 100 | 100 | 100 | 100 | 100 | 100 | 98 |
| UF-10 | | | | | | | | | | | | | | | | |
| 0.05 | 3 | 8 | 6 | 91 | 69 | 47 | 43 | 58 | 49 | 100 | 86 | 60 | 38 | - | - | - |
| 0.10 | 0 | 11 | 94 | 97 | 74 | 69 | 57 | 75 | 91 | 100 | 100 | 98 | 82 | 68 | 64 | 52 |
| 0.50 | 30 | 92 | 100 | 100 | 100 | 100 | 100 | 100 | 100 | 100 | 100 | 100 | 100 | 100 | 100 | 96 |
| PA-5 | | | | | | | | | | | | | | | | |
| 0.05 | 0 | 13 | 44 | 97 | 100 | 100 | 100 | 100 | 100 | 78 | 54 | 0 | - | - | - | - |
| 0.10 | 6 | 77 | 100 | 100 | 100 | 100 | 100 | 100 | 100 | 100 | 100 | 80 | 40 | 0 | - | - |
| 0.50 | 32 | 100 | 100 | 100 | 100 | 100 | 100 | 100 | 100 | 100 | 100 | 100 | 90 | 91 | 76 | 44 |
| PA-10 | | | | | | | | | | | | | | | | |
| 0.05 | 0 | 0 | 12 | 25 | 57 | 19 | 14 | 0 | 6 | 5 | 14 | 0 | - | - | - | - |
| 0.10 | 0 | 15 | 82 | 90 | 100 | 100 | 100 | 100 | 64 | 27 | 40 | 10 | - | - | - | - |
| 0.50 | 29 | 100 | 100 | 100 | 100 | 100 | 100 | 100 | 100 | 100 | 100 | 100 | 98 | 98 | 80 | 38 |
| PU-5 | | | | | | | | | | | | | | | | |
| 0.05 | 3 | 11 | 0 | 0 | 0 | 0 | 0 | 21 | 3 | 13 | 0 | - | - | - | - | - |
| 0.10 | 3 | 17 | 69 | 44 | 0 | 6 | 0 | 24 | 12 | 41 | 0 | - | - | - | - | - |
| 0.50 | 41 | 100 | 100 | 100 | 100 | 100 | 57 | 35 | 29 | 81 | 84 | 48 | 63 | 0 | - | - |

| | | | | | | | | | | | | | | | | |
|---|---|---|---|---|---|---|---|---|---|---|---|---|---|---|---|---|
| PU-10 | | | | | | | | | | | | | | | | |
| 0.05 | 3 | 0 | 0 | 36 | 27 | 3 | 20 | 73 | 29 | 3 | 0 | - | - | - | - | - |
| 0.10 | 6 | 22 | 34 | 100 | 100 | 100 | 100 | 100 | 100 | 63 | 52 | 22 | 0 | - | - | - |
| 0.50 | 100 | 100 | 100 | 100 | 100 | 100 | 100 | 100 | 100 | 100 | 100 | 100 | 93 | 96 | 98 | 38 |
| PVC-5 | | | | | | | | | | | | | | | | |
| 0.05 | 0 | 0 | 6 | 22 | 17 | 5 | 3 | 30 | 11 | 6 | 9 | 0 | - | - | - | - |
| 0.10 | 0 | 0 | 15 | 19 | 20 | 100 | 71 | 25 | 23 | 13 | 28 | 39 | 0 | - | - | - |
| 0.50 | 15 | 100 | 100 | 100 | 100 | 100 | 100 | 100 | 100 | 100 | 100 | 100 | 100 | 95 | 86 | 69 |
| PVC-10 | | | | | | | | | | | | | | | | |
| 0.05 | 6 | 0 | 3 | 8 | 23 | 11 | 11 | 19 | 3 | 3 | 0 | - | - | - | - | - |
| 0.10 | 3 | 0 | 9 | 27 | 29 | 35 | 11 | 5 | 3 | 9 | 0 | - | - | - | - | - |
| 0.50 | 23 | 100 | 100 | 100 | 100 | 100 | 100 | 100 | 66 | 34 | 100 | 100 | 100 | 80 | 79 | 81 |
| CP-50 | | | | | | | | | | | | | | | | |
| 0.05 | 0 | 16 | 8 | 8 | 0 | 0 | 0 | 9 | 23 | 6 | 0 | - | - | - | - | - |
| 0.10 | 0 | 22 | 18 | 11 | 11 | 0 | 9 | 12 | 34 | 12 | 9 | 0 | - | - | - | - |
| 0.50 | 6 | 22 | 42 | 95 | 100 | 100 | 100 | 100 | 100 | 100 | 100 | 100 | 81 | 0 | - | - |
| CP-75 | | | | | | | | | | | | | | | | |
| 0.05 | 6 | 16 | 21 | 22 | 14 | 0 | 5 | 0 | 6 | 32 | 13 | 0 | - | - | - | - |
| 0.10 | 15 | 27 | 26 | 22 | 24 | 6 | 8 | 0 | 11 | 44 | 34 | 0 | - | - | - | - |
| 0.50 | 100 | 100 | 100 | 100 | 100 | 100 | 100 | 100 | 100 | 100 | 100 | 100 | 100 | 18 | 0 | - |

[a]Ten 3rd instar larvae/10 ml treated water solution in each of 4 replications.
[b]Values corrected for natural inhibition of adult emergence (67).

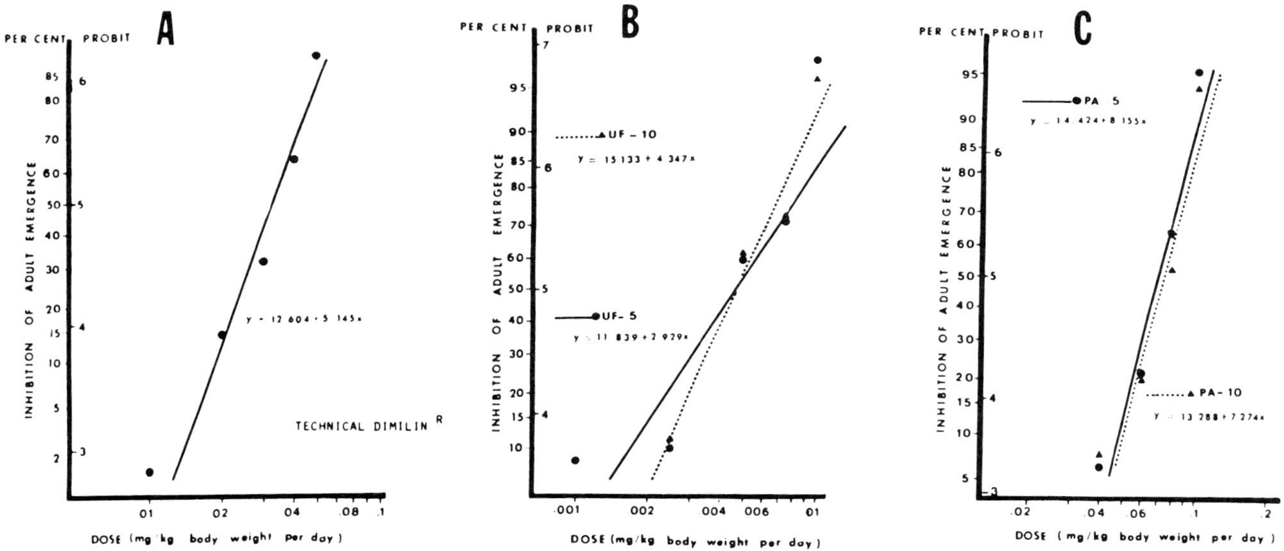


*FIGURE 3. Probit regression lines for effectiveness of technical dimilin and polymer formulations of dimilin against face fly adult emergence from cattle feces when administered as a feed additive to cattle.*

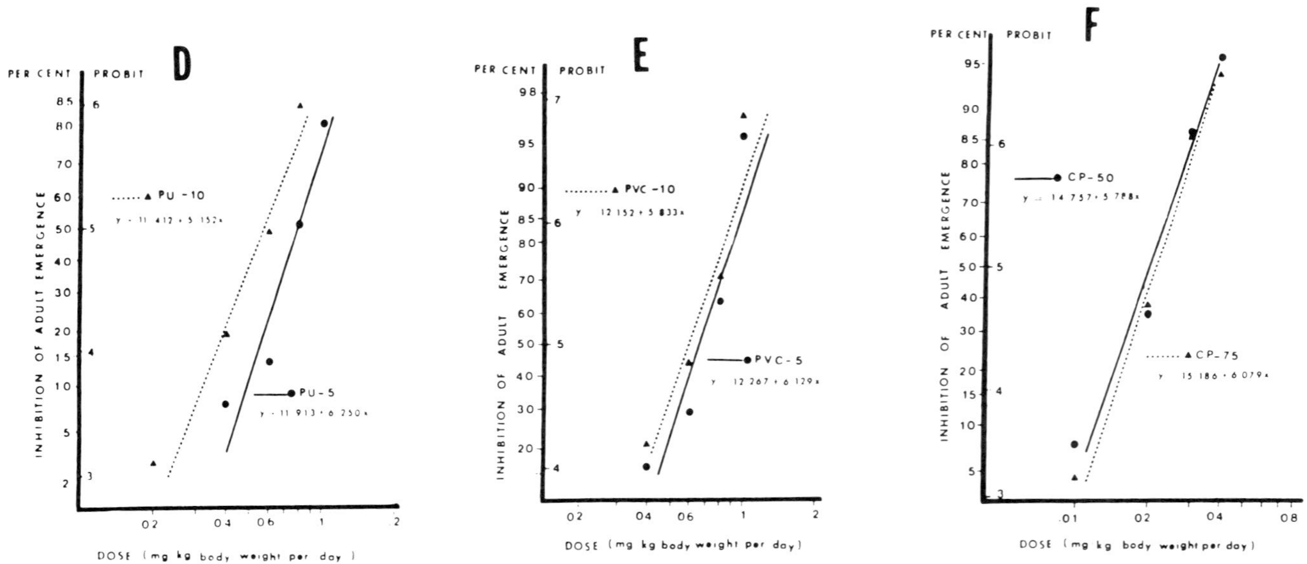


FIGURE 3d, e, and f.

Both PVC-5 and PVC-10 at 0.2 mg/kg of body weight per day showed complete inhibition of adult emergence at day 3 during the treatment. The inhibition of adult emergence declined sharply at day 1 post-treatment. No inhibition of adult emergence was observed at day 3 post-treatment. The $ED_{50}$ of 0.065 mg/kg of body weight per day for PVC-5 was not significantly different at the 5% level from that of 0.059 mg/kg of body weight per day for PVC-10.

Carbopol formulations of Dimilin at dosage rates from 0.01 to 0.05 mg/kg of body weight per day was highly effective in the inhibition of face fly adult emergence (Figure 3F). CP-50 at the rate of 0.01 mg/kg of body weight per day resulted in an 8% inhibition of adult emergence at day 3 during the treatment, whereas 0.05 mg/kg of body weight per day resulted in complete inhibition of adult emergence. The inhibition of adult emergence declined sharply at day 1 post-treatment and very little effect on adult emergence was observed at day 3 post-treatment. Similar results were obtained with CP-75, except that no inhibition of adult emergence could be observed at day 3 post-treatment. The $ED_{50}$ of 0.021 mg/kg of body weight per day for CP-50 was not significantly different at the 5% level from that of 0.021 mg/kg of body weight per day for CP-75.

Based on the $ED_{50}$ values at day 3 during treatment, ureaformaldehyde and Carbopol formulations of Dimilin were superior in the inhibition of face fly adult emergence compared to other formulations of Dimilin. Ureaformaldehyde formulations of Dimilin were 7 times more effective in the inhibition of face fly adult emergence than the technical material.

According to Nye (Thompson-Hayward Co., 1977, personal communication), Dimilin administered in an unprotected form is easily eliminated or metabolized to non-toxic products in an animal's body. He reported that about 42% of AI Dimilin is eliminated from the animal body in the feces. About 15% in urine and 45% in feces is eliminated as non-toxic metabolites. From this information, indications are that Dimilin, when encapsulated, is protected by the polymer carriers from natural breakdown or loss in the animal body and is passed with the feces from the body where it is then released. The ureaformaldehyde polymer is considered to be a "soft" formulation and can be easily broken into small particles by cattle (52). Thus, when ureaformaldehyde formulations of Dimilin are used as feed additive, those formulations will be broken into small particles and easily passed from the body. It is probable that during the passage of these formulations through the body, only small amounts of Dimilin are metabolized or eliminated in the urine, and/or accumulated in the animal's body, thus large amounts are still in the particles of the polymer carriers.

Carbopol formulations of Dimilin are naturally hard formulations but when they are exposed to water, being hydrophilic, they become swollen and soft. These properties should allow the compound to be easily digested as small particles in the animal body, and then act in a manner similar to ureaformaldehyde formulations of Dimilin.

These findings generally agree with those of Breeden *et al.* (54) who found that when methoprene was used as a chicken feed additive, 10 times more technical formulation of methoprene achieved the same level of control of house flies breeding in chicken manure as the encapsulated formulation.

Polyamide, polyurethane foam, and polyvinylchloride formulations of Dimilin have higher $ED_{50}$ values for face files than has the technical formulation. This indicates that these encapsulated formulations released only small amounts of Dimilin into cattle feces within 3 days after treatment, even though large amounts of the encapsulated Dimilin were present in the feces. Thus the Dimilin is slowly released from polymer carriers into the feces. These three types of polymers, considered as "hard" formulations, are very resistant to digestive processes of cattle and are eliminated in the feces in the original sizes. Lloyd and Matthysse (52) reported that the larger size particles released insecticide at lower rates than did the small size particles, thus explaining the release rates of Dimilin into feces at slow rates with the hard large particles. It is probable that these formulations of Dimilin may produce a long residual activity in manure and would be useful for control of flies breeding in animal manure in cattle feed-lots, dairy barns, chicken houses, pig pens, and other similar sites that require long residual control.

Another advantage of encapsulation of Dimilin is that in some animals it may prevent tissue residues. Miller *et al.* (21) for instance, found Dimilin residue in eggs of chickens fed Dimilin. Thus if Dimilin in polymers release only small amounts of Dimilin in the animal's body, the problem of residues in animals tissue can be overcome or at least reduced.

### B. *Residual Activity of Dimilin Formulations Against Adult Emergence*

*1. Residual Activity Against A. aegypti.* The residual activity of Dimilin formulations against 3rd instar *A. aegypti* are presented in Table III.

Samples of a 0.05 ppm technical Dimilin-water solution bioassayed 1 day post-treatment produced 25% inhibition of adult emergence. The inhibition of adult emergence decreased sharply between days 1 and 7 post-treatment, and slowly declined thereafter. No residual activity could be observed at day 60 post-treatment. Water samples of 0.1 ppm of Dimilin bioassayed at days 1, 7, and 15 resulted in complete inhibition of adult emergence. The inhibition declined sharply at day 30 post-treatment and was completely gone at day 60. Dimilin at a concentration of 0.5 ppm produced complete inhibition of adult emergence up to day 45 and declined thereafter. No inhibition of adult emergence could be observed at day 90 post-treatment. This indicated that the higher concentrations of technical Dimilin produced the greater residual activity. However, the residual effective control did not exist longer than 60 days post-treatment even at a concentration of 0.5 ppm.

The ureaformaldehyde formulations of Dimilin showed a low degree of inhibition of *A. aegypti* adult emergence at the first week post-treatment but the inhibition increased with time. The UF-5 and UF-10 at 0.5 ppm showed no difference in the inhibition of adult emergence. Both formulations gave complete inhibition of adult emergence from day 15 post-treatment through the 360 day test. At concentrations of 0.1 and 0.05 ppm, UF-5 was more effective in inhibition of adult emergence than UF-10 at corresponding rates. UF-5 at 0.05 ppm showed no inhibition activity at day 1 post-treatment. The inhibition activity consistently increased and reached complete inhibition of adult emergence at day 45 post-treatment. Effective inhibition of adult emergence (97-100% inhibition) was maintained from day 45 through day 120 post-treatment. However, at day 150 post-treatment the inhibition of adult emergence declined to 92 and to 30% at 180 days. UF-5 at 0.1 ppm showed low level of inhibition of adult emergence (9 to 30%) at the first week post-treatment. The inhibition of adult emergence reached 100% at day 15 post-treatment and maintained complete inhibition of adult emergence for 180 days. The UF-5 at 0.5 ppm gave virtually 100% control for the entire 360 day test. UF-10 at 0.05 and 0.1 ppm demonstrated long residual activity but the level of inhibition was not sufficient to produce effective control up to day 105 post-treatment. It took 120 days post-treatment to produce 100% inhibition of adult emergence.

The residual activity of Polyamide formulations of Dimilin against *A. aegypti* larvae were similar to those of the ureaformaldehyde formulations. PA-5 at 0.05 ppm showed no inhibition of adult emergence at day 1 post-treatment but inhibition increased over time. Complete inhibition was obtained from day 45 through day 105. The inhibition of adult emergence began to drop at day 120 and continued to drop through day 150. PA-5 at 0.1 ppm resulted in a 76% inhibition of adult emergence at 1 week post-treatment, and complete inhibition was maintained from day 15 through day 150 of the test. PA-5 at 0.5 ppm showed complete inhibition from day 7 through the 180 days.

The PA-10 formulation was less effective in residual activity against *A. aegypti* than the PA-5 formulation at corresponding concentrations tested. PA-10 at 0.05 ppm showed a low level of inhibition of adult emergence throughout the test period. PA-10 at 0.1 ppm showed a low level of inhibition of adult emergence at the first week post-treatment. This consistently increased, reaching complete inhibition of adult emergence from day 45 through day 90, then consistently declined from day 105 through day 180. PA-10 at a concentration of 0.5 ppm showed low level of inhibition of adult emergence at day 1, complete inhibition from day 7 through day 180, and 98% control through day 270.

Polyurethane foam formulations produced high levels of inhibition of adult emergence as well as long residual activity at higher concentrations. The PU-5 formulation at concentrations of 0.05 and 0.1 ppm showed a low level of inhibition throughout the test. The PU-5 formulation at 0.5 ppm gave complete inhibition of adult emergence from day 7 through day 60. Thereafter, the inhibition fluctuated between 29 and 84%.

The PU-10 formulation was more effective in the inhibition of *A. aegypti* adult emergence than the PU-5 formulation at corresponding concentrations tested. The PU-10 at 0.05 ppm showed low residual activity up to 120 days post-treatment. At 0.1 ppm, it produced a low level of inhibition for the first 15 days post-treatment and complete inhibition from day 30 through day 105 post-treatment, after which the residual activity began to decline. PU-10 at 0.5 ppm produced complete inhibition of adult emergence for 180 days and 98% inhibition at day 330.

Polyvinylchloride formulations of Dimilin showed similar residual activity to that of polyurethane foam formulations. Both PVC-5 and PVC-10 at concentrations of 0.05 and 0.1 ppm resulted in a low level of inhibition of adult emergence throughout the test. PVC-5 at 0.5 ppm produced complete inhibition of adult emergence from day 7 post-treatment through day 210 of the test and maintained above 95% at day 270. At this concentration, the PVC-10 formulation also gave complete inhibition of adult emergence from day 7 through day 90 post-treatment. Thereafter the inhibition of adult emergence began to slowly decrease but again gave 100% inhibition from day 150 through 210 when it started to slowly decrease.

Carbopol formulations of Dimilin produced similar residual activity to those of polyvinylchloride and polyurethane foam formulations of Dimilin. Both CP-50 and CP-75, at concentrations of 0.05 and 0.1 ppm produced a low percentage of inhibition of *A. aegypti* adult emergence. In both cases the inhibition of adult emergence was never more than 35% throughout the test. The CP-50 formulation was less effective in the inhibition of adult emergence than was the CP-75 formulation. At 0.5 ppm, CP-50 gave low percentages of inhibition at 1 week post-treatment. The inhibition consistently increased over time and reached 100% at day 45 post-treatment. This complete inhibition of adult emergence was maintained through day 180 of the test. The CP-75 formulation at 0.5 ppm produced complete inhibition of adult emergence for 210 days.

The results of these studies on residual effect of Dimilin formulations on *A. aegypti* adult emergence indicate that technical Dimilin persists in water for a shorter period of time than the encapsulated formulations of Dimilin. Technical Dimilin at 0.05, 0.1, and 0.5 ppm, in general, produced an effective inhibition of adult emergence during the first 15 days after application. Thereafter the residual effect on adult emergence decreased rapidly. However, the residual activity was extended as the concentrations increased. At 0.5 ppm, 45 days of complete inhibition of adult emergence was obtained. This confirms other evidence of short residual activity of Dimilin against mosquito species. Mulla *et al.* (75) reported effective control of *Culex tarsalis* for up to 5 days with emulsifiable concentrate and wettable powder formulations of Dimilin at 0.05 lb AI/acre. Schaefer *et al.* (76) obtained effective control of *C. tarsalis* only at the early days after application with 0.04 lb AI/acre, but found no effective residual activity after 10 days post-treatment. Rogers *et al.* (16) reported successful control of *A. taeneorhynchus* with 0.029 lb AI/acre. However, residual activity of this preparation was not sufficient to control the second brood. In one case, residual effective control of *C. tarsalis* and *Anopheles franciscanus* was obtained for 21 days (77).

The studies herein show that encapsulated formulations of Dimilin are superior to the technical material in long residual activity against *A. aegypti*. However, during the first week post-treatment most encapsulated formulations could not produce complete inhibition of adult emergence. This is due to the fact that the encapsulated formulations slowly released Dimilin into the water and it took several days to release enough to be effective against the mosquito. At 0.5 ppm, all encapsulated formulations of Dimilin tested, except PU-5, gave complete inhibition of adult emergence greater than 180 days post-treatment. This indicates that at the rates tested, the encapsulated formulations were effective at least 3 times longer than the technical formulation. At dosage rates of 0.05 and 0.1 ppm, two encapsulated formulations, UF-5 and PA-5, appear to be the most ideal residual mosquito larvicides for further testing. Initially, UF-5 and PA-5 were ineffective in inhibiting adult emergence but reached effective inhibition levels at day 7 to day 30 depending on formulation and concentrations tested. The higher concentrations produced complete inhibition more rapidly as well as retaining longer residual activity than low concentrations.

The results from these studies also indicate that it may be possible to obtain complete inhibition of adult emergence by using a combination of technical Dimilin and the encapsulated formulations. For instance, Dimilin could be applied as a coating to the encapsulated formulation. By the time the unprotected Dimilin is broken down, more Dimilin would be released from the encapsulated formulation, thus a continued 100% inhibition of adult emergence could be obtained. Economical control of mosquitoes also may be obtained by using a combination of the technical Dimilin at 0.1 ppm and the UF-5 or PA-5 formulations at 0.05 ppm. Figure 4 illustrates that technical Dimilin began to loose its effectiveness in the inhibition of adult emergence 15 days post-treatment. At day 20 post-treatment, about 85% inhibition of adult emergence was obtained with technical Dimilin and about 40 and 60% inhibition of adult emergence were produced by UF-5 and PA-5, respectively. At day 30 post-treatment, the inhibition of adult emergence decreased to 69% in 0.1 ppm technical Dimilin treatment, whereas it increased to 86 and 97%, respectively, in the UF-5 and PA-5 treatments. From day 45 post-treatment both UF-5 and PA-5 produced 100% inhibition of adult emergence up to days 105 and 120, respectively, whereas technical Dimilin produced only 22% inhibition of adult emergence at day 45 and had no effect on adult emergence after day 60 post-treatment. By this method, single application may give effective control of mosquitoes up to 3 to 4 months.

2. *Residual Activity Against M. autumnalis.* The residual activity of Dimilin formulations in cattle feces against face fly adult emergence is presented in Table IV.

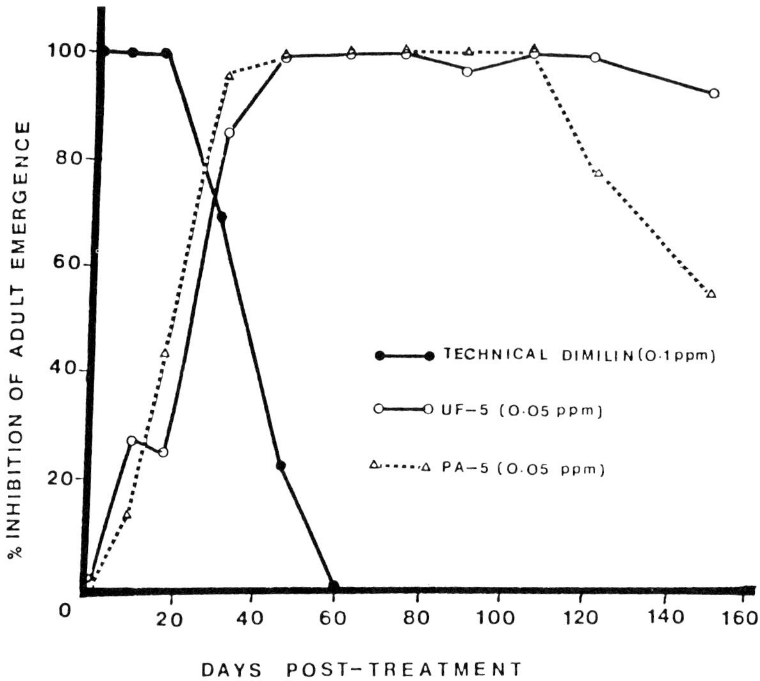


*FIGURE 4. Technical Dimilin versus encapsulated formulations of Dimilin against A. aegypti adult emergence.*

Bioassays of 0.01 ppm AI technical Dimilin-fecal mixture with face fly larvae at day 1 post-treatment resulted in a 90% inhibition of adult emergence. This inhibition decreased sharply to 60% at day 7 and then slowly declined thereafter and no inhibition activity was observed at day 90 post-treatment. Technical Dimilin at the higher concentration of 0.05 and 0.1 ppm produced complete inhibition of face fly adult emergence throughout the 150 day test period.

The UF-5 formulation at 0.05 ppm resulted in low percentage of inhibition of adult emergence at initial days post-treatment. The inhibition of adult emergence sharply increased thereafter. Complete inhibition of adult emergence was obtained from days 15 to 150. UF-5 at the higher concentration of 0.1 ppm resulted in 97% inhibition of adult emergence in 1 day old Dimilin-fecal mixture, and complete inhibition of adult emergence was obtained throughout the remainder of the 150 day test. UF-5 at 0.5 ppm resulted in complete inhibition of adult emergence through the 150 day test period. Similar results were obtained with UF-10 at each corresponding concentration.

Polyamide formulations of Dimilin at the rate of 0.05 and 0.1 ppm had no effect on face fly adult emergence. The PA-5 formulation at 0.5 ppm resulted in low levels of inhibition of adult emergence during the first week post-treatment. The inhibition consistently increased reaching 100% at day 45 post-treatment and maintained this level of inhibition through day 150 post-treatment. PA-10 at 0.5 ppm gave similar results to that of PA-5.

PU-10 had more residual effectiveness in the inhibition of face fly adult emergence than PU-5 at corresponding concentrations. PU-5 at 0.05 ppm had little effect on adult emergence (0 to 3%). The PU-5 formulation at 0.1 ppm fluctuated from 2 to 35% throughout the 150 day test period. PU-5 at 0.5 ppm resulted in complete inhibition of adult emergence throughout the 150 day test. The PU-10 formulation at 0.05 ppm consistently increased with time in the inhibition of adult emergence, reaching the highest inhibition (65%) at day 60 post-treatment. The inhibition of adult emergence began to decline at day 75 and fluctuated between 11 and 57% throughout the balance of the 150 day test. PU-10 at 0.1 ppm slowly increased inhibition of the adult emergence with time reaching 100% inhibition of adult emergence at day 30 and remained at 100% throughout the remainder of the 150 day test. PU-10 at 0.5 ppm showed complete inhibition of adult emergence throughout the 150 day test period.

Polyvinylchloride formulations of Dimilin showed similar results to that of polyurethane foam formulations. The PVC-5 formulation was more effective in the inhibition of face fly adult emergence than the PVC-10 at the low concentrations (0.05 and 0.1 ppm). Both PVC-5 and PVC-10 gave no effective inhibition of adult emergence at a concentration of 0.05 ppm. At 0.1 ppm, PVC-5 showed no effect in the inhibition of adult emergence during the first 30 days post-treatment. However, the effective inhibition of adult emergence (88 to 100%) was obtained from day 45 through day 150 post-treatment. At a 0.1 ppm concentration, PVC-10 had very little effect on the face fly adult emergence. The highest percentage of inhibition was never more than 10% throughout the 150 day test. At 0.5 ppm both PVC-5 and PVC-10 produced complete inhibition of face fly adult emergence from day 45 through the remainder of the 150 day test.

Carbopol formulations of Dimilin showed variation in the inhibition of face fly adult emergence depending upon the composition of Dimilin in the polymer carrier and the concentrations used. The higher percentage of Dimilin composition

TABLE IV. Residual Activity of Dimilin Formulations Against Face Fly Adult Emergence

| Formulation and concentration (ppm)[a] | Corrected percent inhibition of adult emergence from treated cattle feces bioassayed at indicated days post-treatment[b] | | | | | | | | | | |
|---|---|---|---|---|---|---|---|---|---|---|---|
| | 1 | 7 | 15 | 30 | 45 | 60 | 75 | 90 | 105 | 120 | 150 |
| Technical dimilin | | | | | | | | | | | |
| 0.01 | 90 | 60 | 47 | 48 | 28 | 20 | 2 | 0 | 0 | 0 | 0 |
| 0.05 | 100 | 100 | 100 | 100 | 100 | 100 | 100 | 100 | 100 | 100 | 100 |
| 0.10 | 100 | 100 | 100 | 100 | 100 | 100 | 100 | 100 | 100 | 100 | 100 |
| UF-5 | | | | | | | | | | | |
| 0.05 | 23 | 97 | 100 | 100 | 100 | 100 | 100 | 100 | 100 | 100 | 100 |
| 0.10 | 96 | 100 | 100 | 100 | 100 | 100 | 100 | 100 | 100 | 100 | 100 |
| 0.50 | 100 | 100 | 100 | 100 | 100 | 100 | 100 | 100 | 100 | 100 | 100 |
| UF-10 | | | | | | | | | | | |
| 0.05 | 30 | 99 | 100 | 100 | 100 | 100 | 100 | 100 | 100 | 100 | 100 |
| 0.10 | 97 | 100 | 100 | 100 | 100 | 100 | 100 | 100 | 100 | 100 | 100 |
| 0.50 | 100 | 100 | 100 | 100 | 100 | 100 | 100 | 100 | 100 | 100 | 100 |
| PA-5 | | | | | | | | | | | |
| 0.05 | 0 | 0 | 0 | 0 | 0 | 0 | 0 | 0 | 0 | 0 | 0 |
| 0.10 | 0 | 0 | 3 | 2 | 15 | 3 | 3 | 2 | 3 | 13 | 0 |
| 0.50 | 7 | 5 | 74 | 97 | 100 | 100 | 100 | 100 | 100 | 100 | 100 |
| PA-10 | | | | | | | | | | | |
| 0.05 | 0 | 0 | 0 | 0 | 0 | 0 | 0 | 0 | 0 | 0 | 0 |
| 0.10 | 0 | 0 | 0 | 0 | 2 | 3 | 0 | 0 | 0 | 6 | 0 |
| 0.50 | 10 | 14 | 51 | 94 | 92 | 100 | 100 | 100 | 100 | 100 | 100 |

| | | | | | | | | | | | |
|---|---|---|---|---|---|---|---|---|---|---|---|
| PU-5 | | | | | | | | | | | |
| 0.05 | 0 | 0 | 0 | 2 | 3 | 2 | 0 | 0 | - | 0 | 0 |
| 0.10 | 8 | 2 | 3 | 11 | 14 | 33 | 35 | 23 | - | 12 | 29 |
| 0.50 | 100 | 100 | 100 | 100 | 100 | 100 | 100 | 100 | - | 100 | 100 |
| PU-10 | | | | | | | | | | | |
| 0.05 | 6 | 4 | 9 | 45 | 31 | 65 | 33 | 11 | - | 16 | 57 |
| 0.10 | 51 | 63 | 83 | 100 | 100 | 100 | 100 | 100 | - | 100 | 100 |
| 0.50 | 100 | 100 | 100 | 100 | 100 | 100 | 100 | 100 | - | 100 | 100 |
| PVC-5 | | | | | | | | | | | |
| 0.05 | 0 | 0 | 0 | 0 | 2 | 60 | 18 | 6 | 0 | 22 | 6 |
| 0.10 | 0 | 0 | 7 | 8 | 100 | 100 | 97 | 100 | 88 | 100 | 92 |
| 0.50 | 52 | 99 | 100 | 99 | 100 | 100 | 100 | 100 | 100 | 100 | 100 |
| PVC-10 | | | | | | | | | | | |
| 0.05 | 0 | 0 | 0 | 0 | 0 | 0 | 0 | 0 | 0 | 0 | 0 |
| 0.10 | 2 | 0 | 3 | 0 | 2 | 5 | 6 | 0 | 3 | 6 | 9 |
| 0.50 | 12 | 37 | 90 | 97 | 100 | 100 | 100 | 100 | 100 | 100 | 100 |
| CP-50 | | | | | | | | | | | |
| 0.05 | 0 | 0 | 0 | 0 | 0 | 0 | 0 | 0 | - | 0 | 0 |
| 0.10 | 15 | 6 | 5 | 19 | 7 | 26 | 6 | 0 | - | 0 | 8 |
| 0.50 | 100 | 100 | 100 | 100 | 100 | 100 | 100 | 100 | - | 100 | 100 |
| CP-75 | | | | | | | | | | | |
| 0.05 | 24 | 12 | 20 | 22 | 15 | 9 | 19 | 10 | - | 20 | 0 |
| 0.10 | 81 | 82 | 82 | 91 | 83 | 74 | 94 | 68 | - | 100 | 85 |
| 0.50 | 100 | 100 | 100 | 100 | 100 | 100 | 100 | 100 | - | 100 | 100 |

[a]Twenty face fly larvae in each of 4 replications.
[b]Value corrected for natural inhibition of adult emergence in checks by Abbot's formula (67).

showed the greater effective inhibition of adult emergence. The CP-50 at 0.05 ppm gave no effect on face fly emergence whereas CP-75 at this concentration produced inhibition of face fly adult emergence ranging from 9 to 24% from day 1 through day 120 post-treatment. The CP-50 formulation at 0.1 ppm did not give effective inhibition of adult emergence. The CP-75 formulation at 0.1 ppm demonstrated long residual activity against face fly emergence. The inhibition of face fly adult emergence ranged from 68 to 100% throughout the 150 day test. Both CP-50 and CP-75 formulations at the concentration of 0.5 ppm resulted in complete inhibition of adult emergence throughout the 150 day test period.

The results from the studies on residual effect of Dimilin formulations on *M. autumnalis* in cattle feces generally indicate that all Dimilin formulations had a longer residual activity against face flies in feces than against *A. aegypti* in water. After the first week or two the encapsulated formulations at 0.5 ppm gave complete inhibition of adult emergence throughout the 150 day test (Table IV). The residual activity of technical Dimilin was similar except that the 0.01 ppm residual activity decreased at a much slower rate. The unusual behavior of the technical Dimilin in the feces may be due to the fact that the feces were held in an airtight jar and Dimilin was possibly bound in the organic matter and was thus released slowly. The work of Schaefer and Dupras (17) with residue studies of Dimilin in organic matter indicates this may be the reason. However, the lower residual activity reported by Miller *et al.* (18) conducted outdoors, would tend to indicate that the residual activity of Dimilin in cattle feces is much less than in that held in the laboratory under these conditions. Knapp and Ellis (59) found that dichlorvos release rate in an airtight container is dependent upon the air space available. In their studies, when dichlorvos reached the same concentration in the air as in the resin matrix, then the dichlorvos would no longer diffuse from the resin. This concentration equilibrium effect, as they called it, may apply to this situation. More important, however, may be the lack of normal micro flora in the sealed manure that may have helped to prevent degradation of Dimilin.

## IV. SUMMARY AND CONCULSIONS

Laboratory investigations were conducted to determine if the insect growth regulator, Dimilin, could be encapsulated into formulations that would release the toxicant slowly so as to give effective control of target species over a longer period of time than with conventional formulations. *A. aegypti* and *M. autumnalis* larvae were used as the test organisms.

Dimilin was incorporated into 4 types of polymer carriers at the concentrations of 5 and 10% each. These polymer carriers were ureaformaldehyde (UF-5 and UF-10), polyamide (PA-5 and PA-10), polyurethane foam (PU-5 and PU-10), and polyvinylchloride (PVC-5 and PVC-10). Another polymer carrier, Carbopol, was formulated as 50 and 75% concentrations due to the nature of the resin and designated as CP-50 and CP-75, respectively.

Probit analysis was used to calculate the $IC_{50}$s and $ED_{50}$s for Dimilin formulations against the bioassay insects. This was accomplished by using various dosage rates of the formulations in the rearing media and placing the larval form of the respective test organisms in the respective medium for rearing to adults. The criterion for effectiveness of each formulation was based on the concentration needed to inhibit 50% of adult emergence (indicated as $IC_{50}$). When formulations were fed to calves and the feces bioassayed with face fly larvae, the amount that was needed to feed the calves to inhibit 50% adult emergence of the face fly was expressed as the median effective dose of $ED_{50}$.

Bioassays of *A. aegypti* against freshly prepared technical Dimilin resulted in an $IC_{50}$ of 0.0011 ppm. All encapsulated formulations of Dimilin were bioassayed after holding 48 hours in tap water (to allow release of Dimilin). The encapsulated formulations of CP-75, PU-10, UF-5, UF-10, PVC-10, PA-5, CP-50, PU-5, PVC-5, and PA-10 required 62, 210, 230, 240, 390, 410, 500, 500, 530, and 570 times as much Dimilin to acquire the same $IC_{50}$, respectively, as did technical Dimilin. This indicated that the encapsulated formulations slowly release Dimilin during 48 hours after treatment. The amounts of Dimilin left in the polymer was expected to be continuously released into the water. Among polymers of the same type but differing in the Dimilin composition, UF-5 and UF-10 showed no difference in the inhibition of adult emergence. Other encapsulated formulations, CP-75, PU-10, PVC-10, having greater Dimilin composition, released greater amounts of Dimilin than did the same polymers having a lower Dimilin composition. This may be due to the high Dimilin composition in the polymers being more influential than the surface area per volume ratio. However, PA-5 having a lower Dimilin composition, released more Dimilin than did the PA-10. In this case, the surface area per volume may have been responsible for the difference. The polymers having the higher $IC_{50}$

values were expected to have the longer residual longevity against mosquito larvae in water. Thus, residual tests were conducted to prove this theory.

Bioassays of technical Dimilin against face fly larvae in cattle feces resulted in an $IC_{50}$ of 0.0062 ppm. The UF-5, UF-10, CP-75, PU-10, CP-50, PU-5, PVC-5, PVC-10, PA-10, and PA-5 formulations required 8, 9, 10, 16, 31, 35, 69, 130, 140, and 160 times as much Dimilin, respectively, to obtain the same $IC_{50}$ as did technical Dimilin. Among polymers of the same type, UF-5 and UF-10 had no difference in the inhibition of face fly adult emergence. The CP-75, PU-10, PA-10 having a greater Dimilin composition released more Dimilin than did the same polymers having a lower Dimilin composition. The exception of this was the PVC-5, which had a lower Dimilin composition, but released more Dimilin than did the PVC-10. The reasons for these differences can be explained in similar fashion to that in mosquito tests. Again the formulations having the higher $IC_{50}$ values were expected to give the longer residual activity against face flies. Further testing of these formulations as a feed through to calves was conducted to determine if the same results could be obtained against the face fly breeding in the feces of the calves.

Bioassays of the feces from calves fed the Dimilin formulations showed that all formulations were effective against face flies. Technical Dimilin fed to calves at the dosage rate of 0.033 mg/kg of body weight per day was required for an $ED_{50}$ against face flies at day 3 during treatment and 7 times less of the UF-5 and UF-10 was required to obtain the same $ED_{50}$ as with the technical Dimilin. The CP-50 and CP-75 required 1.6 times less and all other formulations required 1.7 to 2.4 times more than was required for the technical Dimilin or from 12 to 17 times more than the UF-5 formulation. Among polymers of the same type, but differing in the Dimilin composition, PU-10 having greater Dimilin composition, released greater amounts of Dimilin. Other polymers of the same type showed no apparent difference in releasing of Dimilin.

These data indicate that the encapsulated formulations of Dimilin may prolong the residual activity of Dimilin against face fly in the feces. Further investigations as a cattle feed additive for residual control of flies breeding in cattle feces are warranted.

The studies on the residual activity of Dimilin formulations against *A. aegypti* adult emergence revealed that technical Dimilin was effective against *A. aegypti* during the first 15 to 45 days post-treatment depending upon the concentrations used. Thereafter, Dimilin in technical form was rapidly degraded in water, whereas various encapsulated formulations of Dimilin continued to be active against mosquitoes throughout the 360 day test. Each formulation of Dimilin varied as to its effectiveness depending upon the concentrations used. However, at the low dosage rate of 0.05 ppm only the UF-5 and PA-5 showed promising residual mosquito larvicides. The UF-5 formulation inhibited adult emergence by 92% for 150 days. The PA-5 at this concentration inhibited adult emergence (78%) for 120 days. Among polymers of the same type, UF-5, PA-5, PVC-5, which had a lower Dimilin composition, showed more residual activity than did the same polymers having greater Dimilin composition. This may be due to the surface area per volume ratio being more influential than the Dimilin composition. The PU-10 and CP-75, which had greater Dimilin composition, produced more residual activity than did the same polymers having lower Dimilin composition. The indication is that the Dimilin composition in these two polymers was more influential than the surface area per volume ratio.

Residual activity of Dimilin formulations in cattle feces against the face fly reacted differently than they did in water against the mosquito. All Dimilin formulations tested were found to produce long residual activity against the face fly in cattle feces. This may be due to the fact that the cattle feces were kept in air tight containers and the Dimilin was probably absorbed into the fibrous materials in the feces, thus releasing slowly into the aqueous portion of the feces.

The results obtained from these studies show that Dimilin can be encapsulated in various polymers to extend its residual longevity. The 5% formulations of UF, PVC, and PA appear to be substantial for residual activity against the mosquito. By the use of encapsulated formulations of Dimilin, long residual effect on the mosquito can be obtained with less Dimilin than would be required with non-encapsulated formulations, also, extended treatment intervals would save labor, time, and money, thus being economically desirable. From an ecological standpoint, because less Dimilin is needed to achieve control of target species, the chances of contamination of the environment would be lessened considerably.

As cattle feed through, these encapsulated formulations of Dimilin not only extend its residual life in the feces, but indications are that they prevent degradation in the animal's body, thus possibly reducing or eliminating tissue residues. These encapsulated formulations may also be useful as feed throughs to livestocks for fly control in cattle feedlots, dairy barns, pig pens, chicken houses, and similar other sites where flies are a problem.

REFERENCES


1. C. A. Edwards, in Environmental Pollution by Pesticides, Plenum Press, London and New York, 1973, 452 pp.
2. G. G. Allan, C. S. Chopra, N. A. Neogi, and R. M. Wilkins, Nature, 234, 349-351 (1971).
3. B. C. Pass and H. W. Dorough, J. Econ. Entomol., 66, 1117-1119 (1973).
4. G. O. Fanger, Proc. Controlled Release Pesticide Symposium, Univ. of Akron, September 16-18, 1974, Akron, Ohio, 20 pp.
5. R. Mulder and A. A. Swennen, Proc. 7th British Insecticide Fungicide Conference, 1973, p. 729-735.
6. J. J. van Daalen, J. Meltzer, R. Mulder, and K. Wellinga, Naturwissenschaften, 59, 312 (1972).
7. R. Mulder and M. J. Gijswijt, Pestic. Sci., 4, 737-745 (1973).
8. K. Wellinga, R. Mulder, and J. J. van Daalen, Naturwissenschaften, 21, 993-998 (1973).
9. L. D. Post and W. R. Vincent, Naturwissenschaften, 60, 431-432 (1973).
10. W. L. Jakob, J. Med. Entomol., 10, 452-455 (1973).
11. M. S. Mulla, H. A. Darwazeh, and R. L. Norland, Proc. 7th British Insecticide Fungicide Conference, 1973, 67, p. 329-338.
12. J. E. Wright, J. Econ. Entomol., 67, 746-747 (1974).
13. T. Miura and R. M. Takahashi, Environ. Entomol., 3, 631-636 (1974).
14. R. M. Takahashi and T. Miura, Proc. California Mosquito Conference Association, 1975, pp. 43:85-89.
15. J. W. Neal Jr., J. Econ. Entomol., 67, 300-301 (1974).
16. A. J. Rogers, C. B. Rathburn Jr., E. J. Reider, G. Dodd, and A. Lafferty, Mosquito News, 36, 273-277 (1976).
17. H. C. Schaefer and E. F. Dupras Jr., J. Agric. Food Chem., 24, 733-739 (1976).
18. R. W. Miller, C. Corley, D. D. Aehles, and L. G. Pickens, J. Agric. Food Chem., 24, 687-688 (1976).
19. C. Corley, R. W. Miller, and K. R. Hill, J. of the AOAC, 57, 1269-1271 (1974).
20. B. A. Sowa and E. P. Marks, Insect Biochem., 5, 855-859 (1975).
21. R. W. Miller, C. Corley, and K. R. Hill, J. Econ. Entomol., 68, 181-182 (1975).
22. R. W. Miller, C. Corley, and S. R. Shufelt, J. Econ. Entomol., 69, 741-743 (1976).
23. T. G. Raley and E. D. Davis, Mosquito News, 9, 68-71 (1949).

24. L. J. Bruce-Chwatt and R. A. Fitz-John, J. Trop. Med. Hyg., 54, 116-121 (1950).
25. R. Elliott, Trans. Roy. Soc. Trop. Med. Hyg., 49, 528-542 (1955).
26. B. R. Evans and E. J. Fink, Mosquito News, 20, 388-389 (1960).
27. W. W. Barnes, A. B. Webb, and L. B. Savage, Mosquito News, 27, 488-490 (1967)
28. W. W. Barnes and A. B. Webb, Mosquito News, 28, 458-541 (1968).
29. L. L. Nelson, W. W. Barnes, F. W. Harris, and M. A. Lawson, J. Econ. Entomol., 63 (6), 1870-1873 (1970).
30. W. Mathis, J. W. Miles, and H. F. Schoof, Bull. World Health Org., 24, 646-649 (1961).
31. J. W. Miles, G. P. Pearce, and J. G. Wochst, J. Econ. Entomol., 10(3), 240-243 (1962).
32. J. T. Whitlaw Jr. and S. E. Evans Jr., J. Econ. Entomol., 61, 889-892 (1968).
33. N. F. Cardarelli and R. K. Mercier, Evaluation of toxic rubber as a mosquito larvicide. B.F. Goodrich Aerospace and Defense Products Report, 1965, 5 pp. Cited by A. H. Boike Jr. and C. B. Rathburn Jr., Mosquito News, 33, 501-505 (1973).
34. N. F. Cardarelli, R. F. Keller, and D. L. Jackson, Biocidal rubber-theory, preparation, and activity, B.F. Goodrich Aerospace and Defense Products Summary Report 1973, 17 pp. Cited by A. H. Boike Jr. and C. B. Rathburn Jr., Mosquito News, 33, 501-505 (1973).
35. H. A. Schultz and A. B. Webb, Mosquito News, 29, 38-41 (1969).
36. A. H. Boike Jr and C. B. Rathburn Jr., Mosquito News, 33, 501-505 (1973).
37. T. A. Parker, Effectiveness of 9.9 Per Cent Dursban in Polyethylene Applied as a Pre-season Larvicide. USAEHA Special Study No. 31-005-71, Defense Documentation Center No. AD 725573, 1970.
38. L. W. Roberts and T. A. Miller, Polymer Formulations of Dursban and Abate as Mosquito Larvicides, USAEHA Special Study No. 31-006-71, Defense Documentation Center No. AD 729344, 1970.
39. R. N. Wilkinson, W. W. Barnes, A. R. Gillogly, and C. D. Minnemeyer, J. Econ. Entomol., 64(1), 1-3 (1971).
40. T. A. Miller, L. L. Nelson, W. W. Young, L. W. Roberts, D. R. Roberts, and R. L. Wilkinson, Mosquito News, 33, 148-155 (1973).
41. T. A. Miller, L. L. Nelson, and W. W. Young, Mosquito News, 33, 172-175 (1973).
42. L. L. Nelson, T. A. Miller, and W. W. Young, Mosquito News, 33, 396-402 (1973).

43. D. R. Roberts, L. W. Roberts, T. A. Miller, L. L. Nelson, and W. W. Young, Mosquito News, 33, 165-171 (1973).
44. E. S. Evans Jr., J. H. Nelson, N. E. Pennington, and W. W. Young, Mosquito News, 35, 343-350 (1975).
45. J. H. Nelson, S. E. Evans Jr., N. E. Pennington, and M. V. Meisch, Mosquito News, 36, 47-51 (1976).
46. L. L. Nelson and T. A. Miller, Laboratory Evaluation of Residues Maintained in Water Treated with Polyvinyl Chloride Formulations of Dursban, USAEHA Special Study No. 31-008-72, 1971.
47. L. L. Nelson, T. A. Miller, I. T. Whitlaw Jr., W. W. Young, and M. A. Lawson, Mosquito News, 33, 403-412 (1973).
48. M. A. Lawson, T. A. Miller, R. J. Oakleaf, and W. W. Young, Mosquito News, 33, 561-567 (1973).
49. T. A. Miller, M. A. Lawson, L. L. Nelson, and W. W. Young, Mosquito News, 33, 413-417 (1973).
50. R. L. Dunn and F. E. Strong, Mosquito News, 33, 110-111 (1973).
51. C. H. Schaefer and W. H. Wilder, J. Econ. Entomol., 65, 1066-1071 (1972).
52. J. E. Lloyd and J. G. Matthysse, J. Econ. Entomol., 59, 363-367 (1966).
53. J. G. Medley and R. O. Drummond, J. Econ. Entomol., 55, 118-121 (1962).
54. G. C. Breeden, E. C. Turner Jr., and W. L. Beane, J. Econ. Entomol., 68, 451-452 (1975).
55. J. Herald, MS Thesis, Univ. of Kentucky, Lexington, Kentucky, 1976, 65 pp.
56. J. W. Kilpatrick, D. R. Maddock, and J. W. Mile, Mosquito News, 55, 951-953 (1962).
57. W. Mathis and H. F. Schoof, J. Econ. Entomol., 61, 1071-1073 (1968).
58. F. W. Knapp, Results of Research in 1966, 79th Annual Report of the Director, Agric, Exp. Stn., Univ. of Kentucky, Lexington, Kentucky, 1966, 106 pp.
59. F. W. Knapp and J. Ellis, J. Econ. Entomol., 68, 130-131 (1975).
60. I. Fox, G. A. Rivera, and I. G. Gayona, J. Econ. Entomol., 62, 1246-1249 (1969).
61. M. D. Price, Pest Control, 28, 47-58 (1960).
62. A. F. Kydonieus, I. K. Smith, and S. Hyman, Proc. Controlled Release Pesticides Symposium, Wright State Univ., Dayton, Ohio, September 8-10, 1975.
63. G. S. Olton, Pest Control, 43, 20, 53 (1975).
64. E. J. Gerberg, AMCA Bull., 5, 109 (1970).
65. S. M. Valder and T. L. Hopkins, Ann. Entomol. Soc. Amer., 61, 827-834 (1968).

66. Anonymous, *Carbopol Water-Soluble Resins*, Service Bulletin GC-36 Revised, B.F. Goodrich Chemical Company, Cleveland, Ohio, 1957.
67. W. S. Abbott, *J. Econ. Entomol.*, *18*, 265-267 (1925).
68. D. J. Finney in *Probit Analysis*, 3rd Edition, Cambridge Univ. Press, London, 1971, 333 pp.
69. A. J. Bar, J. H. Goodnight, J. P. Sall, and J. T. Helwig, in *A User's Guide to SAS*, SAS Institute, Raleigh, North Carolina, 1976, 329 pp.
70. M. Y. G. Hsieh and C. D. Steelman, *Mosquito News*, *34*, 278-282 (1974).
71. C. W. Kruse, G. F. Ludvik, and W. B. Hawkins, *J. Econ. Entomol.*, *45*, 598-601 (1952).
72. J. R. Busvine in *A Critical Review of the Techniques for Testing Insecticides*, The Eastern Press Ltd., London and Reading, 1957, 208 pp.
73. C. H. Stockman, E. H. Bollinger, and R. N. Maranch Jr., *Insect. Biochem.*, *30*, 584-589 (1970).
74. R. W. Miller, *J. Econ. Entomol.*, *67*, 697 (1974).
75. M. S. Mulla and H. A. Darwazeh, *Mosquito News*, *35*, 281-285 (1975).
76. C. H. Schaefer, W. H. Wilder, and F. S. Mulligan, *J. Econ. Entomol.*, *68*, 183-185 (1975).
77. M. S. Mulla and H. A. Darwazeh, *J. Econ. Entomol.*, *69*, 309-312 (1976).

# CONTROLLED ACTIVITY POLYMERS WITH LABILE BONDS TO PENDENT METRIBUZIN

*C. L. McCormick*
*M. M. Fooladi*

Department of Polymer Science
University of Southern Mississippi
Hattiesburg, Mississippi

## I. INTRODUCTION

During the past few years, a growing interest has developed in controlled activity or controlled release polymeric pesticide systems (1-6). These systems offer great promise for enhancing the efficacy of some existing herbicides as well as reducing environmental problems associated with others. Properly formulated, controlled activity systems may significantly reduce losses associated with volatilization, photolytic decomposition, dissolution, and transport from the application area.

Polymeric systems for controlled release of pesticides may be assigned to two general categories based on the mechanism of release. In the first, the pesticide is physically dissolved, adsorbed, entrapped, or dispersed in a polymeric matrix. Release is generally based upon diffusion phenomena; however, chemical or biological degradation of the matrix is also possible. In the second category, the pesticide is chemically bound to the polymer either as a structural unit along the backbone or as a pendent group. Release is largely dependent upon cleavage of chemically labile bonds within the macromolecule.

Polymeric systems with pendent herbicides present the chemist with a number of intriguing possibilities for release rate modification. The nature of the susceptible bond and its distance from the macromolecular backbone can be varied. Additionally, chemical structure, molecular weight, molecular




ISBN 0-12-074450-3

weight distribution, hydrophilicity, and physical form and dimensions may be altered by proper choice of reactants and conditions.

Polymers containing pendent herbicides may be prepared by two major routes (7,8). A herbicide-containing monomer may be synthesized and polymerized or an appropriate herbicide derivative may be directly reacted with a preformed polymer. The advantages offered by the resulting systems are often debated; however, the former method can yield polymers with higher percentages of herbicide per unit weight. In the latter method, a number of biodegradable, natural polymers such as those available as industrial waste (i.e. cellulose, chitin, and lignin) may be utilized.

Neogi (9) synthesized and polymerized vinyl monomers of herbicides including 2-methyl-4-chlorophenoxyacetic acid, pentachlorophenol, and 2,3,5-trichloro-4-hydroxypyridine. Harris and Post (10) prepared homopolymers utilizing the monomers vinyl 2,4-dichlorophenoxyacetate and vinyl 2-(2,4,5-trichlorophenoxy)propionate. Later, in an attempt to increase hydrolysis rates, monomers were synthesized having increased length between the pendent herbicide and the macromolecular backbone (11). Copolymerization (12) of the acryloyloxyethyl or methacryloyloxyethyl esters of 2,3,6-trichlorophenylacetic acid or 2,4-dichlorophenoxyacetic acid with methacrylic acid yielded systems which hydrolyzed only at temperatures above 47$^{o}$C in water buffered to pH 8. Copolymerization of the acryloyloxyethyl ester of 2,4-dichlorophenoxyacetic acid with 2-hydroxyethyl methacrylate or trimethylamine methacrylamide yielded polymers with enhanced hydrolysis rates (13). These rate increases were attributed to intramolecular catalysis and/or increased hydrophilicity arising from inclusion of the comonomer.

Herbicides have been attached to preformed polymers to yield systems with pendent esters, amides, carbamates, or acetals. For example lignin, polyvinyl alcohol, and cellulose have been reacted with 2,4-dichlorophenoxy acetic acid or 2,4-dichlorophenoxybutyric acid derivatives to yield esters (14-22). Cellulose, chitin, amylose, dextran, and polyvinyl alcohol have been reacted with isocyanate derivatives of herbicides to yield systems with hydrolytically labile carbamate bonds (18-22). Hydroxy-containing polymers have been reacted with herbicides containing an aldehyde function to yield biologically active controlled release systems (23). Polymers with hydrolyzable amide bonds have been prepared by reacting aminoethylated resins with 2,4-dichlorophenoxyacetic acid (24).

In our continuing investigations (17-22, 25) of the effect of macromolecular design upon hydrolysis rates of controlled release polymers, we shall in this report describe synthesis, characterization, and hydrolysis studies of polymers with carbamate, urea, and amide bonds to a pendent herbicide, metribuzin. The ultimate objective of this study is to improve the effectiveness of this highly selective, yet often too mobile, herbicide.

## II. EXPERIMENTAL

### A. *Synthesis of Metribuzin-Containing Vinyl Monomers (25)*

Vinyl monomers of 4-amino-(1,1-diemethylethyl)-3-(methylthio)-(1,2,4)-triazin-5(4H)one, metribuzin, were prepared with amide, carbamate, and urea bonds as shown in Figure 1.

The monomers possessing amide bonds were prepared by reacting metribuzin, 1, with acryloyl chloride, 2, or with methacryloyl chloride, 3, under a stream of nitrogen in tetrahydrofuran at 60°C. A small quantity of hydroquinone was added to stabilize the monomer. Elemental analyses, melting points, and percentage yields of the purified products 3 and 5 are given in Table I.

Vinyl monomers of metribuzin-possessing urea bonds were prepared by reacting metribuzin isocyanate, 6,(21) with acrylamide, 7, or with 2-aminoethylacrylate, 9. Similarly monomer 12 was prepared by reacting 6, with N-(2-hydroxyethly)acrylamide. The above reactions were conducted in toluene at 50°C with a tertiary amine catalyst (tributylamine). Additionally, a small amount of hydroquinone was added to prevent monomer polymerization. Melting points, percentage yields, and elemental analyses are given for the purified monomers 8, 10, and 12 in Table I.

### B. *Synthesis of Polymers with Pendent Metribuzin*

The metribuzin-containing monomers 3, 5, 8, 10, and 12 were polymerized in bulk at 150°C or in tetrahydrofuran (THF) at 50°C using benzoyl peroxide as the initiator. The reaction was generally carried out in THF. The resulting polymers were purified by dissolution in THF and precipitation into hexane. Experimental data including number, average molecular weight, inherent viscosity, percent yield,

SCHEME I

SYNTHESIS OF METRIBUZIN VINYL MONOMERS

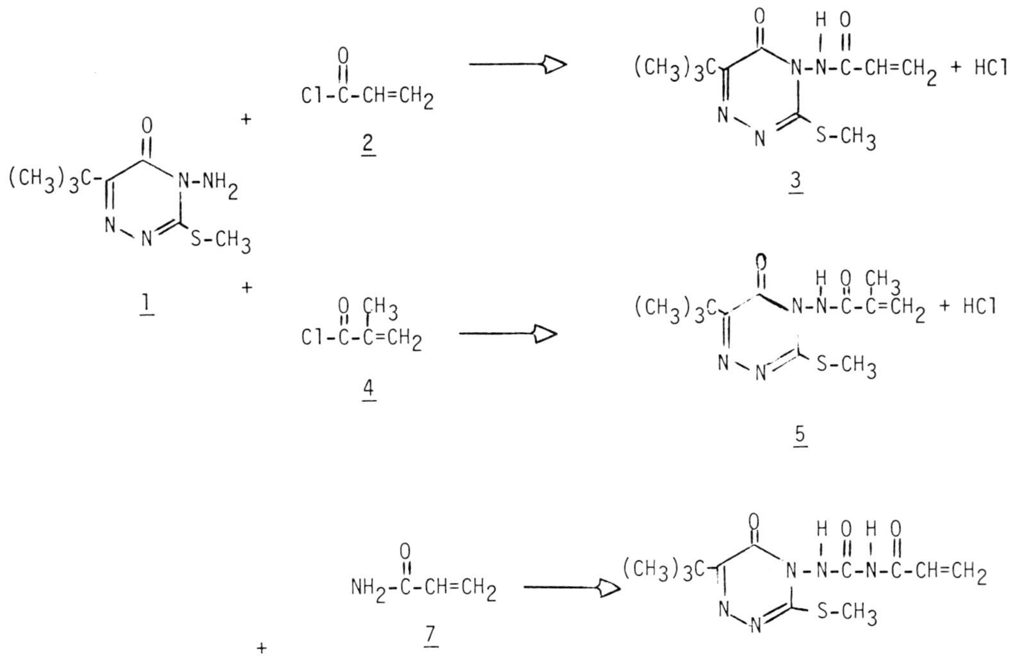

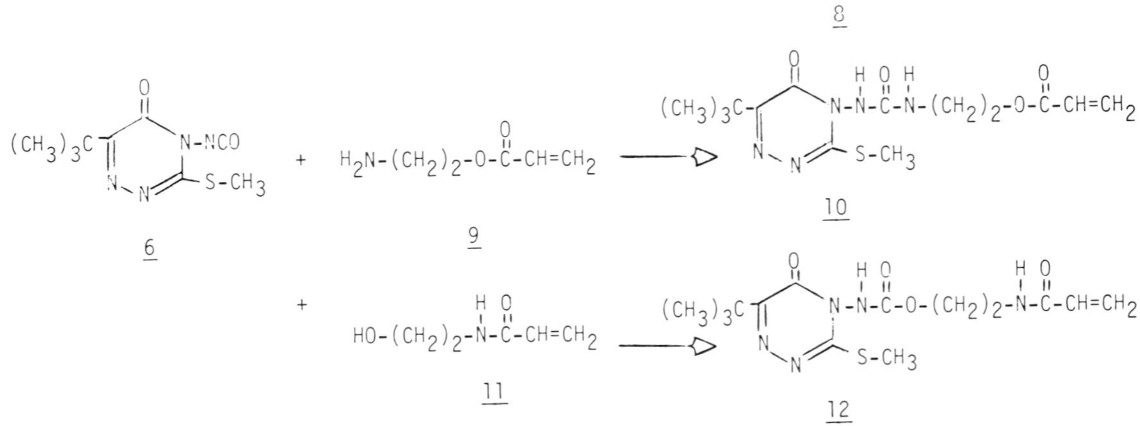

FIGURE 1. *Synthesis of metribuzin vinyl monomers.*

TABLE I. Elemental Analysis, Melting Points, and Percent Yield for Metribuzin Monomers

| Monomer number | Empirical formula | Calculated | | | | Found | | | | M.P. | Yield |
|---|---|---|---|---|---|---|---|---|---|---|---|
| | | C | H | N | S | C | H | N | S | ($C^{o}$) | (%) |
| 3 | $C_{11}H_{16}N_4O_2S$ | 49.16 | 5.96 | 20.86 | 11.92 | 48.80 | 6.22 | 20.12 | 10.97 | 160 | 77 |
| 5 | $C_{12}H_{18}N_4O_2S$ | 51.06 | 6.38 | 19.86 | 11.35 | 50.95 | 6.05 | 20.10 | 10.64 | 165 | 95 |
| 8 | $C_{12}H_{17}N_5O_3S$ | 46.30 | 5.47 | 22.51 | 10.29 | 46.50 | 5.40 | 22.50 | 10.20 | 171 | 86 |
| 10 | $C_{14}H_{21}N_5O_4S$ | 47.32 | 5.91 | 19.72 | 9.01 | 47.10 | 5.80 | 19.70 | 9.15 | ∿50 | 59 |
| 12 | $C_{14}H_{21}N_5O_4S$ | 47.32 | 5.91 | 19.72 | 9.01 | 47.10 | 5.50 | 19.85 | 9.70 | ∿48 | 42 |

and initiator concentration for selected polymers are given in Table II and elemental analyses are shown in Table III.

### *C. Molecular Characterization*

In addition to elemental analyses of monomers and polymers as obtained from Galbraith Laboratories (Tables I and III), infrared spectra (Perkin Elmer R-12B) and nuclear magnetic resonance data (Varian A-60D) were utilized to confirm the monomer and polymer structures. Number average molecular weights, $\bar{M}n$ (Table II), were obtained in ethanol utilizing a Hewlett-Packard Membrane Osmometer 502.

### *D. Release Studies*

Herbicide release in water at room temperature was monitored as a function of time for each polymer and is recorded in Table IV as a percentage of the available herbicide. For these studies, the solid polymer (0.05 g) was submerged in 20 ml of distilled water. Samples (10 μl) were withdrawn at designated intervals and the concentration of released metribuzin was determined by ultraviolet spectoscopy (Cary 17). To rule out possible ultraviolet absorbances of interfering species, a liquid chromatographic technique utilizing a Perkin Elmer HPLC Series 3 with a variable wave length ultraviolet detector was developed.

## III. RESULTS AND DISCUSSION

Vinyl monomers of metribuzin were polymerized to yield controlled release polymers (Table II). Polymers with pendant amide bonds were obtained from N-acryloyl metribuzin, 3, and N-methacryloyl metribuzin, 5. These polymers, F-1 through F-11 and F-12 through F-19, (Table II), exhibited number average molecular weights ranging from 35,000 to 1,000,000 depending upon initiator concentration, reaction time, polymerization method, and temperature. Molecular weights were more controllable in solvent systems than in bulk as expected since higher temperatures were required in the latter.

The polymers F-23 through F-30 and F-31 through F-34 contain extender groups based on ethanol amine. Polymers prepared from monomer 12 contain carbamate bonds capable of hydrolysis near the herbicidal pendent group in addition to

TABLE II. Physical Data and Reaction Conditions for Metribuzin-Containing Polymers

| Polymer number | Monomer number | $\overline{Mn}$ (x $10^3$) | Yield | M.P. (°C) | Initiator conc. (m) | Reaction temp. (°C) | Reaction time (Hrs.) |
|---|---|---|---|---|---|---|---|
| F- 1 | 3 | 81 | 86 | 170-172 | 71. x $10^{-5}$ | 135 | 72 |
| F- 2 | 3 | 317 | 69 | 185 | -0- | 135 | 72 |
| F- 3 | 3 | 42 | 87 | 169 | 1.5 x $10^{-4}$ | 135 | 72 |
| F- 4 | 3 | 270 | 98 | 183 | 4.2 x $10^{-5}$ | 135 | 72 |
| F- 5 | 3 | 200 | 75 | 174-175 | 8.3 x $10^{-6}$ | 150 | 30 |
| F- 6 | 3 | 91 | 60 | 195 | 8.3 x $10^{-6}$ | 150 | 48 |
| F- 7 | 3 | 970 | 60 | 181 | 4.1 x $10^{-6}$ | 30 | 100 |
| F- 8 | 3 | 370 | 58 | 190-192 | -0- | 50 | 44 |
| F- 9 | 3 | 300 | 59 | 189 | 7.5 x $10^{-5}$ | 50 | 44 |
| F-10 | 3 | 730 | 57 | 188 | 1.7 x $10^{-5}$ | 50 | 44 |
| F-11 | 3 | 510 | 57 | 190 | 2.9 x $10^{-5}$ | 50 | 44 |
| F-12 | 5 | 29 | 75 | 190 | 8.3 x $10^{-5}$ | 160 | 44 |
| F-13 | 5 | 130 | 65 | 205 | -0- | 160 | 44 |
| F-14 | 5 | 52 | 70 | 195 | 4.1 x $10^{-5}$ | 160 | 44 |
| F-15 | 5 | 35 | 68 | 210 | 6.2 x $10^{-5}$ | 160 | 44 |
| F-16 | 5 | 1,000 | 96 | 215 | 2.0 x $10^{-5}$ | 50 | 44 |
| F-17 | 5 | 120 | 90 | 220 | -0- | 50 | 44 |
| F-18 | 5 | 150 | 94 | 205 | 4.1 x $10^{-5}$ | 50 | 44 |
| F-19 | 5 | 100 | 93 | 195 | 5.0 x $10^{-5}$ | 50 | 44 |
| F-20 | 5 | 150 | 60 | 245 | 4.1 x $10^{-5}$ | 50 | 20 |
| F-21 | 5 | 65 | 57 | 238 | 4.1 x $10^{-5}$ | 50 | 20 |
| F-22 | 5 | 35 | 64 | 240 | 9.1 x $10^{-5}$ | 50 | 20 |

| | | | | | | | |
|---|---|---|---|---|---|---|---|
| F-23 | 8 | 310 | 87 | 45 | -0- | 135 | 24 |
| F-24 | 8 | 250 | 85 | 40 | $5.6 \times 10^{-5}$ | 135 | 24 |
| F-25 | 8 | 290 | 87 | 42 | $1.5 \times 10^{-5}$ | 135 | 24 |
| F-26 | 8 | 250 | 93 | 46 | $6.0 \times 10^{-5}$ | 135 | 24 |
| F-27 | 10 | 116 | 80 | 35-40 | -0- | 50 | 24 |
| F-28 | 10 | 38 | 80 | 38-39 | $2.1 \times 10^{-5}$ | 50 | 24 |
| F-29 | 10 | 37 | 80 | 39.5 | $7.3 \times 10^{-5}$ | 50 | 24 |
| F-30 | 10 | 44 | 80 | 40 | $1.5 \times 10^{-5}$ | 50 | 24 |
| F-31 | 12 | 110 | 87 | 46 | -0- | 50 | 24 |
| F-32 | 12 | 65 | 75 | 42 | $4.4 \times 10^{-5}$ | 50 | 24 |
| F-33 | 12 | 55 | 76 | 42.5 | $6.6 \times 10^{-5}$ | 50 | 24 |
| F-34 | 12 | 35 | 75 | 40 | $8.9 \times 10^{-5}$ | 50 | 24 |

TABLE III. *Elemental Analysis for Polymers With Pendent Metribuzin*

| Polymer number | Empirical formula | Calculated | | | | Found | | | |
|---|---|---|---|---|---|---|---|---|---|
| | | C | H | N | S | C | H | N | S |
| F-1 | $C_{11}H_{16}N_4O_2S$ | 49.16 | 5.96 | 20.86 | 11.92 | 49.20 | 5.85 | 20.80 | 11.60 |
| F-2 | $C_{11}H_{16}N_4O_2S$ | 49.16 | 5.96 | 20.86 | 11.92 | 49.31 | 5.90 | 20.75 | 11.81 |
| F-5 | $C_{11}H_{16}N_4O_2S$ | 49.16 | 5.96 | 20.86 | 11.92 | 48.60 | 6.30 | 19.65 | 10.98 |
| F-8 | $C_{11}H_{16}N_4O_2S$ | 49.16 | 5.96 | 20.86 | 11.92 | 49.10 | 5.71 | 20.60 | 11.50 |
| F-12 | $C_{12}H_{18}N_4O_2S$ | 51.06 | 6.38 | 19.86 | 11.35 | 50.98 | 6.10 | 19.15 | 11.12 |
| F-16 | $C_{12}H_{18}N_4O_2S$ | 51.06 | 6.38 | 19.86 | 11.35 | 51.10 | 6.42 | 19.71 | 11.50 |
| F-20 | $C_{12}H_{17}N_5O_3S$ | 46.30 | 5.47 | 22.51 | 10.29 | 46.10 | 5.30 | 22.65 | 10.50 |
| F-23 | $C_{14}H_{21}N_5O_4S$ | 47.32 | 5.92 | 19.72 | 9.01 | 47.21 | 5.68 | 20.00 | 9.10 |
| F-24 | $C_{14}H_{21}N_5O_4S$ | 47.32 | 5.92 | 19.72 | 9.01 | 47.10 | 5.81 | 19.60 | 9.12 |
| F-27 | $C_{14}H_{21}N_5O_4S$ | 47.32 | 5.92 | 19.72 | 9.01 | 47.25 | 5.85 | 19.65 | 9.20 |
| F-31 | $C_{14}H_{21}N_5O_4S$ | 47.32 | 5.92 | 19.72 | 9.01 | 47.60 | 6.10 | 20.00 | 8.85 |

TABLE IV. Release of Pendent Metribuzin from Vinyl Polymers in $H_2O$ at Room Temperature

| | | | Percent release | | | |
|---|---|---|---|---|---|---|
| Polymer number | Conc. mg/25 ml | % Pesticide in polymer | 5-min | 1-day | 2-days | 7-days |
| F- 1 | 2.0 | 80 | -0- | 5 | 9 | 20 |
| F-10 | 3.4 | 80 | 4 | 1 | 15 | 30 |
| F-11 | 5.0 | 80 | 5 | 10 | 13 | 33 |
| F-13 | 5.2 | 76 | 3 | 11 | 13 | 40 |
| F-19 | 4.1 | 76 | 6 | 11 | 14 | 32 |
| F-23 | 2.8 | 60 | 4 | 16 | 22 | 27 |
| F-25 | 4.0 | 60 | 6 | 7 | 11 | 28 |
| F-26 | 4.6 | 60 | -0- | 12 | 19 | 43 |
| F-29 | 3.0 | 60 | 3 | 13 | 23 | 28 |

the amide group close to the macromolecular backbone. Polymers prepared from monomer 8, on the other hand, possess urea and ester bonds. Molecular weights of the urea- and carbamate-containing polymers ranged from 35,000 to 310,000.

Preliminary release studies are reported in Table IV for polymers which were immersed in water at 25°C. In these experiments, no attempts were made to control particle size or pH. Additional problems making interpretation of comparative data difficult include hydrolysis of the released metribuzin (20) to products with no active chromophores, varying glass transition temperatures of each polymer type, and the changing hydrophilic nature of the polymer as pendent groups are hydrolyzed.

Despite these limitations, a number of interesting observations may be made. The electronic nature of the triazine ring and its proximity to the urea or amide bond lead to suprisingly rapid rates of hydrolysis in distilled water. The influence of spacer groups and molecular weight in these systems on hydrolysis rates has not been fully elucidated.

The polymers prepared from the methacrylic monomer, 5, show similar rates of release when compared to those prepared from the acrylic monomer, 3. Interestingly, however, phytotoxicity studies (18,20) show much greater activity for the methacrylic polymers (F-12 through F-19). $C^{13}$ nuclear magnetic resonance studies have shown that these polymers tend to depolymerize at extended times to a monomer-polymer mixture. This phenomenon may be due to a low ceiling temperature of the methacrylic polymers. The monomer itself could then serve as a controlled release molecule; its additional mobility could account for its high biological activity.

Rates of metribuzin release from the polymers reported in this work apparently do not diminish with time over the first seven days as do the rates reported for a number of other pendant polymeric systems. These results may be due to the changing functionality pendant to the macromolecular backbone as hydrolysis proceeds. Some enhancement of release rate may be brought about by the increasing hydrophilicity expected as the amide bonds are hydrolyzed to yield polyacrylic acid units in the copolymer. Additionally, anchimeric assistance by the resulting groups may promote rapid hydrolysis of the remaining amide bonds. Kinetic experiments under controlled conditions of particle size, pH, temperature, and molecular weight are in progress to confirm these preliminary conclusions.

ACKNOWLEDGMENTS

Research support for this project by Hopkins Agricultural Chemical Company of Madison, Wisconsin, is gratefully acknowledged.

REFERENCES

1. N. F. Cardarelli, (Ed.), Proc. Controlled Release Pesticide Symposium, Univ. of Akron, Akron, Ohio, September 1974.
2. F. W. Harris, (Ed.), Proc. Controlled Release Pesticide Symposium, Wright State Univ., Dayton, Ohio, September 1975.
3. N. F. Cardarelli, (Ed.), Proc. Controlled Release Pesticide Symposium, Univ. of Akron, Akron, Ohio, September 1976.
4. H. B. Scher, (Ed.), Controlled Release Pesticides, ACS Symp. Series 53, Washington, D.C. 1977.
5. G. A. Montemarrano, (Ed.), Proc. Controlled Release Symposium, Gaithersburg, Maryland, August 1978.
6. R. F. Goulding, (Ed.), Proc. Controlled Release Pesticide Symposium, Oregon State Univ., Corvallis, Oregon, August 1977.
7. G. G. Allan, *et al.*, Chem Tech, 171, March 1973.
8. F. W. Harris, Proc. Controlled Release Pesticide Symposium, 1.33-1.43, Univ. of Akron, Akron, Ohio, September 1976.
9. A. N. Negoi, Ph.D. Dissertation, Univ. of Washington, Seattle, Washington, 1970.
10. F. W. Harris and L. K. Post, J. Poly. Sci., A-1, 13, 225 (1975).
11. F. W. Harris and L. K. Post, Polymer Preprints, 16 (2), 622 (1975).
12. F. W. Harris *et al.*, Controlled Release Polymeric Formulations, ACS Symp. Series 33, Washington, D.C., 222 (1976).
13. F. W. Harris *et al.*, Controlled Release Pesticides, H. B. Scher, (Ed.), ACS Symp. Series 53, Washington, D.C., 112 (1977).
14. G. G. Allan, in Lignins, K. U. Sarkanen and C. H. Ludwig, (Eds.), Wiley, New York, p. 612 (1971).
15. G. G. Allan *et al.*, Int. Pest Control, 14(2), 15 (1972).
16. G. G. Allen *et al.*, Int. Pest Control, March/April 17 (1975).

17. C. L. McCormick and M. M. Fooladi, Controlled Release Pesticides, ACS Symp. Series 53, 112-125, H. B. Scher, (Ed.), Washington, D.C. 1977.
18. C. L. McCormick and K. E. Savage, Proc. Controlled Release Symposium, 28-40, Corvallis, Oregon, 1977.
19. C. L. McCormick, D. K. Lichatowich, and M. M. Fooladi, Proc. Controlled Release Symposium, 3.6-3.17, Gaithersburg, Maryland, August 1978.
20. K. E. Savage, C. L. McCormick, and B. H. Hutchinson, Proc. Controlled Release Symposium, 3.18-3.28, Gaithersburg, Maryland, August 1978.
21. C. L. McCormick *et al.*, Organic Coatings and Plastics Preprints, ACS 40, 555-559 (1979).
22. C. L. McCormick and D. K. Lichatowich, Polymer Letters, August 1979.
23. Schacht, *et al.*, Macromol. Chem., 179, 837-840 (1978).
24. H. D. Jakubka and E. Busch, Z. Chem., 13(3), 105 (1973).
25. M. M. Fooladi, Ph.D. Dissertation, Univ. of Southern Mississippi, May 1979.

# CHLORPYRIFOS RELEASE FROM MOSQUITO LARVICIDE FORMULATIONS

*K. G. Seymour*

Agricultural Products Department
The Dow Chemical Company
Midland, Michigan

Various authors (1-5) have mathematically examined the release of active ingredient from controlled release devices. Most treatments deal with amounts or rates of release to the environment. However, many pesticides are simultaneously decomposed in the surrounding soil or water. DURSBAN[1] 10CR is one such insecticide. DURSBAN 10CR is registered in the U.S. for the control of mosquito larvae in natural bodies of water. This function is dependant upon a balance between the rate of release into the water and the simultaneous degradation of the insecticide after entering the water phase.

Early screening by the Edgewood Arsenal Environmental Hygiene group identified particulate plastic formulations of O,O-diethyl-O-(3,5,6-trichloro-2-pyridyl) phosphorothioate, common name chlorpyrifos, as a promising approach to controlling mosquito larvae in aqueous habitats (7). An extensive R&D program by that group and co-workers (8-18), with collaboration by The Dow Chemical Company Agricultural Products Department, resulted in a large array of data concerning the selection and functionality of the product. This report considers some key parameters affecting the product functionality and compares initial active ingredient release rates for various preparations.

Before considering the chlorpyrifos controlled release data, it is instructive to review the relevant chemical, physical and biological data. Chlorpyrifos is very insoluble

[1] *Trademark of The Dow Chemical Company.*




ISBN 0-12-074450-3

TABLE I. Chlorpyrifos Properties

| | |
|---|---|
| Structure: | 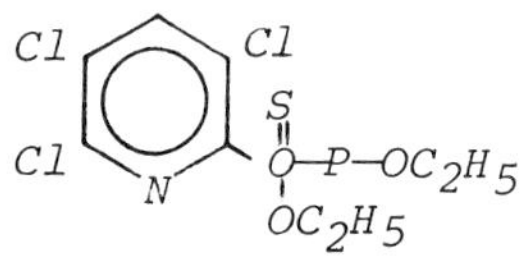 |
| Molecular Wt: | 350.5 |
| Melting Point: | 41.5 - 43.5°C |
| Vapor Pressure: | $1.87 \times 10^{-5}$ mm Hg, 25°C<br>$8.15 \times 10^{-5}$ mm Hg, 35°C |
| Solubility, 25°C: | |
| Water | 0.4 ppm |
| Xylene | 400 g/100 g solvent |
| Methanol | 45 g/100 g solvent |
| Acetone | 650 g/100 g solvent |
| Aliphatic Oils | 5-15 g/100 g solvent |
| Acute Toxicity: | |
| Oral $LD_{50}$, male rat | 163 mg/Kg |
| duck (mallard) | 76 mg/Kg |
| Solubility Ratio, 25°C: | |
| Isooctane/water | $2 \times 10^6$ |
| Benzene/water | $20 \times 10^6$ |

in water (19), soluble in organic solvents, shows moderate oral toxicity and is relatively non-volatile (Table I). Toxicity to aquatic life varies widely for various species (Table II).

Chlorpyrifos is hydrolyzed by water. While the precise sequence of the degradation may vary, the end result is cleavage of the pyridyl-O-P linkage to form 3,5,6-trichloropyridinol. The hydrolysis reaction is first order (20). Meikle and Youngston (20) also found that the rate in buffered distilled water increased about three times for each $10^{o}C$ temperature increase. Changing the pH from 5 to 7 to 8 changed the half life from 23 to 35 to 63 days. However, trace contaminants in water have by far the greatest effect on degradation rate (Table III). Schaefer and Dupras (21) also observed rapid loss in natural water, but they observed extended efficacy in polluted waters, as did Wade (22) in cattle dipping vat studies.

The DURSBAN 10CR product is a cylindrical pellet, with a variable particle length because of mechanical factors during the extrusion-cutting process. A data reduction procedure to allow comparison of various formulation lots, or of different formulations, would be helpful in interpreting the accumulated data. It was also desired to estimate whether significant release rate variation might result from various preparative procedures, or particle size variations.

The accumulated field data (12-16, 18) suggest that DURSBAN 10CR maintains a steady-state chlorpyrifos concentration in water such that constant delivery rate is a

*TABLE II. Aquatic Organism Toxicity*[a]

| | *Water temp. $^{o}F$* | *Hours exposure* | *$LC_{50}$ ppb* |
|---|---|---|---|
| *Bluegill* | *80* | *96* | *3* |
| *Channel catfish* | *80* | *96* | *13* |
| *Mosquito fish* | *72* | *36* | *230* |
| *Grass shrimp* | *36* | *48* | *2.4* |
| *Blue crab* | *--* | *24* | *120* |

[a]*From DURSBAN Insecticide Technical Data Sheet, (19).*

*TABLE III. Chlorpyrifos Hydrolysis (First Order Reaction)*

| *Water* | *Rate constant ($25^oC$) $Days^{-1}$* | |
|---|---|---|
| *Buffered distilled water, pH 6.9* | *0.02* | *reference 20* |
| *Ibid plus Cu(II) 0.6 ppm* | *0.23* | *reference 20* |
| *Tap water, pH 7.7* | *0.47* | *reference 20* |
| *"Canal" water, pH 8* | *0.45* | *reference 20* |
| *Pond water* | *0.46* | *reference 21* |

$$\frac{k \text{ natural water}}{k \text{ distilled water}} = 23$$

possibility. Consider also that the plastic matrix is plasticized by the chlorpyrifos, that the partition coefficient for chlorpyrifos in organic solvents greatly favors the solvent (Table I) and that Nelson *et al.* (13) found release rate to be independent of active ingredient loading, although theory and results with several other plastic formulations indicated release rate should increase with loading. If, however, internal diffusion is not limiting, then dissolution rate will initially be limiting and the particles will provide a nearly constant rate until the internal concentration decreases enough to affect either diffusion, or the instantaneous solubility ratio. After that point, the release rate may decrease.

The Edgewood group had observed that distilled water in contact with DURSBAN 10CR (or similar preparations) increased in chlorpyrifos content for several weeks (12-14). Distilled water tests had produced higher concentrations than had natural waters (Table IV) but the (lower) concentration in natural water was reached more rapidly. In view of these points, a constant rate derivation appears reasonable.

Baker and Lonsdale (6) considered the general case discussed here, proceeding generally as follows: Consider a pesticide, dispensing device or particle releasing active

*TABLE IV. Chlorpyrifos Concentrations Maintained in Water by CPE + PE Pellet[a] Formulations (13)*

| | *M-3640*[b] | *1-4758-114*[b] |
|---|---|---|
| *Chlorpyrifos content* | *9.4%* | *15.0%* |
| *Pellet size* | *1/16" x 1/8"* | *1/16" x 1/8"* |
| *Dosage* | *1 ppm* | *1 ppm* |
| *Distilled water-lab (8-week concn)* | *38 ppb* | *32 ppb* |
| *Hydrant water-pools (avg. conc.)* | *1.4* | *1.3* |
| *Ratio* $\frac{\text{distilled water}}{\text{hydrant water}}$ | *27* | *25* |

[a]*Chlorinated polyethylene plus polyethylene.*
[b]*Dow Chemical Company experimental sample designations.*

ingredient (A.I.) at a constant rate into a fixed volume

$$\frac{dM}{dt} = A \tag{1}$$

where M = amount released, t = time and A = release rate. Simultaneously, the active ingredient is degraded in the environment in a first order reaction.

$$\frac{dM}{dt} = -kM \tag{2}$$

By mass balance, the net accumulation in the fixed volume environment is release minus degradation. Then

$$\frac{dM}{dt} = A - kM \tag{3}$$

and

$$\ln \frac{A}{A-kM} = kt\ . \tag{4}$$

Expressing Baker and Lonsdale's derivations in exponential form for convenience,

$$M_t = \frac{A}{K} (1-e^{-kt}). \tag{5}$$

If k is known, a plot of $M_t$ (or concentration) vs. $(1-e^{-kt})$ yields an experimental value for A. Here k = 0.14/week (20). The calculations reported below were made with k = 0.14/week (20) and measurements were made at 25°C.

Results from a shelf-life stability experiment[2] provide a test of the above relation. A single lot of DURSBAN 10CR had been subdivided and stored 30 days at 90°F and 85% RH, 6 and 12 months at 100°F, and 12 months at outside ambient temperature. Initial and final release rates were measured into triple-distilled water, generally following procedures employed by the Edgewood group. The hexane extracts of the water were analyzed[3] using gas chromatographic conditions as listed in Table V. No change in release rate was found. The data (Figure 1) follow the postulated model quite well through seven weeks, after which the points diverge and the rate appears to decrease, as postulated above. 1.5 ppm chlorpyrifos was initially added as the dose rate. The least squares slope (A) of Figure 1 was 33.8 ppb/week. Therefore, release per unit added was 33.8/1500 = .0225, or 2.25% of the amount added was released per week over the initial seven weeks of apparent constant release rate.

Another example of the $M_t$ vs. $(1-e^{-kt})$ relation is shown in Figure 2. In this experiment (by G.L. Boswell) a different mechanical chopping procedure was being evaluated. The particles were not cylindrical, but more nearly oblate spheroids. The plastic matrix was the same as for the Figure 1 experiment, and for regular production. It is apparent that, through the eight week test, accumulation was linear with $(1-e^{-kt})$.

[2]*R.E. Krause, Agricultural Products Department, Dow Chemical U.S.A., unpublished report.*
[3]*By P.K. Gilmore with supervision by R.L. McKellar, Agricultural Products Department, The Dow Chemical Company, Midland, Michigan.*

*TABLE V. Gas Chromatographic Conditions for the Determination of Chlorpyrifos*[a]

| | |
|---|---|
| *Gas chromatograph* | *Hewlett-Packard Model MT-222, equipped with a flame photometric detector.* |
| *Flame photometric detector* | *Melpar, Model FPD 100AT with a 526 nm interference filter (selective, sensitive response to phosphorus) TRACOR Inc.* |
| *Column* | *11% OV-17+QF-1 on Gas-Chrom.Q 100/120 mesh 6' x 3 mm.* |
| *Strip chart recorder* | *1 mv imput* |
| *Automatic sampler* | *Model 7671A Hewlett-Packard* |
| *Column temperature* | *215$^{o}$C* |
| *Injector block temperature* | *220$^{o}$C* |
| *Detector temperature* | *225$^{o}$C* |
| *Carrier gas, nitrogen* | *200 ml/min* |
| *Burner gas, oxygen* | *30 ml/min* |
| *Burner gas, hydrogen* | *200 ml/min* |
| *Attenuator* | *16 x $10^{4}$* |
| *Chart speed* | *0.25 in/min* |
| *10-μl injection* | |

[a]*Courtesy of R.L. McKellar.*

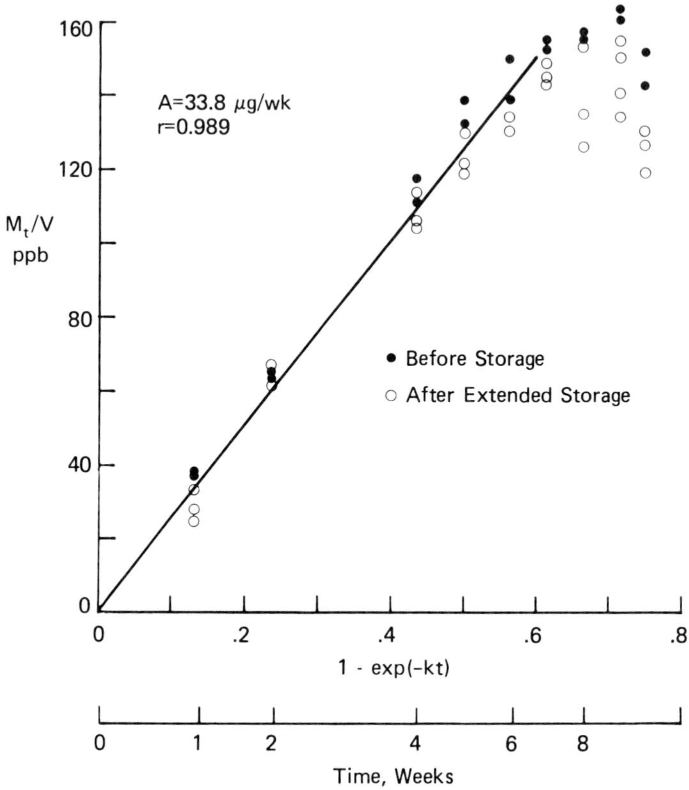


*FIGURE 1. Chlorpyrifos release rate from DURSBAN 10CR before and after extended storage at 38°C or ambient temperature. Reservoir dose = 1.5 ppm.*

Repeatability of the procedure has not been rigorously studied. However, different workers measuring release rate from the same lot of DURSBAN 10CR insecticide obtained similar results (Table VI). Other evaluations confirmed the repeatability of the comparison (lots A and B, Table VII).

It is unfortunate that the published data by the Edgewood group (10-13, 15) generally do not allow detailed re-calculation. The very large number of excellent observations and experiments they reported were necessarily condensed or presented graphically for reasons of space and clarity. One publication (18) does present detailed data. Calculations from those data are included in Table VII. Their release rates were 2/3 - 3/4 of that observed for the other

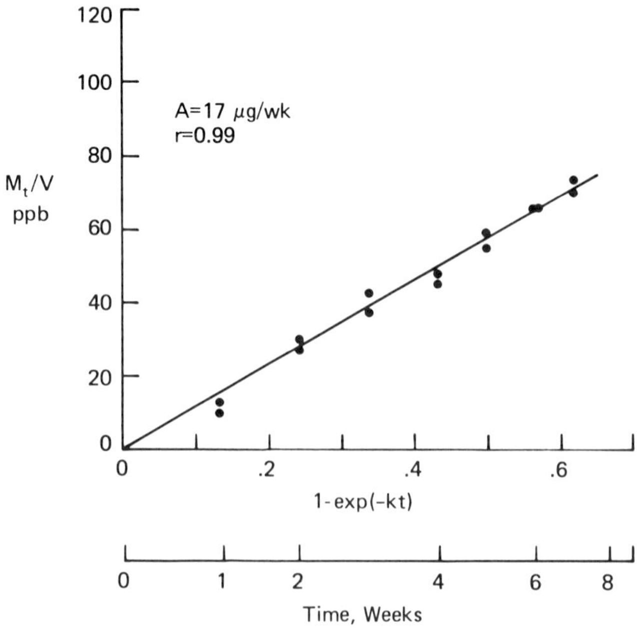


*FIGURE 2. Chlorpyrifos release rate from experimental, non-cylindrical particles. Reservoir dose = 1.0 ppm.*

preparations reported on here. However, the Edgewood group noted the need to increase release somewhat, and the preparations reported in this paper more nearly represent the formulations utilized in the later, extensive experiments that established efficacy, safety, etc. under actual field conditions.

Comparison of the release rates for the experimental shapes (Table VII) show lower release rates for two of the preparations. The next two released faster, but still not equal to the "standard" of 2.0 - 2.3% per week, even though the surface/volume ratio was higher.

Release rate values appear useful for comparing formulations and for interpreting the effect of formulation variables. The constant release rate model chosen was found to fit an assortment of chlorpyrifos—polyethylene/chlorinated polyethylene formulation preparations during the early portion of the dissolution process. Studies extending over the entire release period, although valuable in their own right, are much too extended in time for normal laboratory comparison of preparations. The studies reported here were

TABLE VI. Release Rate from a Single "DURSBAN 10CR" Insecticide Lot, as Measured by Different Researchers at Different Times

| Test | 1977 Correlation | | 1978 Shelf-life (Figure 1) |
|---|---|---|---|
| Reservoir dose added, ppm | 1.0 | 1.5 | 1.5 |
| A, ppb/wk | 24 | 35 | 34 |
| Correlation coefficient, r | 0.993 | 0.997 | 0.989 |
| Relative release, percentage released per week | 2.4 | 2.3 | 2.2 |

TABLE VII. Release Rates from Various Preparations of Chlorpyrifos/PE + CPE[a] Formulations

| Sample | Release Rate %/week | Surface/Volume $mm^{-1}$ |
|---|---|---|
| Nelson, et al. (18) | | |
| (1000 ppm dose) | 1.4 | |
| (500 ppm dose) | 1.7 | |
| Dow results | | |
| Lot A | 2.4 | |
| Table 6 | 2.2 | |
| Lot B | 2.0 | 3.5 |
| | 2.3 | |
| Experimental shapes | 1.1 | 2.4 |
| | 1.1 | 2.1 |
| | 1.8 | 5 |
| | 1.7 | 5 |

[a]Polyethylene plus chlorinated polyethylene.

conducted in distilled water at or near 25°C. It should again be emphasized that natural waters accelerate the chlorpyrifos hydrolysis rate. Concentrations found after treatment of natural water with DURSBAN 10CR insecticide will, therefore, be substantially lower than for distilled water.

## REFERENCES


1. T. Higuchi, J. Pharm. Sci., 50, 874 (1961).
2. T. Higuchi, J. Pharm. Sci., 52, 1145 (1963).
3. R. W. Baker and H. K. Lonsdale in Adv. Experimental Medicine and Biology, A. C. Tanquary and R. E. Lacey, (Eds.), Plenum Press, New York 1974.
4. D. H. Lewis and D. R. Cowsar, ACS Symp. Series 53, 1, H. B. Sher (Ed.), American Chemical Society, Washington, D.C. 1977.
5. R. W. Baker and H. K. Lonsdale, Chem. Tech., 5, 668 (1975).
6. R. W. Baker and H. K. Lonsdale, Proc. Controlled Release Pesticide Symposium, Wright State Univ., Dayton, Ohio, pp. 9-39 (1975).
7. J. T. Whitlaw Jr. and E. S. Evans Jr., Econ. Ent., 61, 889 (1968).
8. T. A. Miller, L. L. Nelson, W. W. Young, L. W. Roberts, and R. N. Wilkinson, Mosquito News, 33(2), 148 (1973).
9. L. W. Roberts, D. R. Roberts, T. A. Miller, L. L. Nelson, and W. W. Young, Mosquito News, 33(2), 155 (1973).
10. D. R. Roberts, L. W. Roberts, T. A. Miller, L. L. Nelson, and W. W. Young, Mosquito News, 33(2), 165 (1973).
11. T. A. Miller, L. L. Nelson, and W. W. Young, Mosquito News, 33(2), 172 (1973).
12. L. L. Nelson, T. A. Miller, and W. W. Young, Mosquito News, 33(3), 396 (1973).
13. L. L. Nelson, T. A. Miller, J. T. Whitlaw Jr., W. W. Young, and M. A. Lawson, Mosquito News, 33(3), 403 (1973).
14. M. A. Lawson, T. A. Miller, R. J. Oakleaf, and W. W. Young, Mosquito News, 33(4), 561 (1973).
15. E. S. Evans Jr., J. H. Nelson, N. E. Pennington, and M. V. Meisch, Mosquito News, 36(1), 47 (1976).
16. J. H. Nelson, E. S. Evans Jr., N. E. Pennington, and M. V. Meisch, Mosquito News, 36(1), 47 (1976).

17. J. N. Nelson, D. L. Stoneburner, E. S. Evans Jr., N. E. Pennington, and M. V. Meisch, Bul. Env. Contamin. and Toxicology, 15, 630 (1976).
18. J. H. Nelson and E. S. Evans Jr., U.S. Army Environmental Hygiene Agency Report 44-022-73/75, Aberdeen Proving Ground, Maryland (1973).
19. Anon. DURSBAN Insecticide Technical Data Sheet, Agricultural Products Department, Dow Chemical U.S.A., Midland, Michigan (1977).
20. W. R. Meikle and C. R. Youngston, Arch. Env. Contamin. and Toxicology, 7, 13 (1978).
21. C. H. Schaefer and E. F. Dupras Jr., J. Econ. Ent., 63, 701 (1970).
22. L. L. Wade, J. Econ. Ent., 61, 908 (1968).

# EVALUATION OF PLANT GROWTH REGULATORS USING THE SIMULATED CONTROLLED RELEASE APPROACH

*Richard M. Wilkins*

Department of Agricultural Biology
The University of Newcastle Upon Tyne

*The steady development of formulations based on the controlled release of biologically active compounds has revealed a substantial number of applications in plant protection and related fields. Many of these applications are based on existing herbicides, insecticides, nematocides, etc., and also on fertilizers. A new area has been the development of insect pheromones for which the potential for pest control is now only being realized through controlled release technology. Controlled release can improve on existing commercial products and make them safer and more efficient, but new compounds will be devised that depend on controlled action to be successful.*

*This paper concerns the application of controlled release techniques to new areas in crop production. One such area is the use of plant growth regulators. From the wide range of plant regulators available those that stimulate growth could have the greatest potential for use in controlled release formulations. For initial investigation compounds that have cytokinin-like activity such as leaf senescence inhibition have been chosen.*

*The approaches for identifying the new areas for controlled release are discussed. Simulation of the continuous action in the field by repeated conventional applications allows these new areas to be quickly evaluated without recourse to formulation. Finally, results from preliminary field studies, in which the simulated approach is used to assess the potential for benzimidazole derivatives as controlled release senescence inhibitors in wheat, are given.*




ISBN 0-12-074450-3

## I. NEW AREAS OF APPLICATION OF CONTROLLED RELEASE SYSTEMS

Many new areas where the controlled release approach can give benefits exist, especially in plant protection. Benefits resulting from controlled release of insecticides or herbicides have been fully recognized and many formulations have been developed and have been tested in the field. However, little attention has been paid to the area of pathogen control, especially where fungal diseases are concerned. Opportunities exist for the development of formulations for extended control of soil-borne pathogens (1) for diseases of trees and also for inhibiting the decomposition of materials and fibers. In many situations, however, there is the danger of fungicide resistance arising and durable success will depend on careful selection of fungicide and target pathogen.

An area that has not received attention is that of plant growth regulation. Natural plant growth is controlled by an interaction of endogenous regulators. As these normal plant hormones have insufficient stability outside of the plant the use of synthetic analogs has been developed for manipulation of growth. The search for chemicals to modify and improve plant growth is now considered important by most major agrochemicals companies and considerable research effort is devoted to their discovery. However, this process is more difficult than pesticide screening. This is because selective action on a particular physiological process, at a low concentration and without any undesirable side-effects on the plant, is required. The restrictions of conventional pesticide application (i.e. foliar sprays or soil-applied granules) in which a large dosage is applied and is then rapidly lost could be limiting the effectiveness of many plant growth regulators. Thus the advantages of the controlled release of plant growth regulators would be first to maintain active amounts in the plant and second to minimize adverse side effects from overly high concentrations. The controlled release approach could thus mimic the natural continuous production and movement of the endogenous hormones within the plant.

## II. THE POTENTIAL FOR CONTROLLED ACTION IN PLANT GROWTH REGULATION

A wide range of substances have been found, that, in small concentrations, affect the physiological processes of plants (2). The exogenous regulators are classified according to the properties of the natural plant hormones as auxins, gibberellins, cytokinins or inhibitors (3). Auxins cause elongation of plant cells. Gibberellins are compounds that stimulate cell division, cell enlargement, or both. Cytokinins stimulate cell division and differentiation. Inhibitors are a diverse group of plant growth regulators that inhibit or retard physiological or biochemical processes in plants. The distinction between plant growth regulators and herbicides is often only a matter of dose rate as many herbicides (e.g. 2,4-D) also have auxin-like properties.

Many of the limited range of currently commercially available plant growth regulators (4) are intended to modify the crop plant in ways similar to those achieved by plant breeding. Effects induced include shortening of plant height, encouraging branching, initiation or termination of the dormancy of seeds, buds and tubers, promotion or delay of flowering and plant maturation, leaf and fruit drop and enhancement of rooting and propagation. These plant modifications are designed to adapt the crop to modern cultural techniques where many of the requirements of traditional varieties (ability to compete against weeds and to assimilate nutrients from poor soils to give optimum yields) are not so important. In addition many of the uses of plant growth regulators are to modify the crop development to fit mechanized systems of operations, especially in harvesting. Defoliants and fruit abscission agents fall into this category. One of the features of many of the current commercial uses of plant growth regulators is that the effect is over a short period and thus a single spray application of the chemical is usually adequate. Thus the use of controlled release approaches in these situations is unlikely to lead to any improvements. Certain uses of current plant growth regulators could be extended by continued action. The use of inhibitors in controlled release form could be used for long-term maintenance of turf and ornamentals (5) although the present use of maleic hydrazide and daminozide is already economical. Similarly, in agricultural situations growth inhibition such as stem shortening in cereals, by chlormequat, provides season-long effects with a single application and controlled release delivery systems could not improve on present formulations.

Greater potential for controlled release is likely to exist through plant stimulation rather than inhibition. Substances for this purpose could be existing auxins, gibberellins or cytokinins, including both natural hormones or synthetic analogs. Examples of the wide range of structures available in this latter group are given in Figure 1, many of which offer opportunity for formulation by covalent bonds.

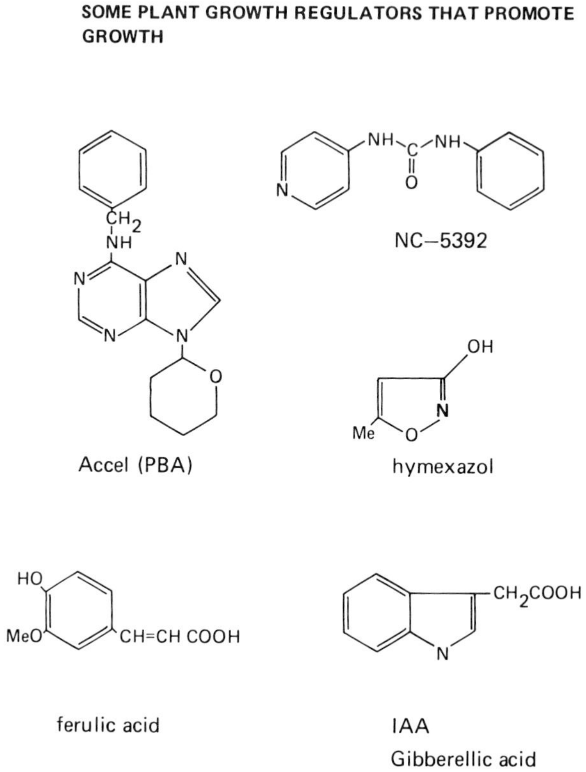


*FIGURE 1. Some plant growth regulators that stimulate growth.*

Auxins, gibberellins and cytokinins can promote growth, either individually or in concert but the cytokinins are likely to have a less deletarious effect on the plant should the concentration of released chemical exceed the safe limit. However, field evidence has demonstrated that long-term effects of 2,4-DB, an auxin-like herbicide, can have quite considerable and beneficial stimulation on growth of Douglas fir (5).

Cytokinins have a range of regulatory effects on plants (4) including:

1. stimulation of the doubling of nuclear DNA and cell division;
2. elongation of stem sections;
3. leaf enlargement;
4. inhibition of root growth;
5. differentiation of buds in callus tissue, leaves, roots, cotyledons or stem pieces;

6. chloroplast development;
7. germination and breaking of seed dormancy, and
8. delay of leaf senescence and nutrient mobilization.

Possibly the most useful effect for continuous action is the delay in the senescence of leaves. The breakdown of chlorophyll is inhibited and the photosynthetic activity of the leaves is maintained. The potential for cytokinins as senescence inhibitors has been recognized for some time. The synthetic cytokinins Accel (see Figure 1) and 6-benzyl-aminopurine (benzyladenine (BA)) have been used to delay senescence in fruit and vegetables in storage (7). In the field, spray applications of kinetin (Figure 2) and BA to rice plants at different stages of growth have retarded leaf senescence and increased grain yields (8). This use of senescence inhibitors in the field could be optimized if a controlled action of the chemical could be obtained.

As a result of the limited stability of the natural cytokinins when sprayed on the plant the related synthetic compounds based on the purine nucleus have been developed. It is doubtful whether these natural compounds would have the required soil properties for uptake by plants from the root-zone. Controlled delivery to plants through the roots would mimic the natural production of cytokinins in the roots followed by movement to the leaves (9). However, chemically bonded combinations of BA and other cytokinins with starch and cellulose have been prepared with very low rates of

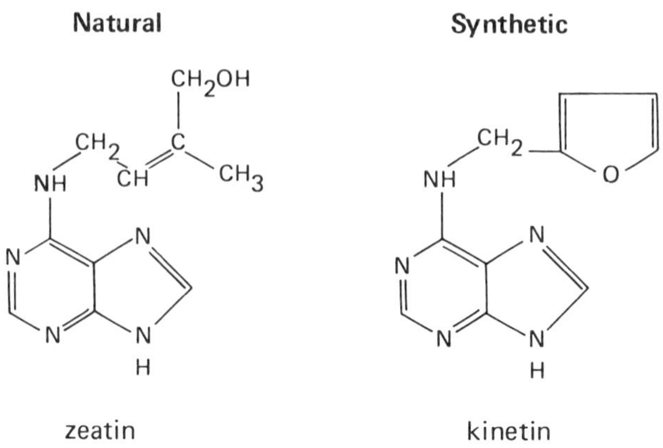


*FIGURE 2. The structure of cytokinins.*

substitution (10). Release of the active ingredient was followed using a soybean callus bioassay but no indication was given of release rates under biological conditions nor in the field.

An alternative approach is to exploit the anti-senescent properties of non-purine compounds such as the examples given in Figure 3. These have all been found to prevent senescence in detached leaves and at the same time may not possess all the other properties of the natural cytokinins. Many of these compounds have potential for controlled release formulation.

## III. BENZIMIDAZOLES AS SENESCENCE INHIBITORS

The structural similarity between benzylaminobenzimidadole (Figure 3) and 6-benzylaminopurine (BA) is clear and accounts for the cytokinin-like activity of the former compound. This activity is conferred on many other benzimidazole derivatives including those compounds that find current commercial use as broad spectrum systemic fungicides. The anti-senescent activity of the benzimidazole fungicides was observed in the field from the 'greening' or 'cosmetic'

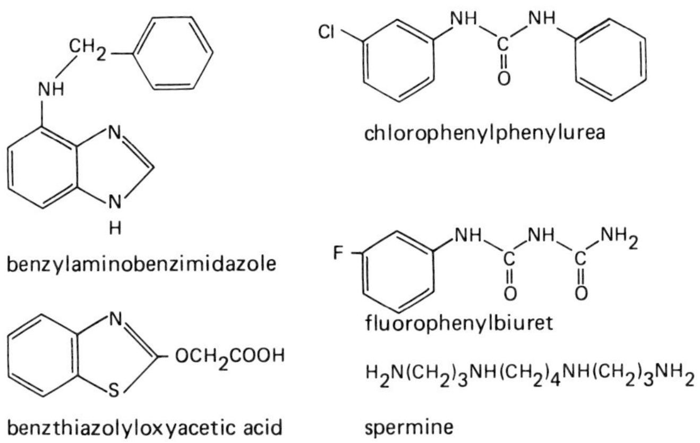


*FIGURE 3. Non-purine compounds possessing cytokinin-like activity.*

effect in treated crops. Sometimes yield increases were noted in the absence of disease symptoms (11). These effects were especially notable with benomyl (methyl 1-[(butylamino) carbonyl]-1 H-benzimidazol-2-ylcarbamate) and carbendazim or MBC (methyl 1H-benzimidazol-2-yl carbamate) from their use for disease control in cereals (see Figure 4). As other fungicides (e.g. zineb) have also increased the yields in wheat in the absence of diseases, this effect could be due to the suppression of saprophytic and weakly parasitic leaf fungi (12). However, carbendazim and benomyl have demonstrated anti-senescent activity in a range of laboratory tests, specifically the inhibition of chlorophyll decomposition in leaves (13-15).

In addition to their systemic movement in plants, fungicidal and anti-senescent activity benzimidazole derivatives can control plant mites, repel aphids and effect plant viral diseases. Some derivatives (thiophanate-ethyl) have anti-helminthic properties in mammals. Benomyl and carbendazim (and also the benzimidazole fungicides such as thiabendazole and thiophanate methyl) have been approved for agricultural use and are commercially available. These systemic fungicides can be taken up by the roots or leaves of plants and thus should be appropriate candidates for controlled release

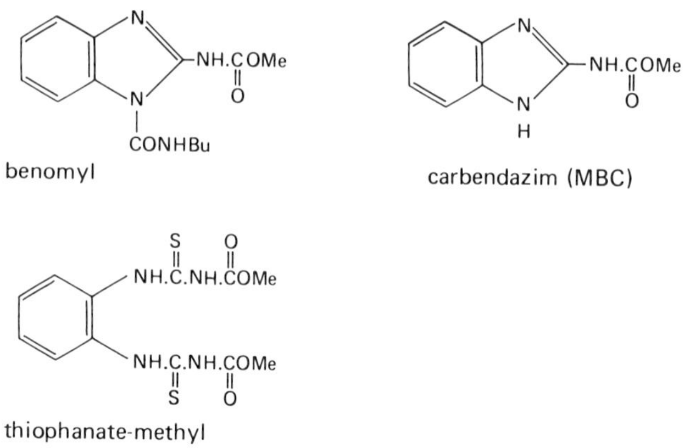


*FIGURE 4. Benzimidazole fungicides. Benomyl and thiophanatemethyl are converted to MBC in water and in plant tissue.*

delivery systems. With these factors being favorable, it was decided to investigate the potential of controlled release forms of carbendazim and benomyl for long-term senescence by simulated field trials.

## IV. SIMULATION OF CONTROLLED ACTION IN THE FIELD

The advantages of new controlled release applications have often only been realized following the development of a suitable formulation. Thus, the existence of a formulation with the appropriate release characteristics has been a prerequisite to proving the effectiveness of the continued action of the bioactive agent. The sequence in this mode of problem solving is typically to:

i. identify a problem that may be amenable to a controlled release approach;

ii. select appropriate bioactive materials;

iii. develop a suitable controlled release formulation;

iv. test the formulation under simulated field conditions, e.g. greenhouse, wind-tunnel or beaker, and to

v. finally field test.

Frequently the different steps in this procedure are carried out by different research groups and the use of this sequence has been encouraged by the availability of controlled release systems (e.g. microencapsulation) that can have a wide range of applications.

In order to establish that a benefit exists by a continuous or controlled action of a bioactive material, it can be advantageous to simulate the effect of a controlled release formulation under field conditions. Such a simulation under experimental conditions could comprise repeated applications of an appropriate dose rate in order that the threshold concentration at the target organism is maintained. Having established the optimum dosage and period of control then suitable formulations with appropriate release characteristics can be prepared and tested. In situations where the biological response to long-term release is not known, this simulated approach can be more effective than the sequence of steps as outlined before, where the formulation is designed and prepared prior to any field evaluation. However, if a particular controlled release formulation is available then it can be field tested in as wide a range of situations as possible.

The benefit of controlled release formulation of agrochemicals in improving crop-pest selectivity, can also be tested by simulation in the field. Lack of selectivity of a pesticide may proscribe its use in single conventional applications. For example, the herbicide ethofumesate at normal dosages of 1-2 kg a.i./ha gives excellent control of weed grasses (*Poa annua*) but severely damages some fine turf grasses (*Agrostis spp.* and *Festuca rubra ssp. commuta*) in New Zealand (16). Three sequential treatments of ethofumesate at 0.25 kg a.i./ha gave safe and aesthetically acceptable control of the weed grass. The continuous action of a low selective dose is not only more economical in herbicide, but allows the desired grasses to suppress the weed grass. In this type of situation, the preliminary field experiment simulating continued action must be a prerequisite to development of the controlled release formulation.

## V. SIMULATED FIELD TESTING OF BENZIMIDAZOLES FOR CONTROLLED RELEASE SENESCENCE INHIBITION

To assess the anti-senescent effect of the fungicidal benzimidazoles, benomyl and carbendazim, in a field situation the crop should be substantially free of disease. In these conditions, the fungicidal action will not contribute to any increased yields. For a preliminary investigation, winter wheat was chosen. In wheat the components of yield largely depend on the flag leaf (17) and through continuous delay of senescence of leaves leading to the flag leaf, a grain yield increase may be obtained. The low disease levels in winter wheat in northern Britain during 1976 provided an excellent opportunity to assess the concept of controlled release senescence delay.

## VI. EXPERIMENTAL

Field experiments were conducted at the Cockle Park Experimental Farm of the University of Newcastle upon Tyne. A range of varieties of winter wheat were used to assess the general applicability of the concept. The varieties used are given in Table I.

The simulation of the controlled release regime was achieved by repeating the spraying of benomyl and carbendazim at 2-weekly intervals throughout late spring and summer of 1976. Details of the treatments are in Table II.

Both the early standard and late standard treatments are the current commercial timings for disease control.

*TABLE I. Winter Wheat Varieties Used*

| | | |
|---|---|---|
| *Anouk* | *Feldkrone* | *B6221/9/2* |
| *Armada* | *Flanders* | *CM 221* |
| *Atou* | *Guilleman 107* | *TJB 259/83* |
| *Benoist 5308* | *Maris Hobbit* | *TJB 370/491* |
| *Cappelle Desprez* | *Maris Huntsman* | *UCT 39/3* |

The treatments were allocated in a random block design covering 21 variety strips. The plot size was 2 m x 5 m with a 0.5 m discard between adjacent plots. All treatments were made at 0.56 kg a.i./ha (the recommended commercial rate) in 280 l/ha water (for the first 3 sequential applications) or 550 l/ha using the 50% wettable powder formulations Benlate and Bavistin. A backpack sprayer fitted with a 2 m boom carrying 6 fan-jet nozzles was used.

### *A. Assessments*

Disease levels in the crop stand were low throughout the season. The principal diseases, powdery mildew, Septoria and eye-spot, occurred only at very low incidence levels. Visual assessments were made of powdery mildew and Septoria while eye-spot counts were made by pulling tillers at random. The low incidence of eye-spot was confirmed prior to harvesting by a count of white-heads. Visual assessments were also made (8 June and 22 July) on the appearance of the crop stand.

*TABLE II. Treatments and Timing*

| *Treatment* | *Crop growth state* | *Date* |
|---|---|---|
| *Early standard* | *2 node* | *13 May* |
| *Late standard* | *flag leaf erect* | *16 July* |
| *Repeated spray* | *7 x 2-weekly intervals 2 node to flowering* | *13 May-6 August* |
| *Untreated* | --- | --- |

The plots were harvested using a Claas Compact combine harvester with a 2 m swath and fitted with an on-board balance. Discards were cut and removed beforehand and individual plots were harvested taking the usual precautions. Samples of grain from each plot were taken to assess moisture content, thousand grain weight, and grain appearance.

## VII. RESULTS AND DISCUSSION

Assessment of the appearance of the wheat at the later date did show that the plots receiving the repeated spraying were 'greener' than the standard single treatments or the untreated checks. The visual assessment scores in Table III indicate the 'cosmetic' effect on the crop. A contribution to the appearance of the crops may also have been made by cereal aphids which were affected by the treatments (18).

The grain yields converted to a standard moisture content of 15% are summarized in Table IV.

A yield increase for the simulated controlled release approach is indicated and when compared to the mean of all the other treatments (both standards and untreated), the repeated benomyl application gave a yield increase of 8% and the repeated carbendazim treatment an increase of 7%. These are mean increases for a wide range of varieties, and could be considered indicative of a larger number of trials. Certain varieties gave high responses while others responded very little. Although this experiment is limited to one growing season, and thus must be treated with caution, the results are promising for the potential value of a controlled release senescence inhibitor. Where diseases are a problem, of course, yield increases due to disease suppression alone will be important and useful.

*TABLE III. Crop Stand Appearance: Visual Scoring*

| *Treatment* | *Mean Scores*[a] *benomyl* | *carbendazim* |
|---|---|---|
| *Early standard* | *3.0* | *3.4* |
| *Late standard* | *3.3* | *3.3* |
| *Repeated spray* | *4.7* | *4.1* |
| *Untreated* | *3.4* | *3.4* |

[a] *Mean of scores based on 1-5 scale: 1=dull, 5=green.*

*TABLE IV. Harvest Grain Yields*

| *Treatment* | *Yield t/ha* | *% Yield compared to untreated* | *% Yield compared to all other treatments* |
|---|---|---|---|
| *Benomyl:* | | | |
| *Early standard* | *7.70* | *102* | *--* |
| *Late standard* | *7.52* | *99* | *--* |
| *Repeated spray* | *8.21* | *108* | *108* |
| *Untreated* | *7.58* | *--* | *--* |
| *Carbendazim:* | | | |
| *Early standard* | *7.31* | *96* | *--* |
| *Late standard* | *7.42* | *97* | *--* |
| *Repeated spray* | *8.00* | *105* | *107* |
| *Untreated* | *7.64* | *--* | *--* |

For this experiment, the dose rate for the repeated treatment was set at the same as for the single treatments, i.e. the commercial rate of 0.56 kg a.i/ha. Thus, the cumulative total for the repeated treatment was 3.92 kg a.i/ha. This has established the safety of the treatment to the crop but the increase in crop yield certainly does not offset the costs. Further trials are needed to determine a lower optimum dose rate and if yield increases are consistent, then appropriate controlled release formulations could be prepared.

All plots treated with carbendazim showed a slight increase (6%) in mean moisture contents when compared to the mean of untreated and all the benomyl treatments. The moisture contents of grain treated with benomyl were not significantly different from the untreated grain. Thousand grain weights showed no appreciable change according to treatment. This provides additional support for the lack of disease during the late growing period.

## VIII. CONCLUSIONS

In order to identify new areas of application of controlled release to agriculture and related fields, it may be advantageous to test crop response by simulating the effects of the proposed formulation. This can be done by repeated conventional applications such as spraying. This approach is the most suitable means of identifying potential advantages of controlled release plant growth regulators, in addition to pesticides. To test this approach and to evaluate the use of controlled release senescence inhibitors, repeated spray treatments of winter wheat with the benzimidazole fungicides, benomyl and carbendazim, have given yield increases when compared to single applications. Although these yield increases may not be economically justified in the present context, this approach has demonstrated that controlled release formulations for plant growth regulators could have considerable potential.

## ACKNOWLEDGMENTS

The author wishes to thank Professor G.E. Russell and also Mr. David Rivett of the National Institute of Agricultural Botany for their considerable help.

## REFERENCES

1. H. T. Dellicolli, in Controlled Release Pesticides, H. B. Scher, (Ed.), American Chemical Society, Washington, D.C., 1977, p. 84.
2. P. J. Kemp, PANS Plant Growth Regulator Index, PANS, 25, (2) 211 (1979).
3. R. J. Weaver, in Plant Growth Substances in Agriculture, Freeman, San Francisco, 1972.
4. L. G. Nickell, Chem. & Engr. News, October 9, 18 (1978).
5. A. E. Eimert, Hort. Science, 11 (32), 374 (1976).
6. G. G. Allen, *et al.*, in Proc. Controlled Release Symposium, Controlled Release Society, New Orleans, Louisiana, August 6-8, 1979, p. 3.20.
7. D. T. Dennis, M. Stubbs, and T. P. Coultate, Can. J. Bot., 45, 1019 (1967).
8. A. K. Biswas and M. A. Choudhuri, J. Nucl. Agric. Biol., 7 (1), 14-18 (1978).
9. J. Van Staden and J. E. Davey, Plant, Cell and Environ., 2, 93 (1979).
10. S. Bittner, I. Perry, and Y. Knobler, Phytochem., 16, 305 (1977).
11. J. E. E. Jenkins, Proc. 7th Brit. Insecticide and Fungicide Conf., 1973, p. 781.

12. C. H. Dickinson and P. R. Walpole, Expl. Husb., 29, 23 (1975).
13. B. Staskawicz, Pestic. Biochem. Physiol., 8, 106 (1978).
14. R. G. M. Skene, J. Hort. Sci., 47, 179 (1972).
15. P. S. Weaver, Rep. Agric. Hort. Res. Stn. Univ. Bristol, 1971, p. 149.
16. J. R. F. McLean, Proc. Weed and Pest Control Conference, 1977, p. 192.
17. G. N. Thorne, in The Growth of Cereals and Grasses, F. L. Milthorpe and J. D. Ivins, (Eds.), Butterworths, London, 1966, p. 88.
18. K. W. Bailiss, G. A. Partis, C. J. Hodgson, and E. V. Stone, Ann. App. Biol., 89 (9), 443 (1978).

# CHEMODYNAMICS AND ENVIRONMENTAL TOXICOLOGY OF CONTROLLED RELEASE ORGANOTIN MOLLUSCICIDES

*Nate F. Cardarelli*
*William Evans*

Environmental Management Laboratory
University of Akron
Akron, Ohio

*The efficacy of controlled release trialkyl organotin molluscicides has been amply demonstrated under field conditions at over 40 test sites in Africa, Asia and South America (1,2). Cost/benefit ratios appear to be extremely favorable. The major impediment to large scale usage of the three commercially available materials is the need to establish benefit/risk ratios for various environmental conditions. Basically the need is to assess the degree of environmental hazard involved. It is believed that the organotins, under foreseeable use conditions, impose less ecological burden than conventional molluscicides now in extensive use. Our present task is to supply convincing evidence to the World Health Organization and a multitude of public health agencies that: 1. Controlled release molluscicides are safe in toxicological terms, and 2. That the destruction of the snail host is a superior approach to disease intervention than the use of highly hazardous drugs in medical therapy.*

*Since 1965 the number of Schistosomiasis victims has increased from around 200 million to present estimates of over 350 million humans with several million deaths annually and an incalculable economic burden. Obviously medical treatment and conventional usage of molluscicides is ineffective.*

*This report summarizes the known chemodynamics and toxicological information regarding controlled release bis (tri-n-butyltin oxide, "TBTO") and tributyltin fluoride "TBTF", molluscicides and indicates areas where further information is desirable.*




ISBN 0-12-074450-3

## I. INTRODUCTION

Controlled release molluscicides consist of an agent monolithically dispersed in a polymeric matrix. TBTO and TBTF exhibit considerable solubility in select elastomeric materials—natural rubber being a solvent of choice due to its biodegradability. Upon immersion in water solute molecules are slowly emitted from the dispenser surface through a diffusion-dissolution mechanism described elsewhere (2). Dispenser geometry may take any form normal to rubber and plastic commodities—pellets, granules, powders, strips, sheets, cords, tubes, etc., thus providing far greater versatility than presently available with conventional molluscicides. Release rate at the dispenser surface is moderated by additives that vary the diffusion path length. Commercially available materials of this type include BioMet$^{tm}$ SRM, 6% TBTO in natural rubber; a product of the M & T Chemical Co. (Rahway, New Jersey) and CBL-9B, 20.3% TBTF in natural rubber, available from the Creative Biology Laboratory (Barberton, Ohio).

In contrast Ecopro$^{R}$ 1330, 30% TBTF in a plastic base emits it's agent through a leaching mechanism, long term efficacy being developed through the use of a secondary polymer that increases free volume and a porosigen coleachant. This material is available in quantity from the Environmental Chemicals Co. (Barrington, Illinois). The principles underlying the use of a porosigen are described by Cardarelli and Walker (3), and the use of a secondary polymer to modulate free volume is discussed by Cardarelli (4).

Conventional mollusciciding operations rely upon usage of a relatively massive agent dosage. This "overkill" is essential in that a lethal quantity of the toxicant must be imparted to each target snail prior to natural detoxification processes reducing the active concentration below the 100% mortality threshold. It is not practical in most situations to maintain a continuous toxicant concentration in the treated water higher than about 0.05 ppm through controlled-release techniques—assuming that several years of protection are expected with one application. Consequently, controlled release methodology is keyed to the continuous exposure of the target species to ultralow organotin concentrations and the consequent development of a terminal chronic intoxification. Obviously, non-target hydrobionts are exposed for a prolonged duration as well.

## II. ENVIRONMENTAL CHEMODYNAMICS

### A. *Transport Dynamics*

Destruction of the target snail requires that toxicant molecules migrate through the polymeric matrix of the dispensing unit, pass into the interfacial zone, traverse a water path, reach the site of snail contact, enter the snail and accumulate at sites of cidal activity. Physical and chemical barriers interpose at each segment of the trajectory. Watercourse constituents may permeate the dispenser and/or accumulate on its surface thus chemically or physically blocking agent egress. Once released the organotin molecule may chemically react with water constituents such as inorganic ions, or be absorbed on suspended matter and thus detoxified or removed from snail contact. Upon reaching the target the toxic molecule must overcome both active and passive defense mechanisms inherent in molluscan physiology, traverse snail tissue and reach, as a toxicant, specific sites where interruption of one or more vital processes can occur.

### B. *Organotin Emission*

The rate of organotin release from a given formulation under varying water quality conditions has not been determined with accuracy by strictly chemical means at the ultralow concentrations involved. TBTO solubility does not exceed 3 ppm, although colloidal suspension results in the 3 to 36 ppm range, so that water concentrations as high as 36 ppm are possible. TBTF solubility is probably less than 1 ppm. Pellet loss based upon Sn determination after a given immersion period is not more than 0.6 $\mu g/cm^2$-day for TBTO (BioMet$^{tm}$ SRM) and 0.4 $\mu g/cm^2$-day for TBTF (CBL-9B). Conventional analytical techniques used with organotins at low concentrations do not speciate - but rather are based upon inorganic tin analysis. Thus no separation has been made between the active agent and various degradants of less toxic activity. In most natural conditions, distinction between the xenobiotic and the normal tin background would be impossible at the molluscicidal concentration range using known analytical techniques. Recently Sherman has demonstrated that a modified phenylfluorone technique can be used with reliability to 50 parts-per-trillion and a ± 5% error at 1 part-per-billion (5). Positive separation of inorganic from organic tin is achieved. The merit in this technique was not confirmed until June 1979 and consequently insufficient time was available to establish release rate curves.

In an earlier study[1], 1-$C^{14}$ labelled TBTO was used in the preparation of BioMet SRM pellets (6). Sample pellets were immersed in water and the TBTO emission rate (assuming that there is no decomposition of the agent within the emitter) determined by liquid scintillation chromatography as depicted in Figure 1. Release equilibrium is achieved in about 14 days, the value being 0.58 ± 0.03 micrograms/$cm^2$-day in demineralized water of pH 6.8. The initial peak loss and rapid drop to a constant release rate is typical of controlled release formulations (2).

### *C. Water Quality Effects*

It is well known that water constituents significantly influence the activity of conventional molluscicides such as the salts of niclosamide, trifenmorph, copper sulfate and the pentachlorophenates. In order to assess the effect of water quality on the efficacy of BioMet SRM and CBL-9B, a series of bioassay studies were performed[2]: *Biomphalaria glabrata* snails were exposed to varying concentrations of organotins arising from dispenser emission in water containing set levels of inorganic ions and suspendents. Table I depicts "standard" mortality rates in demineralized water. Results of replicated tests under varying water quality conditions are shown in Table II.

Inorganic ion content was adjusted using the sodium or potassium salt of the given anion. pH was maintained using NaOH or $HNO_3$, the nitrate ion showing little influence on mortality. It was recognized that in the standard bioassay using 10 snails per replicate in 1 liter of water, the pH gradually changes with metabolic exudates, decomposing lettuce (used as food) and the like. Typical methods of buffering, such as using mixed phosphates, fails due to toxic effects on the test snail. The presence of *B. glabrata* gradually lowers alkaline pH and raises acidic pH towards an apparent preferred 6.6 value. The change is slow and seldom exceeds ± .04 pH units in a given 7 to 10 day study.

[1]*Project sponsored by the Edna McConnel Clark Foundation, New York, NY, U.S.A., Grant No. 276-0091.*

[2]*Work sponsored by the U.S. National Institutes of Health, Bethesda, MD, Grant No. AI11861-01A1 1975/1766 (7).*

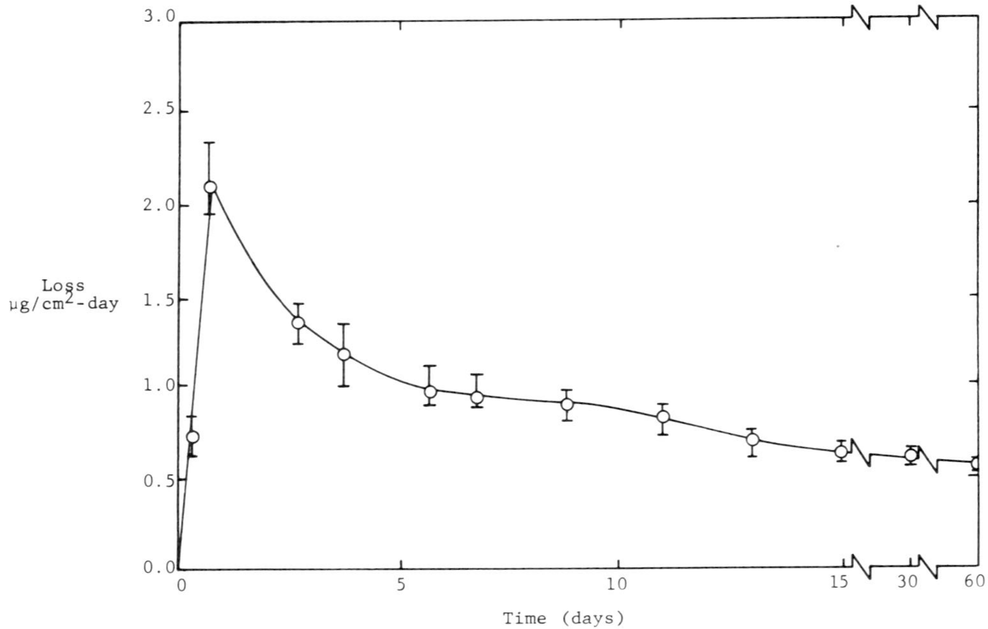


*FIGURE 1.* $C^{14}$ *labelled TBTO emission from BioMet*® *SRM (replicated samples).*

*TABLE I. LT Values for Several Snail Species Using Demineralized Water*

| | | BioMet SRM[a] | | CBL-9B[a] | |
|---|---|---|---|---|---|
| | | $LT_{50}$ | $LT_{100}$ | $LT_{50}$ | $LT_{100}$ |
| Species | pH | days | days | days | days |
| *Biomphalaria glabrata* | 6.0 | 4.5[b] | 6[b] | 1.5[b] | 3[b] |
| | 7.0 | 2.5 | 3 | 2.5 | 4 |
| | 8.0 | 1.5 | 3 | 4 | 6 |
| *Bluinus globosus* | 6.0 | 4 | 8 | 3.5 | 7 |
| | 7.0 | 3.5 | 7 | 3 | 8 |
| | 8.0 | 3 | 6 | 4 | 8 |

[a]100 ppm pellet (6 ppm total active agent BioMet SRM, 20 ppm total active agent CBL-9B).
[b]Average of three replicates of ten snails each.

The inorganic ion concentrations used were at the maximum level expected in natural waters.

Preliminary bioassay indicated that the materials used in preparing the various water systems were not in themselves toxic.

It was observed that the effect of the various water constituents examined was to accelerate or decrease the rate of kill - the final result being the same - 100% mortality. Since the agents involved are non-toxic, the shifting of the $LT_{100}$ value could be due to chemical interference, purgation by organic matter so that effective concentration is reduced, or changes in molluscan physiology and thus physiological response to a xenobiotic. Note that BioMet SRM activity is greater in alkaline water and CBL-9B in acidic.

Organic materials such as yeast and humic acid over a 1 to 100 ppm concentration range are not toxic to snails, and in systems treated with organotins do not appreciably effect LT values. This is not indicative of a lack of chemical interaction - in that yeast may absorb the agent, but snail contact is maintained by ingestion of the resulting ligands. CBL-9B and BioMet SRM were also examined in systems containing 100 ppm of several suspended colloidal clays. Kaolin

*TABLE II. The Effect of Various Inorganic ions on the $LT_{100}$ for Biomphalaria glabrata Exposed to Controlled Release Molluscicides*[a]

| | | $LT_{100}$ (days) | | | | | |
|---|---|---|---|---|---|---|---|
| | | BioMet[R] SMR[b] | | | CBL-9B[b] | | |
| Test materials | Concentrations (ppm) | pH = 6.0 | 7.0 | 8.0 | 6.0 | 7.0 | 8.0 |
| $K^+$ | 30 | 6 | 7 | 11 | 5 | 4 | 7 |
| $Na^+$ | 300 | 6 | 5 | 6 | 3 | 5 | 5 |
| $Ca^{++}$ | 600 | 3 | 4 | 5 | 1 | 3 | 5 |
| $Mg^{++}$ | 125 | 6 | 7 | 12 | 5 | 4 | 9 |
| $Sr^{++}$ | 300 | 4 | 6 | 8 | 5 | 4 | 8 |
| $Fe^{++}$ | 10 | 6[c] | - | - | 5[c] | - | - |
| $SO_4^{=}$ | 570 | 8 | 4 | 5 | 4 | 4 | 5 |
| $Cl^-$ | 150 | 6 | 8 | 12 | 5 | 4 | 7 |
| $SO_3^{=}$ | 10 | 5 | 7 | 6 | 5 | 6 | 6 |
| $CO_3^{=}$ | 350 | 6 | 6 | 6 | 4 | 4 | 5 |
| $NO_3^-$ | 300 | 6 | 5 | 7 | 4 | 4 | 6 |
| $NO_2^-$ | 0.5 | 5 | 7 | 7 | 5 | 6 | 6 |
| $PO_2^{\equiv}$ | 100 | 8 | 5 | 7 | 6 | 6 | 9 |
| $SiO_3^{=}$ | 30 | 9 | 6 | 10 | 6 | 5 | 8 |
| $F^-$ | 7 | 5 | 6 | 7 | 5 | 5 | 6 |

[a]Each experiment was replicated five times using ten adult snails per replicate.
[b]100 ppm pellet conc. (6 ppm TBTO total active agent and 20 ppm TBTF total active agent).
[c]Ferrous ion can only be maintained in solution in the acidic pH range.

and four montmorillonites, varying in their respective cation exchange capacities, were used. The presence of relatively high levels of extremely absorptive surfaces does not significantly alter the mortality curve. It is likely that the snail ingests such materials during browsing. Table III summarizes the $LT_{100}$ values for suspendents and organic materials.

Results of water quality studies are summarized in Table IV.

TABLE III. Effect of Inorganic Suspendents and Organic Matter on the Mortality of Biomphalaria glabrata Exposed to Controlled Release Organotins[a]

| Material | Conc. (ppm) | LT$_{100}$ (days) BioMet SRM pH = 6.0 | 7.0 | 8.0 | CBL-9B 6.0 | 7.0 | 8.0 |
|---|---|---|---|---|---|---|---|
| Kaolin | 100 | 6 | 6 | 5 | 5 | 6 | 4 |
| Montmorillonite A[b] | 100 | 4 | 5 | 5 | 5 | 6 | 4 |
| Montmorillonite B[c] | 100 | 5 | 6 | 6 | 4 | 6 | 4 |
| Montmorillonite C[d] | 100 | 4 | 5 | 7 | 5 | 5 | 5 |
| Montmorillonite D[e] | 100 | 5 | 6 | 6 | 4 | 6 | 4 |
| Yeast[f] | 100 | 5 | 7 | 5 | 6 | 7 | 6 |
| Humic acid[g] | 100 | - | - | 10 | - | - | 6 |

[a]10 ppm pellet conc. all tests performed in replicates of five with 10 adult snails per test.
[b]Bentolite L. (Georgia Kaolin Co., Elizabeth, NJ) 80 meg/100g, 71.7% $SiO_2$, 15.7% $Al_2O_3$.
[c]Gelwhite G.P. (Georgia Kaolin Co.), 100 meg/100 g, 58.6% $SiO_2$, 21.2% $Al_2O_3$.
[d]Thixo-Jel Bentonite (Georgia Kaolin Co.), 100 meg/100 g, 66.8% $SiO_2$, 16.4% $Al_2O_3$.
[e]Mineral Colloid B.P. (Georgia Kaolin Co.), 90 meg/100 g, 55.7% $SiO_2$, 24.4% $Al_2O_3$.
[f]Brewers yeast, 100 ppm initial concentration. Since yeast is a living organism, concentration increases with time under normal circumstances. However, organotins are moderately toxic to yeast cells and thus a complex concentration schema is present. Photometric determination of true concentration was not possible due to interference from snail by-products.
[g]Humic acid can only be solubilized in the alkaline range.

*TABLE IV. Variation in $LT_{100}$ Value with Changes in Water Constituents*

| | | $LT_{100}$ Variance from standard | | | |
|---|---|---|---|---|---|
| | | BioMet SRM | | CBL-9B | |
| Category | pH = | acidic | alkaline | acidic | alkaline |
| Inorganic ions | | No effect | Increase | No effect | Increase |
| Suspended organic | | No effect | No effect | No effect | Decrease |
| Soluble organic | | -- | Increase | -- | No effect |
| Suspended inorganic | | No effect | Increase | No effect | No effect |
| pH | | Increase | Decrease | Decrease | Increase |

The effect of water quality on organotin induced snail mortality is apparently significantly less than that reported for conventional molluscicides. The dramatic detoxification effect of $CaCO_3$, mud and yeast cells on the ethanolamine salt of niclosamide has been noted (8,9). Meredith (10) further observes that the decrease in niclosamide activity is related to the total exchange capacity of the soil in question. At $CaCO_3$ concentrations of 0.1%, Da Souza *et al.* (8) reports that ethanolamine niclosamide activity decreases 23% and trifenmorph drops by 60%. Similarly copper ion dosages used for snail control are dependent upon water quality factors (11). Fulvic, humic and amino acids complex copper. Turbid matter and water hardness detoxify copper ion, both predominating over the also significant pH effect.

### D. Absorption of Organotins

Organotins are extremely hydrophobic and rapidly absorbed or adsorbed from the water course by bottom soil, algae, vascular plants and the like. Trialkyl and triaryl tins absorb on soil particles forming complex ligands. The Freudlich isotherm K values for TBTO and TBTF range from 2000 to 12,000 depending upon soil type, indicating extremely strong bonding (12,13). Typical values for other pesticides absorbed by similar type soils range from 10 to several hundred.

Lysimetry using radiolabelled organotins indicated very small organotin concentrations in the leachates, 0.01% to 0.2% over a 12 to 18 inch leach path (13). High organic content soils display the greatest absorbency followed by clay loams and sandy loams in that order. Schatzberg and Harris report that TBTO movement away from the point of ejection is mostly in the vertical direction in sandy soils whereas horizontal migration predominates in clay and organic top soils (14). Soil absorption is dramatically evident in BioMet SRM pellets and other materials returned after seven months immersion in natural test sites in Rhodesia (15). It was found that:

1. Mud deposits on BioMet SRM pellets contained 0.24 to 0.58 ppm tin.
2. Mud samples taken just below such pellets contained 0.09 to 5.07 ppm tin.
3. Both clear bottom water samples and near surface water samples contained no detectable tin.
4. Mud contaminated bottom waters contained 0.18 ppm tin.

In order to determine the relative amounts of organotin in the water and bottom soil, BioMet SRM dispensers releasing $C^{14}$ labelled TBTO were placed in a water-soil system and water aliquots examined for Beta emission. After pellet removal, the water $\beta$ count dropped to background levels in 4 days, indicating a complete loss of soluble organotin into the bottom soil (or perhaps the glass walls of the test containers used). Using more sensitive techniques with the same method, Christie found that label completely disappeared from water in 16 days (16).

It is likely that an equilibrium condition exists between the rate of ejection from the dispenser and absorption in the soil with considerable dependence upon water quality and the type of soil involved.

In a number of similar experiments, the soil to water ratio for TBTO or TBTF appears to be approximately 15:1 to 19:1. Bioassay confirms this finding.

In laboratory systems where flowing water passed over BioMet SRM treated bottom soil, Shiff *et al.* (17) noted that browsing snails succumbed while snails isolated in the effluent were not affected. Two implications were deduced from these findings: 1) In a treated system, organotin toxicants mainly reside in the soil and intoxication occurs as target molluscs ingest soil particles during browsing activities, and 2) If the mode of contact is during snail-soil interaction, then dosage requirements for effective molluscicidal operations might be based upon bottom area rather than total water volume. Flow rate may be of minor importance. Evans (18) found that snails free to browse the bottom soil in 4 liter BioMet SRM treated aquaria succumbed at twice the rate of caged snails having no direct soil contact. Compartmentalized large aquaria were used in more elaborate studies. A two inch thick high organic content soil substrate was used with a test pellet of CBL-9B or BioMet SRM placed at one end. Placement was either on the water surface, midway in the water column or on top of the soil. Water circulation was continuously maintained. By placing target snails at various average distances from the dispenser site and observing mortality, the movement of a hypothetical toxicant front could be tracked. Furthermore, by allowing some snails in a given compartment to freely browse the soil while isolating others in the same compartment from soil contact, via siene type caging, the relative amounts of organotin in the water and the soil might be charted through both bioassay and periodic analysis of the soil and water. Caged snails were fed lettuce ad libitum. Table V illustrates one run of a series. Snail mortality and new ovipositing were observed daily.

A number of variables were evaluated in a series of similar experiments, including water depth, location of the dispenser, toxicity of the system after removal of the organotin pellet, soil toxicity and the like. It was apparent that the organotin, though emitted into the water, was rapidly absorbed by the bottom soil. Whether the point of emission was on the water surface, midway in the water or resting on the soil was not significant.

Analysis of soil and water using the Trachman *et al.* (19) method provided results shown in Table VI.

Aquaria bottom soils remain toxic to snails for relatively short periods of time - perhaps four weeks. This is not a question of continual downward percolation removing the toxicant from the snail browsing zone. Snails have been exposed to total soil columns taken as cores from aquaria and known to contain tin. Such aliquots are toxic for a short period of time and usually are non-toxic by day 12. Every test to date has confirmed the hypothesis that the organotin molecules contacting soil particles are absorbed and slowly form ligands of a non-toxic nature. It is also possible that the organotin is chemically altered to a non-toxic form.

Evidence strongly supports the contention that the predominant mode of snail contact with the toxic agent is through soil ingestion - provided that such particulate matter has been freshly exposed to organotin. Organotin solutes are present at ultralow concentrations within a treated water body over the release life of the dispensing pellet.

TABLE V. Bioassay of *B. glabrata* at Varying Distances from a Controlled Release Molluscicide Pellet

| Agent | Active conc. | Average distance pellet to snail | Soil browsing snails[a] | | Isolated snails[b] | |
|---|---|---|---|---|---|---|
| | | | $LT_{100}$ (days) | No. egg masses | $LT_{100}$ (days) | No. egg masses |
| CBL-9B | 2 ppm | 12 cm | 9 | 0 | 12 | 0 |
| | | 24 cm | 10 | 0 | 13 | 0 |
| | | 36 cm | 13 | 0 | 14 | 0 |
| | | 48 cm | 13 | 0 | 14 | 0 |
| | | 60 cm | 14 | 0 | 20 | 23 |
| BioMet® SRM | 0.6 ppm | 12 cm | 17 | 0 | 22 | 3 |
| | | 24 cm | 25 | 0 | 22 | 5 |
| | | 36 cm | 22 | 0 | 25 | 10 |
| | | 48 cm | 20 | 0 | 25 | 15 |
| | | 60 cm | 22 | 0 | 25 | 19 |

[a] 20 adult *B. glabrata* per compartment.
[b] 10 adult *b. glabrata* per compartment.

*TABLE VI. Analysis for Residual Tin in Soil and Water from Aquaria Systems, Containing Two Inches of Bottom Soil, Exposed for 28 Days to CBL-9B*[a]

| Pellet location | Water depth (cm) | Pellet dosage (ppm) | Residual tin: soil (ppm) | Residual tin: water (ppm) |
|---|---|---|---|---|
| Soil surface | 10.5 | 90.3 | 0.236 | 0.009 |
| | 21.1 | 45.2 | 0.136 | 0.012 |
| | 24.4 | 39.0 | 0.154 | 0.013 |
| Water surface | 10.5 | 90.3 | 0.181 | 0.014 |
| | 21.1 | 45.2 | 0.342 | 0.111 |
| | 24.4 | 39.0 | 0.436 | 0.014 |

[a]*Analytical work was performed by M & T Chemicals Inc., Rahway, New Jersey.*

Chu (20) examined BioMet SRM against *B. rohlfsi* in repeat challenge tests under various laboratory environments. Using $LD_{100}$ (24 hr.) criteria he notes that in water alone, 389 days longevity is achieved whereas with water + silt, 328 days and in water + mud, only 284 days.

## E. *Area Versus Volume Dosage*

If snail ingestion of organotin laden bottom soils is the primary mode of contact, then the area application dosage should correlate with the LT value whereas a volume dosage should not. In order to investigate this hypothesis several studies were performed[3]. In the initial test, BioMet SRM pellets were applied to the surface of a standard soil mixture, 2.5 cm depth, in a 40 liter aquaria with varying water depths. Fifty adult *B. glabrata* were used in each aquaria. Table VII depicts results. BioMet SRM dosage, based on pellet weight, was 0.072g TBTO total toxicant in each aquaria. Note that #3, having the greater water volume but the lesser mud surface

[3]*Work sponsored by the Edna McConnell Clark Foundation, New York, NY.*

*TABLE VII. BioMet SRM Bioassay B. glabrata at Constant Surface Dosage (0.056 mg/cm² TBTO)*

| | #1 | #2 | #3 | #4 |
|---|---|---|---|---|
| Total available toxicant | 5.4 ppm | 2.7 ppm | 2.3 ppm | 0.0 ppm |
| Mud surface area | 1257 $cm^2$ | 1257 $cm^2$ | 1257 $cm^2$ | 1257 $cm^2$ |
| Water depth | 10.4 cm | 20.8 cm | 24.1 cm | 14.5 cm |
| Water volume | 13.3L | 26.6L | 30.7L | 18.4L |
| $LT_{10}$ | 3.2 days | 3.2 days | 3.2 days | 10+ days[a] |
| $LT_{50}$ | 6.0 | 5.7 | 5.2 | -- |
| $LT_{90}$ | 9.0 | 7.0 | 7.1 | -- |
| $LT_{100}$ | 10.0 | 8.0 | 10.0 | -- |

[a]Pellets removed day 10 and water changed.

area, had a higher mortality rate whereas, #1 and #2, having the same surface area but different water volume, had nearly identical LT values.

Organotin pellets were removed on day 10 and 50 additional snails added to each aquaria to assess residual toxicity. The $LT_{100}$ values were 52 days for both aquaria #1 and #2 and 40 days for #3.

Water samples were taken from each aquaria on day 10 and used in replicated snail bioassay without the presence of soil. After 68 days, mortality was 43% in #1, 20% in #2, 27% in #3 and 18% in the control.

Fifty-two days after pellet removal, 90 g aliquots of mud from aquaria #3 were placed in 1000 ml of DM water and bioassayed. Snail mortality was essentially identical to that of the controls.

After 18 days pre-wash, BioMet SRM pellets were placed in fresh soil-water systems using identical dosages, based on area, but differing water depths. Fifty adult *B. glabrata* added and the results reported in Table VII were observed.

Repeated experiments showed no significant difference. Water is essentially non-toxic within two weeks, the bottom soil is toxic, but loses toxicity rapidly when the TBTO emitter is removed. It was thus indicated that dosage requirements may be predicated on the bottom surface area of a given water body rather than water volume, provided that the target species do indeed browse the soil.

### *F. Organotin Transmission Through Soil*

Organotins absorb on the mineral particulates of soil, Slesinger (13) noting that clay loam having 4.93% organic carbon content shows little distinction in adsorption coefficients over high organic content (33.3%) muck. He further notes that soils immobilize organotins. In lysimetry tests, downward movement of TBTO and other organotins was very low, less than 0.1% of the applied material appearing in the leachate. Schatzberg *et al.* (21) reports that all soils exhibit a high affinity to organotin. Once absorbed, such materials are not easily removed. Shaking organotin-bound soils in water for 24 hours and bioassay of the supernatent showed no toxic activity. In-house studies indicated that as little as 0.5 cm of silt over the dispensing pellet reduced snail mortality to 50% or less under standard test conditions.

### *G. Environmental Degradation of Organotins*

Trialkyl and triaryl organotins degrade readily under various environmental influences. Half-life values, for TBTO, depicted in Table VIII, vary in accordance with use conditions. Ritchie *et al.* (22) determined the chemical stability of TBTO as being over 8 and under 16 days.

TBTO, TBTF, and to a lesser extent, TPTF, are readily prone to degradation through hydrolytic and photolytic cleavage, microbial decomposition, and chemical activity with soil constituents. The mode of degradation will vary in accordance with environmental parameters.

Slesinger, *et al.* (23) measured the half-lives of these materials under various environmental conditions. Values are given in Table IX.

Mazaev *et al.* (24) notes that the hydrolysis of organotins at very low concentrations is in a practical sense a first order reaction following the general expression:

(1) $C_t = C_o e^{-bt}$ where $C_t$ is the organotin concentration at time t, $C_o$ the initial concentration and b a decomposition coefficient. Solving for half life

(2) $T_{\frac{1}{2}} = 0.693/b$

Values for b were measured and found to be 0.038 (TBTO in distilled water), 0.043 (TBTO in pond water) and 0.167 (TBTO in pond water with sand).

TABLE VIII. Half-Life of Bis (tri-n-butyltin) Oxide (24)

| Agent | Environment | $T\frac{1}{2}$ |
|---|---|---|
| TBTO | Distilled water | 18.2 days |
| | Pond water | 16.1 days |
| | Pond water + sand | 4.15 days |

Degradation rate and initial degradants formed depend upon the environmental situation over the biologically active life of the parent compound. Table X summarizes the initial degradant under several environmental conditions.

The rapid decay under natural conditions of the trialkyl and triaryl tins of molluscicidal importance has not been challenged in the literature. The basic questions raised concern the elucidation of the degradation pathways under various environmental conditions and the toxicology of each degradant. It is well recognized that trialkyltins lose biological activity as the organic groups are removed (32-34). With the exception of tributyltin carbonate, the sequential degradants become significantly less toxic. The dialkyltins show low level activity while the monoalkyl materials are essentially biologically inert.

The hypothesized environmental fate of TBTF and TBTO in an aqueous medium is shown in Figure 2.

In a consideration of TBTF degradation, Monaghan *et al.* (26) advances the following scheme based upon the very low concentrations involved and LeChateliers principle:

$$(C_4H_9)_3SnF + H_2O \longrightarrow (C_4H_9)_3SnOH + HF$$

$$2(C_4H_9)_3SnOH \longrightarrow (C_4H_9)_2O + H_2O$$

Work by Kochkin *et al.* (35) using polarographic techniques confirms that TBTO decomposes to monobutyltin hydroxide in water.

TBTO readily reacts with $CO_2$ in the air or in water to form the carbonate. Aqueous destannylation is slow. Slesinger and Dressler (23) note that TBTO conversion to dibutyltin - was 2 to 6% by post-immersion day 35 and TBTF 3 to 7%. Rates of dealkylination under varying environmental parameters are unknown, but very with pH. The Sn-C bond is relatively weak and readily cleaved through hydrolysis alone, or at an increased rate in the presence of ultraviolet

TABLE IX. Organotin Half-Lives Under Various Environmental Stresses (23)

| Chemical | Environment | Half-life (days) | Amount Degraded |
|---|---|---|---|
| TBTO | U.V. Photolysis (water) | 18.5 | -- |
| TBTF | U.V. Photolysis (water) | 2.0 | -- |
| TPTF[a] | U.V. Photolysis (water) | 54 | -- |
| TBTO | U.V. Photolysis (soil) | 21 | 68% (35 days) |
| TBTF | U.V. Photolysis (soil) | 17 | -- |
| TPTF | U.V. Photolysis (soil) | 37 | -- |
| TBTO (1 ppm) | Microbe culture | -- | 37% (126 days) |
| TBTO (10 ppm) | Microbe culture | -- | 23% (126 days) |
| TBTF (1 ppm) | Microbe culture | -- | 27% (126 days) |
| TBTF (10 ppm) | Microbe culture | -- | 18% (126 days) |
| TPTF (0.1 ppm) | Microbe culture | -- | 48% (126 days) |
| TPTF (0.5 ppm) | Microbe culture | -- | 58% (126 days) |
| TBTO | Soil | 70 | -- |
| TBTF | Soil | 50 | -- |
| TPTF | Soil | 15 | -- |

[a]TPTF = Triphenyltin fluoride.

*TABLE X. Primary Environmental Degradants*

| Parent | Environment | Primary Degradant | Reference |
|---|---|---|---|
| TBTO | Water | Dibutyltin oxide | 23 |
| | | Tributyltin carbonate | 35 |
| | | Dibutyltin hydroxide | 26 |
| | Soil | Dibutyltin oxicd | 23 |
| | U.V./water | Dibutyltin oxide | 23 |
| | Animal tissue | Dibutyltin-x | 27 |
| | | Tributyltin hydroxide | 28 |
| | Plant material | Tributyltin carbonate | 29 |
| TBTF | Water | Tributyltin hydroxide | 25 |
| | | Tributyltin oxide | 25 |
| TPTA[a] | Water | Diphenyltin acetate | 30 |
| | Soil | Diphenyltin-derivatives | 31 |
| | U.V./water | Diphenyltin acetate | 31 |
| TPTF | U.V./water | Diphenyltin fluoride | 23 |
| | Soil | Diphenyltin fluoride | 23 |

[a]Triphenyltin acetate

radiation. The ultimate decomposition product, $SnO_2$, is non-toxic and ubiquitous. Natural abundance varies from 1 to 250 ppm in soil, 2-3 ppm in igneous rock, and as Sn, 1 to 5 ppm in ocean sediments.

TBTO and TBTF are readily absorbed by organic matter. Mossbauer spectroscopy shows high substantivity of TBTO for cellulose bearing materials, possibly due to an in situ reaction with $CO_2$ to form polymeric organotin carbonates (25). U.V. radiation accelerates the reaction. Also organotins, such as TBTO, react readily with amino acids and fatty acids under laboratory conditions. It is thus expected that reactions would occur with proteinaceous material and lipids in the use environment. Such reactions tend to detoxify the organotin moiety (to be published). However, selective compounds may be somewhat molluscicidal - tributyltin laurate for example.

Analysis of the work of a number of researchers (13, 22, 24, 29, 33, 35-37) and in-house studies have led to the following composite degradation scheme, Table XI, under natural aquatic conditions.

Molluscicidal organotins would not follow a strictly chemical decomposition scheme in the use environment. TBTO, TBTF, TPTA and TPTF are subjected to biodegradation via microbial activity. Gram negative bacterial soil microbes and fungi act as degradants.

Biochemical degradation organotins applied to plants: triphenyltin acetate, triphenyltin chloride and triphenyltin hydroxide fungicides have been extensively studied (37). It is generally concluded that organotins do not penetrate the plant species involved, no translocation occurs and degradation is rapid, resulting in inorganic tin residues (38).

In vitro organotin degradation activity arises from successive dealkylation through hydrolytic cleavage of the fairly weak Sn-C bond via electrophilic or nucleophilic attack. Biochemical reactions in living organisms differ significantly. Studies indicate that destannylation occur through hydroxylation of carbon-hydrogen bonds (39-41). Organotin compounds are converted through a tetra-, tri-, di-, mono-sequence under biological oxidation conditions (42). With tributyltin derivatives, the α and β hydroxyl metabolites gave destannylated products. Fish (42) notes that tributyltin decomposes to dibutyltin acetate, butyl alcohol and 1-butene. Stable metabolites included γ-hydroxybutyldibutyltin chloride and δ-hydroxylbutyldibutyltin chloride, both being only slightly less toxic to mice than tributyltin chloride.

Alkyltin (IV) cations react strongly with N and O donors in living systems. At physiological pH $(C_4H_9)_3Sn^+$ and $(C_4H_9)_2Sn^{++}$ ions are hydrolyzed to $(C_4H_9)_3SnOH$ and $(C_4H_9)_2Sn(OH)_2$ with little tendency towards formation of polynuclear systems (43).

Concern exists as to the possible methylation of organic or inorganic tin in biological systems. In vitro methylation is not considered likely (44). In vivo methylation requires a complex series of reactions. $-SnX_3$ must homolytically cleave methylcobalamin and a succession of reductions occur (45,46). Pseudomonas bacteria will methylate inorganic tin. Since the amount of inorganic tin entering the biosphere from the degradation of organotin molluscicides would be extremely small compared to the naturally occuring Sn; this process should not be of concern.

A biomagnification factor of 170 has been observed for inorganic tin in aquatic plants exposed to industrial discharges (47). No evidence of similar organotin accumulations in plant life or animals has been reported.

$$(C_4H_9)_3SnF \xrightarrow{H_2O} (C_4H_9)_3SnOSn(C_4H_9)_3 \xrightarrow{H_2O/Cl^-} (C_4H_9)_3SnCl$$

$$(C_4H_9)_3SnOSn(C_4H_9)_3 \xrightarrow{H_2O/CO_2} (C_4H_9)_3SnOC(=O)OSn(C_4H_9)_3 \longrightarrow (C_4H_9)_3SnOH$$

$$(C_4H_9)_3SnOC(=O)OSn(C_4H_9)_3 \xrightarrow{UV/H_2O,O_2} (C_4H_9)_2SnO$$

$$(C_4H_9)_3SnOH \xrightarrow{UV/H_2O,O_2} (C_4H_9)_2SnO \underset{}{\overset{H_2O}{\rightleftharpoons}} (C_4H_9)_2Sn(OH)_2 \xrightarrow{UV/H_2O} C_4H_9Sn(=O)OH \xrightarrow{UV/H_2O} SnO_2 + C_4H_{10}$$

*FIGURE 2. Hypothetical degradation pathways: TBTF and TBTO in an aqueous medium.*

*TABLE XI. Serial Degradants and Degradation Scheme (Aquatic System)*

| *Material* | *Degradant* | *Half-life* | *General toxicity of degradant* |
|---|---|---|---|
| *TBTF* | *TBTO* | *12-24 hrs.* | *Toxic* |
| *TBTO* | *TBT-CO*<br>*TBT-OH*$^{3}$ | *1-7 days(?)*<br>*4-18 days* | *Sl. toxic*<br>*Sl. toxic* |
| *TBT-CO*$_3$ | *TBT-OH* | *14-35 days(?)* | *Sl. toxic* |
| *TBTO-OH* | *Dibutyltin oxide (DBTO)* | *4-12 hrs.* | *Sl. toxic* |
| *DBTO* | *DBT(OH)* | *3-15 days* | *Non-toxic* |
| *DBT(OH)*$_2$ | $(C_4H_9)_2\text{-}\overset{O}{\overset{\parallel}{Sn}}(OH)$ | *12-18 hrs* | *Non-toxic* |
| $(C_4H_9)_2\text{-}\overset{O}{\overset{\parallel}{Sn}}(OH)$ | $SnO_2 + C_4H_{10}$ | *4-11 days* | *Non-toxic* |

## *H. Organotin Molluscicidal Mechanism*

Microscopical examination of snail tissue exposed to massive TBTO concentrations indicates proteolytic attack against specific tissue. In-house studies based on a single acute dose time profile indicate massive damage to connective tissue and various membranes. Gross pathological examination shows accumulative destruction of cell walls and loss of impermeability to hemolymph. Massive hemorrhage preceeds mortality. At the lower concentrations practical to snail control via controlled release organotin molluscicide a chronic intoxication exists. The nature of the kill mechanism is not known. TBTO and TBTF react readily with amino acids in vitro. Smith *et al.* (48) postulates that the lower trialkyl tins inhibit oxidative phosphorylation perhaps through a pentacoordinate trialkyltin-histidine blocking mechanism.

The dialkyltins, in contrast, are known to react readily with sulfhydryl groups thus interfering with α-ketoacid oxidation in biological systems (48).

In-house work on the elucidation of the molluscicidal mechanism has been presented by Allen *et al.* (49).

## III. ENVIRONMENTAL TOXICOLOGY

Due to the mechanism of snail contact microenvironmental studies have shown that fish, crayfish, aquatic plants and other members of the non-target biota are unaffected in treated soil/water systems lethal to snails (1,2). This has been confirmed in field bioassays regarding piscine species, phytoplankton, insecta, oligochaetes and the like (50-52).

Acute toxicity data is available for TBTO and TBTF and a number of the degradants. However, such data is of relatively little use in establishing the environmental hazard in that non-target organisms are continuously exposed to ultra-low organotin levels for prolonged periods of time. (Materials returned after 7 to 18 months field immersion were analyzed for residual organotin and emission lives of 4 to 22 years extrapolated). It is generally acknowledged that stannic oxide, the end product of any conceivable degradation pathway, is non-toxic. Analysis of human tissue indicates all organs possess tin, and tin is a known vital trace element in rodents, if not man (53).

Acute toxocity in rodents is presented in Table XII. Once the agent is bound within a non-digestable polymeric matrix, the $LD_{50}$ value increases dramatically; non-formulated TBTF is ca 200 mg/kg while Ecopro 1330 (TBTF/plastic) is over 2500 (54).

Soil fertility is unaffected at relatively high organotin content (55). Some fungi and gram positive bacteria are sensitive to low levels of TBTO. Marine algae and some fresh water species are destroyed at 0.1 to 0.01 ppm TBTO or TBTF concentration. Molluscicidal dosages are in the 0.003 to 0.007 ppm range). Phytotoxicity as regards vascular plants is not evident save at relatively high dosages, above 10 ppm. Insecticidal dosages are generally in the 1 to 10 ppm range (56). However, mosquito and chironomid larva are highly susceptible to TBTF intoxication (57). Crustacea will survive high TBTO concentrations for 6 months or longer (6). Fish toxicity, acute or chronic, is not observed at molluscicidal dosages.

Microenvironmental studies have indicated that in an ecological system containing soil, water, plant materials, snails and a representative fish, *Lebistes reticulatus*, snails are destroyed without piscicidal activity noted (58).

*TABLE XII. Acute Toxicity of Organotin Materials*

| *Material* | *Species* | *Administration* | $LD_{50}$ *(mg/kg)* |
|---|---|---|---|
| *TBTO* | *Rat* | *Oral* | *148 to 234* |
| | *Mouse* | *Oral* | *122 to 203* |
| | *Rat* | *Inhalation* | *56* |
| *TBTO* | *Rabbit* | *Dermal* | *11,700* |
| *TBTF* | *Rat* | *Oral* | *200* |
| | *Rabbit* | *Dermal* | *680* |
| *Ecopro 1330 (30% TBTF)* | *Rabbit* | *Oral* | *2500-5000* |
| | *Rabbit* | *Dermal* | *(skin irritation index 4.7)* |
| *DBTO*[a] | *Rat* | *Oral* | *487 to 520* |
| *CBL-9B (20% TBTF)* | *Rat* | *Oral* | *Over 1000* |

[a]*Dibutyltin oxide*

In over 40 field test sites containing fish during and after treatment with controlled release organotins, no fish intoxication or mortality was observed. Fish do avoid treated areas and application in a lake margin tends to drive fish outward to organotin free waters. In the conventional bioassay test one can demonstrate fish toxicity at relatively low organotin concentrations, 0.1 to 1 ppm. However, absorption on added soil or plant matter tends to purge the water resulting in no mortality at the stated levels. The possible bioaccumulation of the subject organotins by piscine life has yet to be investigated.

Studies with rodents over 4 generations have indicated that TBTO and TBTF at 100X molluscicidal dosages and the breakdown products cause no gross physiological alterations or teratogenicity (6). Higher levels showed some liver damage with TBTO, TBTF, TBT-$CO_3$ and DBTO. Mice ingesting radio labelled TBTO over a 15 to 30 day period excreted 92 to 96% of the label unaltered mainly via the feces. The remainder accumulated mostly in the kidneys, indicative of metabolism, with lesser amounts in the liver, spleen and fatty tissue (59). Clearance, once $^{14}C$-TBTO was removed from the diet, was rapid; most organs being free of label within a few days.

Due to the mechanism of kill, proteolysis and inhibition of transamination, development of resistance in molluscs to TBTO or TBTF is considered unlikely. In-house studies indicate a distinctive drop in tolerance of the S1 and S1 generations wherein the parent snail generation was exposed to sublethal dosages (to be published).

Importantly, the subject organotins are non-carcinogenic (60) and probably non-mutagenic (61), at least for short term (6 mo.) exposures. In fact, TBTO complexes have shown significant antitumorigenic properties (62,63). It is interesting to note that the three major drugs used in schistosomian therapy are carcinogenic (64,65) and two are mutagenic (66).

Recent studies with insecta indicate that species response may be selective. Unpublished data[4] provides some indication that controlled release TBTO and TBTF will repel or destroy Simulium, the chironomidae, and Caddis flies (*Hydropsyche, Limnephilis and Neothremna)* larvae. Ephemeroptera, Plecoptera, excepting *Nemoura*, and *Coleoptera* do not appear to be affected.

## IV. SUMMARY

Efficacy of the several organotin controlled release molluscicides has been demonstrated in laboratory and limited field investigations (1,2,6,15,20,30,52,66-68). Field and analytical studies to date point to an expected pellet life of three or more years. Focal control, wherein snail breeding sites only are treated, appears to be a superior approach over area treatment. This technique may be applicable to transmission sites on manmade lakes where a total volume treatment would be economically impossible. A lethal barrier might also be feasible using controlled release molluscicides.

In order to establish a condition of chronic and ultimately lethal intoxication, the target snail must be exposed over some period of time to the organotin content of the water course/soil bottom. Of course, all elements of the general biota are similarly exposed.

[4]*Andrews, D.A., personal correspondence, Idaho State University, July 3, 1979.*

It should be shown that the presence of inorganic ions, soluble and insoluble organic matter and inorganic suspendants will cause a variation in the lethal time to complete population mortality. However, the magnitude of this change is so slight as to have no practical effect on intervention of the schistosomiasis transmission cycle. pH is a major determinant of the $LT_{100}$ under varying water quality conditions.

The molluscicidal trialkyl organotins degrade readily through hydrolytic and photolytic cleavage so that half-lives in water vary from a few days to 18 days. Several degradation pathways are discussed, all resulting in the innocuous stannic oxide as the ultimate product. The rate of decay varies with water parameters, but is fairly rapid so that the agents of interest, TBTO and TBTF are properly classified as environmentally non-persistent. Molluscicidal organotins are readily absorbed by soil particles and detoxified within a few weeks by unknown processes.

Snail contact is essentially through ingestion of organotin contaminated soil particles and possibly organic matter. It is believed that most agent ingestion occurs during browsing activity on the bottom soil. Experiments have demonstrated that snails isolated from soil contact show a much longer $LT_{100}$, measurements indicating that at any given time the soil-to-water organotin ratio ranges from 15 to 1 to 19 to 1. Laboratory study confirms that dosage requirements can be based on bottom surface area and total water volume is not a pertinent consideration. Organotins show only a small transport distance in soil so that pellets buried in mud or silt will not be effective against snails.

The kill mechanism appears to be through proteolysis of select membranes when dosage is sufficient for actute intoxication. Under ultralow concentrations whereby a chronic intoxication is induced, the mechanism is unknown. However, the propensity for TBTO and TBTF to react readily with amino acids may lead to a blocking of peptide formation and/or oxidative phosphorylation involving a histidine complex.

Microenvironmental studies indicate that TBTO and TBTF dosages exist where snails are destroyed without significant loss of fish life.

Mammalian ingestion of the subject organotins at 100x or more of the molluscicidal dosages is apparently not harmful, with the majority excreted unchanged. Biochemical reactions lead to rapid loss of toxicity and tissue clearance is rapid once the emitting source is removed. Although data is limited, it is reasonable to assume that tetraogenic, carcinogenic and other mutagenic effects are absent. However, mutagenic effects may be present in snails and are certainly evident in mosquito larvae.

In a natural environment non-target biota, vascular plants, most algae species, oligochaetes, crustacea, amphibia, most insecta, fish, and phytoplankton show no visible evidence of intoxication.

It is believed that the environmental hazard arising from the use of controlled release organotin molluscicides is dramatically less than that resulting from the present conventional snail control methodology and medicinals used in therapy.

## REFERENCES

1. N. F. Cardarelli, Controlled Release Molluscicide, Monogr. Publ. Univ. Akron, Ohio, 136 pp., May 1977.
2. N. F. Cardarelli, Controlled Release Pesticide Formulations, CRC Press, Cleveland, Ohio, 210 pp. 1976.
3. N. F. Cardarelli and K. E. Walker, U.S. Patent 4,012,221, March, 1977.
4. N. F. Cardarelli, U.S. Patent 4,166,111, August, 1979.
5. R. L. Sherman and T. L. Carlson, A Modified Phenylfluorone Method for Determining Organotin Compounds in ppb and Sub-ppb Range, July 1979 (submitted for publication).
6. N. F. Cardarelli, Laboratory and Field Evaluation of Controlled Release Molluscicide and Schistolarvicides, Annual Report, Edna McConnell Clark Foundation, New York, Grant No. 276-0091, July 1, 1977 (unpublished).
7. N. F. Cardarelli and K. E. Walker, Chronic Effects of Ultralow Toxicant Concentrations, unpublished report, U.S. Natl. Inst. Hlth., Bethesda, Maryland, Grant AI11861-01A1, 1975/1976, July 1, 1976.
8. C. P. DaSouza and E. Paulini, Rev. Brasil, Malariol, Doencas Trop., 21, 799-818 (1969).
9. C. P. DaSouza and E. Paulini, Rev. Brazil. Malariol. Doencas Trop., 20, 155-160 (1968).
10. R. Meredith, WHO Rep. Ser. WHO/SCHISTO/71.14, 1971.
11. D. J. McLeay, Int. Copper Res. Assoc. Rep. Proj. 126B, January 1 - December 31, 1970.
12. R. R. Wilkenson *et al.*, Organotins: Production, usage, environmental fate and effects, and human health effects, in Report Organotin Workshop, M. Good (Ed.), University of New Orleans, New Orleans, Louisiana, February 17-19, 1978, pp. 229-241.
13. A. E. Slesinger, The Safe Disposal of Organotins in Soil. Unpublished report, M & T Chemicals, Inc., Rahway, New Jersey, 1977.

14. P. Schatzberg and L. Harris, Organotin detoxification and soil migration in Report Organotin Workshop, M. Good (Ed.), University of New Orleans, New Orleans, Louisiana, February 17-19, 1978, pp. 95-107.
15. C. J. Shiff and C. Yiannakis, Controlled release molluscicide studies. Unpublished report, Blair Research Laboratory, Salisbury, Rhodesia 1976.
16. J. D. Christie, in Laboratory and field evaluation of controlled release molluscicides and schistolarvicides, Progress Report XIV, unpublished, N.F. Cardarelli, (Ed.), Edna McConnell Clark Foundation Grant 276-0091, April 1978.
17. C. J. Shiff *et al.*, in Proc. Controlled Release Pesticide Symp., Wright State Univ., Dayton, Ohio, September 8-10, 1975, pp. 177-186.
18. W. Evans, Proc. Controlled Release Pesticide Symp., Univ. of Akron, Akron, Ohio, 5.48-5.51, September 13-15, 1976.
19. H. L. Trachman *et al.*, Anal. Chem., 49, 1090-1093 (1977).
20. K. Y. Chu, Bull. WHO, 54, 417-420 (1976).
21. P. Schatzberg and L. Harris, Organotin detoxification in Report Organotin Workshop, M. Good (Ed.), New Orleans Univ., New Orleans, Louisiana, February 17-19, 1978, pp. 95-107.
22. L. S. Ritchie and L. A. Berrios-Duran, Bull. WHO, 40, 471-473 (1969).
23. A. E. Slesinger and I. Dressler, The Environmental Chemistry of Three Organotin Chemicals, in Report Organotin Workshop, M. Good (Ed.), Univ. of New Orleans, New Orleans, Louisiana, February 17-19, 1978, pp. 115-162.
24. V. T. Mazaev *et al.*, Gig. Sanit. (Russian), 3, 17-20 (1976).
25. A. W. Sheldon, J. Paint Tech., 47, 54-58 (1975).
26. C. P. Monaghan *et al.*, An Evaluation of Leaching Mechanisms for Organotin Containing Antifouling Coatings, unpublished report, Intl. Controlled Release Pesticide Symp., Oregon State Univ., Corvallis, Oregon, August 22-24, 1977.
27. R. H. Fish, Review: Bioorganotin Chemistry, in Report Organotin Workshop, M. Good, (Ed.), Univ. of New Orleans, New Orleans, Louisiana, February 17-19, 1978, pp. 179-183.
28. R. S. Tobias, Organotin (IV) ions in aqueous media, in Report Organotin Workshop, M. Good, (Ed.), Univ. of New Orleans, New Orleans, Louisiana, February 17-19, 1978, pp. 21-27.
29. P. J. Smith, Chemicals & Industry, 5, 874-875 (1977).
30. K. R. S. Ascher and S. Nissim, World Review of Pest Control, 3, 188-193 (1964).

31. R. D. Barnes *et al.*, Pesticide Science, 5, 304-307 (1974).
32. R. K. Ingham *et al.*, Chem. Rev., 60, 520-522 (1960).
33. J. M. Barnes and H. B. Stoner, Pharmacol. Rev, 11, 211-231 (1959).
34. R. J. Vizgirda, Paint and Varnish Production, 62(12), 25-28 (1972).
35. D. A. Kochkin *et al.*, J. Gen. Chem. U.S.S.R., 39, 1740-1744 (1969).
36. R. C. Poller and A. Parkin, ACS Symp. Series No. 41, Sucrochemistry Symposium, American Chemical Society, Washington, D.C. 1977, pp. 145-152.
37. R. R. Wilkinson, Assessment of the Need for, the Character of, and the Impact Resulting from Limitations on Selected Organotins: Phase I, Draft Final Report, Midwest Research Institute to the Environmental Protection Agency, July 1977, 360 pp.
38. K. Freitag and R. Bock, Pest. Sci., 5, 731-739 (1974).
39. J. J. Zuckerman, Advan. in Chem. Series No. 157, American Chemical Society, Washington, D.C., 1976, Chap. 14.
40. R. H. Fish *et al.*, J. Organo-Metallic Chem., 93, C1-C5 (1975).
41. E. C. Kimmel *et al.*, J. Agr. Food Chem., 25(1), 1-8 (1977).
42. R. H. Fish, Review: Bioorganotin Chemistry in Report Organotin Workshop, M. Good, (Ed.), Univ. of New Orleans, New Orleans, Louisiana, February 17-19, 1978, pp. 179-183.
43. R. S. Tobias, Organotin (IV) Ions in Aqueous Media, in Report Organotin Workshop, M. Good, (Ed.), Univ. of New Orleans, New Orleans, Louisiana, February 17-19, 1978, pp. 21-27.
44. J. S. Thayer, Proc. of Controlled Release Pesticide Symp., 2.25-2.30, Gaithersburg, Maryland, August 14-16, 1978.
45. W. P. Ridley *et al.*, Science, 197, 329-332 (1977).
46. L. J. Dizikes *et al.*, J. Amer. Chem. Soc., 100, 1010-1012 (1978).
47. P. Scramel *et al.*, Int. J. Environ. Studies, 5, 37-40 (1973).
48. P. Smith and L. Smith, Chem. Brit., 11(6), 208-212, 226 (1975).
49. A. J. Allen, B. M. Quitter, and C. Radick, in Controlled Release of Bioactive Materials, R. W. Baker, (Ed.), Academic Press, New York, 1980.
50. C. J. Shiff and A. C. Evans, Central African J. Med., 23(11), Suppl. 6-11 (1977).

51. C. J. Shiff, Proc. Controlled Release Pesticide Symp., 5.18-5.26, Univ. of Akron, September 13-15, 1976.
52. B. Gilbert *et al.*, Bull. WHO, 49, 633-636 (1973).
53. K. Schwarz, in Newer Trace Elements in Nutrition, pp. 313-325, W. Mertz and W. E. Cornatzer, (Eds.), Marcel Dekker, New York, 1971.
54. J. Biesemeir, Ecopro 1330, Wisconsin Alumni Foundation Report No. 8010432 (unpublished), March 1978.
55. W. B. Bollen and C. M. Tu, Tin and Its Uses, 94, 13-15 (1972).
56. H. Floch *et al.*, Bull. Soc. Pathol. Exot., 57, 454-465 (1964).
57. N. F. Cardarelli, Mosquito News, 38(3), 328-333 (1978).
58. N. F. Cardarelli, in Molluscicides in Schistosomiasis Control, T. C. Cheng, (Ed.), Academic Press, New York, 1974, pp. 177-240.
59. W. Evans *et al.*, J. Tox. Environ. Health, (to be published 1979).
60. J. R. M. Innes *et al.*, J. Natl. Cancer Inst., 42, 1104-1114 (1969).
61. S. S. Epstein *et al.*, J. Tox. Appl. Pharmacol., 23, 288-325 (1972).
62. S. Ozaki and S. Hideaki, Japan Patent 76,100,089, September 1976.
63. P. J. Smith, Ann. Rep., Internatl. Tin Res. Inst., p. 18, Great Britain, 1977.
64. W. H. Haese *et al.*, J. Pharmacol. Exp. Ther., 186, 430-440 (1973).
65. Abdallah *et al.*, Report of a WHO Consultant Group on the Comparative Evaluation of New Schistosomicidal Drugs for use in Treatment Campaigns, unpublished report, WHO/SCHISTO/72.19, 1972.
66. L. S. Ritchie *et al.*, Prolonged Application of an Organotin Against *Bromphalaria glabrata* and *Schistosoma mansoni*, Rep. Puerto Rico Nuclear Center, San Juan, PR, 1974.
67. A. T. Santos *et al.*, Proc. Controlled Release Pesticide Symp., 114-123, Oregon State Univ., Corvallis, Oregon, Autust 22-24, 1977.
68. A. Mansoori, Proc. Controlled Release Pesticide Symp., 5.54, Univ. of Akron, Akron, Ohio, September 13-15, 1976.

# BIOASSAYS AND ENVIRONMENTAL EFFECTS OF ORGANOTIN MARINE ANTIFOULANTS

*Mary L. Good*
*Dee S. Dundee*
*Gary Swindler*

Departments of Chemistry
and Biological Sciences
University of New Orleans
Lakefront, New Orleans, Louisiana

*Coatings containing various organotins as marine antifoulants have been proven to provide effective, long-lived protection of marine vessels and marine installation. Their efficacy as compared to other antifoulants and other antifouling methodologies has been the focus of a number of both fundamental studies and technological investigation. In addition, the relative environmental impact of their widespread use as compared to alternate antifouling procedures has become an urgent question. This paper reports our initial studies of the effects of tributyltin acetate on both invertebrates and vertebrates from a local, south Louisiana estuary. Summary results are: (1) at concentrations of organotin of 0.15 ppm and above, all vertebrates (fishes) died within three hours (most died within twenty minutes of exposure); (2) at an organotin concentration of 0.10 ppm all vertebrates died within twenty-four hours; (3) at an organotin concentration of 0.15 ppm all invertebrates (mollusc) died between five and eight days; (4) all controls remained healthy and active throughout the experiments; and (5) some habitat substrates were found to detoxify organotin solutions by absorbing (or adsorbing) the toxicant.*




ISBN 0-12-074450-3

## I. INTRODUCTION

"Fouling" of solid objects in marine waters is a well-known phenomenon. Ships, pilings and piers all serve as attachments for various marine invertebrates. When one considers our energy shortage, this fouling becomes critical. "The drag by a six-month accumulation of marine organisms can force a vessel to burn 40% more fuel just to maintain normal cruising speed" (1). Thus any attempt at fuel economy in the shipping industry and by the Navy must emphasize the development of effective antifoulant procedures.

The only practical antifoulants over the years have been paints containing toxic chemicals. The most widely used toxicant has been cuprous oxide but organocompounds of arsenic, mercury, lead, and tin have also been evaluated. The cuprous oxide coatings are good fouling inhibitors, but they have short effective life times and the current price for copper makes them economically unattractive. The coatings containing organoarsenic, organomercury or organolead are known to have serious toxicological effects on vertebrates and have been either restricted in use or completely removed from the commercial market.

The organotin coatings have been evaluated by a number of groups, particularly by investigators at the Naval Ship Research and Development Center at Annapolis, MD. The organotin toxicants, particularly the tributyl tin derivatives, provide superior antifouling coatings for relatively long time periods. In addition, when applied on conductive substrates they do not promote corrosion as do copper-based coatings. Thus, these organotin containing coatings appear to be the most promising antifoulants yet developed.

If these organotins are to be widely used in the shipping industry and by the Navy, they will very quickly be distributed in all marine waters. Current research here at the University of New Orleans and at Louisiana State University is concentrated on determining accurate leach rates for the toxicants and the leaching mechanisms for the organotins (2-8). These basic studies are expected to provide definitive information about the chemical form of the toxicant released to the environment and the mass transfer of toxicant from the coating surface to the surroundings. The work reported here represents our initial investigation of the direct relationship between the efficacy of the organotin toxicants and their environmental impact.

## II. EXPERIMENTAL

A small travel trailer was converted to a mobile laboratory and fitted with external water pumps, fish tanks, aerators, etc. The trailer could be parked on the shore of a body of water where organisms could be collected and maintained in "natural" habitats. The organisms used in these experiments were estuarine forms taken from Lake Pontchartrain and maintained in lake water. A series of all-glass tanks (6 liter capacity) were used for each experiment (Table I). Each tank contained four liters of lake water with one half of the tanks containing one liter of substrate taken from the shallow lake bottom. Each tank was aerated with a glass diffusion tube connected to rubber hoses to minimize lab-induced metal contaminants. Two different habitat substrates were investigated. The first was a silt-clay mixture of high organic content which was collected just west of the Lake Pontchartrain Causeway on the south shore of the lake. The second substrate was obtained from the lake bottom in the vicinity of the trailer (across the street from the University of New Orleans Campus) and was largely sand with little organic matter.

Five to eight invertebrate filter-feeders, *Rangia cuneata* Sowerby (1831) (Mollusca: Pelecypoda), were maintained in each tank. To establish comparative baseline data on non-filter-feeders, vertebrates were also tested. Three to five fishes (a mixture of *Gambusia affinis, Poecilla latipinna* and *Fundulus grandis*) from the mouth of a canal emptying into the estuary were added to each tank. In early experiments the fish were tested separately from the clams. In later tests, the fish and clams were maintained together in the same tank. No differences in toxicity were noted. Several other organisms including grass shrimp and blue crabs were casually tested as the above described experiments were in progress. Preliminary data are reported for them although the results are not to be considered final since no controls were maintained.

The toxicant stock solution was prepared by dissolving tributyltin acetate (TBTA) in a relatively small volume of habitat lake water. Appropriate dilutions were then made to provide organotin solutions of varying concentrations. Each experimental tank was filled with the habitat lake water containing the toxicant and each control tank was filled with lake water. Organisms were introduced into the experimental and control tanks from holding tanks used for maintaining the organisms after they are collected. The analysis of the original stock solution and its subsequent dilutions for

TABLE I. Outline of Experiments Completed

| Tank No. | 4 Liters Lake Water | 1 Liter Substrate | Clams | Fishes | Cont. of TBTA (ppm) |
|---|---|---|---|---|---|
| 1 control | x | - | x | x | - |
| 2 control | x | clay | x | x | - |
| 3 | x | - | x | x | 1.00 |
| 4 | x | clay | x | x | 1.00 |
| 5 | x | - | x | x | 0.75 |
| 6 | x | clay | x | x | 0.75 |
| 7 | x | - | x | x | 0.50 |
| 8 | x | clay | x | x | 0.50 |
| 9 | x | - | x | x | 0.25 |
| 10 | x | clay | x | x | 0.25 |
| 11 control | x | - | x | x | - |
| 12 control | x | clay | x | x | - |
| 13 | x | - | x | x | 0.20 |
| 14 | x | clay | x | x | 0.20 |
| 15 | x | - | x | x | 0.15 |
| 16 | x | clay | x | x | 0.15 |
| 17 | x | - | x | x | 0.10 |
| 18 | x | clay | x | x | 0.10 |
| 19 | x | - | x | x | 0.05 |
| 20 | x | clay | x | x | 0.05 |
| 21 | 3 liters | - | x | - | 1.33 |
| 22 | 3 liters | sand | x | - | 1.33 |
| 23 | 3 liters | - | x | - | 1.00 |
| 24 | 3 liters | sand | x | - | 1.00 |
| 25 | 3 liters | - | x | - | 0.67 |
| 26 | 3 liters | sand | x | - | 0.67 |
| 27 | 3 liters | - | x | - | 0.33 |
| 28 | 3 liters | sand | x | - | 0.33 |
| 29 | x | - | x | x | 0.09 |
| 30 | x | sand | x | x | 0.09 |
| 31 | x | - | x | x | 0.06 |
| 32 | x | sand | x | x | 0.06 |
| 33 | x | - | x | x | 0.04 |
| 34 | x | sand | x | x | 0.04 |
| 35 | x | - | x | x | 0.02 |
| 36 | x | sand | x | x | 0.02 |

tributyltin acetate were carried out spectrophotometrically using dithizone (9-11). The TBTA was extracted from lake-water with chloroform. The chloroform solution was equilibrated with an aqueous buffer solution of pH 8.1 and dithizone was added. After the development of the TBTA-dithizone complex color, the absorbance was determined at 450 nm. on a Beckman DU spectrophotometer. The complex obeys Beer's law over a wide range of concentrations and the method provides simultaneously a concentration step and an analysis step for the direct determination of organotins.

## III. RESULTS AND DISCUSSION

Table I indicates the breadth of experiments performed as a function of organotin concentration. Tanks 1 through 10 contained TBTA in the range of 1.0 ppm to 0.25 ppm. A second run using only the *Rangia cuneta* (clams) was carried out using 3 liters of lake water and is shown in the table as experiments No. 21-28 with organotin concentrations ranging from 1.33 ppm to 0.33 ppm. It became apparent that the adult *Rangia cuneata* were dying rapidly even at the lowest concentration (0.25 ppm) of organotin. Subsequent experiments were then carried out with organotin concentrations of 0.20 ppm to 0.05 ppm (tanks 13-20) and 0.09 ppm to 0.02 ppm (tanks 29-36). During trial runs without substrate it became apparent that the lethal dosage for adult *Rangia cuneata* (3-6 cm. in length) is near 0.10 ppm. At 1.0 ppm the clams all died within 5 days. At the lowest concentration of 0.02 ppm all of the adult clams survived the full length of the experiments of 17 days. The in-between concentration of 0.10 ppm produced a mixed survival rate of 8 days to experiment termination at 17 days. Figure 1 portrays the survival rate of clams versus organotin concentration in a three-dimensional display which clearly indicates the concentration effects.

The distress symptoms for the clams began to appear several hours after exposure to organotin at concentrations of 1.00 ppm to 0.04 ppm. Symptoms include the following: (1) siphons barely protrude from the shells, in contrast to the controls where the siphons extend normally from the shells; (2) excess mucus is released and the water becomes cloudy; (3) siphons and mantle edges appear swollen and rubbery; and (4) the clams become insensitive to touch. Ultimate death ensues whether the valves are agape or closed. Different fatality rates were observed for the tanks with

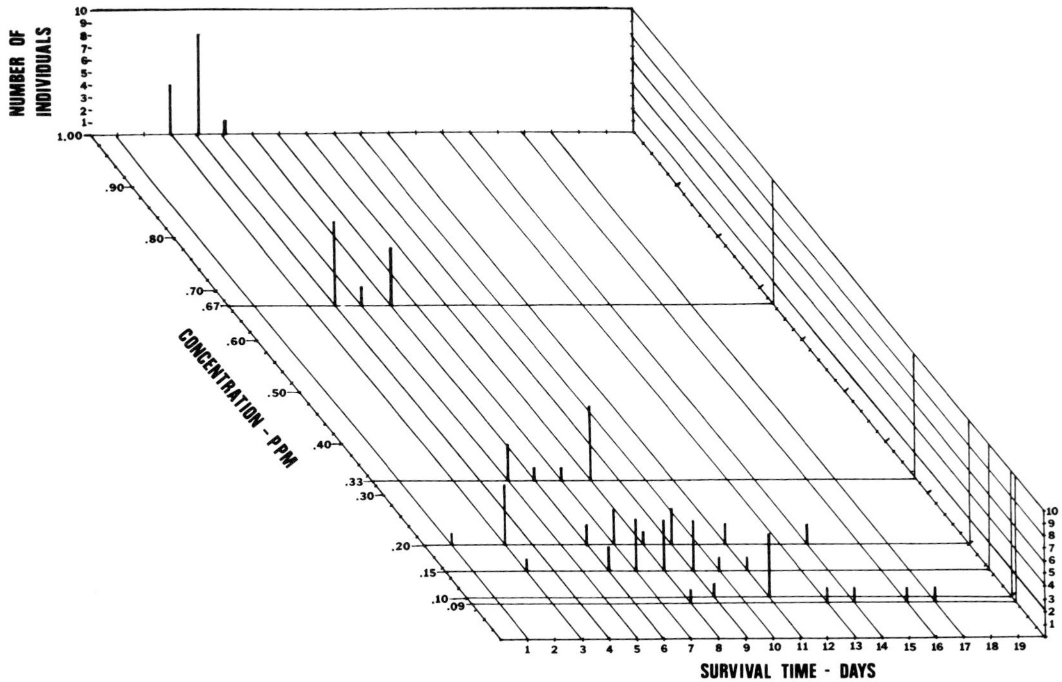


*FIGURE 1. Plot of adult Rangia Cuneata: organotin concentration vs. survival time vs. Number of individuals dead.*

substrates and differences were discernable between the clay and sand substrates. A detailed experiment for investigating this effect is outlined below.

Juvenile *Rangia cuneata* (1-2 cm.) were exposed to organotin concentrations of 0.09 ppm to 0.02 ppm both with and without the sandy substrate obtained where they were collected. At concentrations between 0.09 ppm and 0.04 ppm all clams died within 6 days while others survived longer periods of time. All had died after 20 days. The lethal concentration for juvenile *Rangia cuneata* is below 0.02 ppm, the lowest concentration checked in these experiments. The distress symptoms for the juvenile clams were more difficult to define but in general they follow those described above for adults, as shown in Figure 2.

The fishes were initially tested separate from the clams in organotin concentrations ranging from 1.00 ppm to 0.25 ppm. In later experiments ranging down to 0.02 ppm organotin they were tested together with the clams. No appreciable differences were found between the two types of experiments. At concentrations of 1.00 ppm to 0.20 ppm the fishes died within the time span of 20 minutes to 3 hours. At concentrations of 0.10 ppm all died within 24 hours and at 0.02 ppm total fatalities were obtained within 12 days. The distress symptoms preceeding the death of fishes were the same at all concentrations tested and consisted of: (1) erratic swimming movements; (2) loss of equilibrium and orientation; (3) convulsions; and (4) finally loss of buoyancy.

Several other organisms were casually tested (without controls) in the TBTA solutions and preliminary results indicate that the threshold tolerance for grass shrimp (*Paleomonetes* sp.) is about 0.25 ppm and that blue crabs (*Callinectes sapidus*) survived a three day experiment at 0.25 ppm but appeared distressed. Further definitive work with these species is planned.

In order to further define the apparent differences in life spans observed in the experiments where substrates were added, a new experiment was designed. Two of the glass tanks were set up, each with one liter of a specific substrate (silt-clay or sand) and four liters of lake water containing 0.20 ppm TBTA. The tanks were left for 20 days and two portions of the water were then taken out of each tank, one portion was filtered and one was not. These four water samples were placed in four new tanks without substrate and five clams and five fish were added to each. The experiment was continued for 20 days. During that time no deaths of clams or fish occurred in the tanks containing water previously in contact with the silt-clay substrate. In the filtered water from the sand substrate, three fish deaths

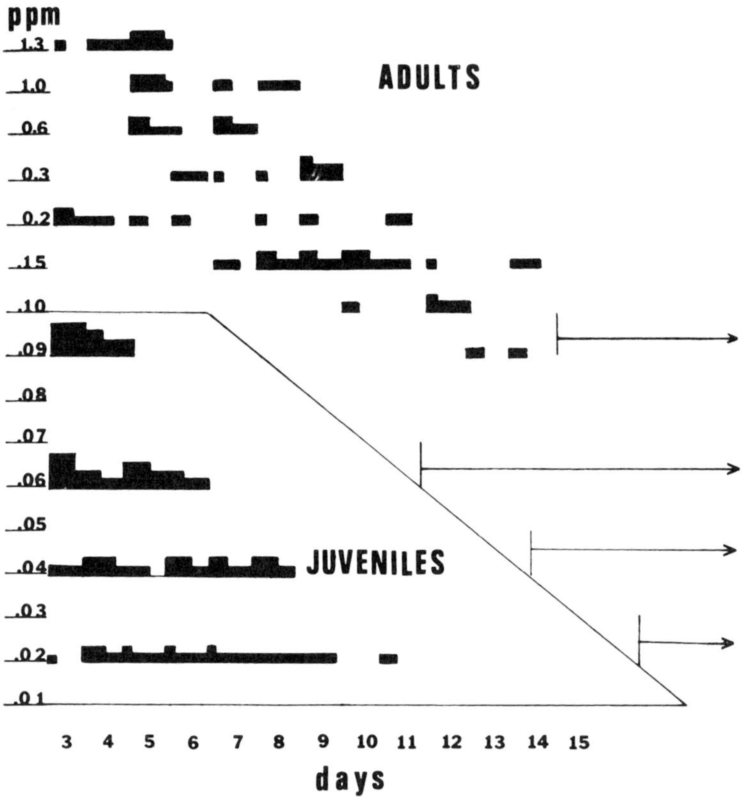


*FIGURE 2. Comparison of death rates of adult and juvenile Rangia Cuneata at various concentrations of tributyltin acetate.*

occurred (but no clam deaths) during the 20 days. The unfiltered water from the sand substrate produced an environment where four out of five of the fish and clams were dead after 20 days. These results indicate that the clay bottoms (or the organic matter) effectively absorb and/or adsorb the organotin and make it unavailable both to fish and bottom feeders. The sand bottoms are significantly less effective as organotin "sinks" and allow some of the toxicant to remain in solution and adsorbed on small particles which are suspended in the water but unfilterable. A consideration of the long-term environmental effects would indicate that this is a most significant finding, in that organotins released from marine coatings may be de-activated by certain types of harbor bottoms. Detailed work on the adsorptivity of various organotins of different substrates is now underway in our laboratories (chemical work supported by the Office of Naval Research and Louisiana Sea Grant).

The approximate lethal concentrations of TBTA for the estuarine fish determined in this study compares well with the $LC_{100}$ = 0.75 ppm found by Flack and co-workers (12) for goldfish and guppys and the tributyltin oxide (an organotin dimer) $EC_{50}$ = 0.031 ppm found for trout by Chliamovitch and Kuhn (13). The distress symptoms reported here for fish are also compatible with Chliamovitch and Kuhn's (13) observation that tributyltin oxide produced damage in the gill epithelium and interfered with the respiratory process in fishes.

Taken collectively, all of the bioassays reported in the literature (and these results) of organotin compounds of the form $R_3SnX$ indicate that the actual toxicant species released in aqueous solution is the same regardless of the nature of X. Thus in the tributyl series, the $EC_{50}$, LD, and $LC_{100}$ values appear to be essentially the same for any given tributyltin derivative. This observation is consistent with our preliminary conclusion that all of the tributyltin molecular species hydrolyze in water by first forming the tributyltin cation (4).

## IV. CONCLUSIONS

These experiments indicate clearly the toxicity of triorganotins to estuarine organisms and the variation in toxicity from one species to another. It also points out the possible roles of environmental substrates on the de-toxification of antifoulant toxicants released to the environment. This preliminary study demonstrates the need for further basic experimentation to determine actual leaching rates for antifouling coatings, the final environmental fate of anitfoulants, and the definitive mode of action by which the toxicants kill fouling organisms and non-involved environmental species. Our own plans include bioassay studies using salt water organisms and panels coated with currently used organotin antifoulants. We also expect to investigate in detail the adsorption characteristics of the organotin species on a number of solid materials including clays, sands and naturally occurring mixed substrates. Such studies are mandated by these initial results before wide-spread use of the organotin materials could be encouraged.

## ACKNOWLEDGMENTS

This project was supported primarily from funds provided by the Maritime Administration, U.S. Department of Commerce.

## REFERENCES

1. E. A. Starbird and R. F. Sisson, National Geographic, November, 623 (1973).
2. E. O'Brien, C. Monaghan, and M. Good, in Organometallic Polymers, C. E. Carraher and C. Pittman, (Eds.), Academic Press, New York, 1978, p. 207.
3. J. F. Hoffman, K. C. Kappel, L. M. Frenzel, and M. L. Good, in Organometallic Polymers, C. E. Carraher and C. Pittman, (Eds.), Academic Press, New York, 1978, p. 195.
4. C. P. Monaghan, J. F. Hoffman, E. J. O'Brien, L. M. Frenzel, and M. L. Good, in Proc. Controlled Release Pesticide Symposium, Corvallis, Oregon, 1977.
5. V. H. Kulkarni and M. L. Good, Anal. Chem., 50, 973 (1978).
6. M. L. Good, V. H. Kulkarni, C. P. Monaghan, and J. F. Hoffman, in Proc. of Third Coastal Marsh and Estuary Management Symposium: Environmental Conditions in the Louisiana Coastal Zone, J. W. Day Jr., D. D. Culley Jr., A. J. Mumphrey, and R. E. Turner, (Eds.), Louisiana State Univ., Div. of Continuing Education, Baton Rouge, Louisiana, 1979, p. 19.
7. C. P. Monaghan, V. H. Kulkarni, and M. L. Good, in Organometals and Organometalloids: Occurance and Fate in Environment, ACS Symp. Series No. 82, F. E.

Brinckman and J. M. Bellama, (Eds.), American Chemical Society, Washington, D.C., 1978, p. 539.
8. C. P. Monaghan, V. H. Kulkarni, and M. L. Good, in Proc. Controlled Release Symposium, National Bureau of Standards, Washington, D.C., August 1978.
9. W. N. Aldridge and J. E. Cremer, Analyst, 82, 37 (1957).
10. L. Chromy and K. Uhacz, J. Oil Col. Chem. Assoc., 51, 494 (1968).
11. V. H. Kulkarni and M. L. Good, manuscript in preparation.
12. H. Flock, R. Deschiens, and T. Flock, Bull. Soc. Pathol. Exot., 57, 454 (1964).
13. Y. P. Chliamovitch and C. Kuhn, J. Fish Biol., 10, 575 (1977).

# THE BIOCIDAL MECHANISM OF CONTROLLED RELEASE BIS (TRI-N-BUTYLTIN) OXIDE IN *BIOMPHALARIA GLABRATA*

*Anthony J. Allen*
*Bernadette M. Quitter*
*Colleen M. Radick*

Environmental Management Laboratory
University of Akron
Akron, Ohio

## I. INTRODUCTION

Tributyl organotin chemicals have long been recognized as exhibiting molluscicidal potential. When properly formulated in polymeric matrices, tributyltin fluoride (TBTF) and bis (tri-n-butyltin) oxide (TBTO) release slowly upon immersion in water. Agent release from polymeric dispersions ranges from 4 years to over 20 years based upon analysis of residual organotin in specimens returned after field trials (1,2). Effective release levels vary from 1 ppb/day to 7 ppb/day depending upon water quality. Laboratory and field evaluation of controlled release organotin formulations have indicated long term activity at ultra low dosage levels against various host species of *S. mansoni*, *S. hematobium* and *S. japonicum* (3). Thus the use of controlled release organotin compounds has been proposed to provide long term effective control of snail vectors of schistosomiasis.

Organotins such as TBTO and TBTF are particularly advantageous because they are environmentally non-persistent and extremely low concentrations can be used to achieve effective snail control with minimal adverse effects on other life forms. Teratogenicity studies in our laboratories have shown no effect over four generations of mice continuously exposed to TBTO and TBTF in drinking water at >100X the molluscicidal dosage (2). To date there have been no reports




ISBN 0-12-074450-3

of adverse effects on fish, crayfish, marine and freshwater shellfish, insects, aquatic plants, birds and mammals. Toxicity is noted with gram positive bacteria and possibly several monocellular fungi species. An additional advantage of the organotin compounds is that there is no apparent immunological resistance built up by the snail. Examination of the progeny of snails exposed to sublethal concentrations of TBTO and TBTF shows decreasing tolerance to both materials and dramatic changes in growth rate, fecundity, and ova mortality through at least two successive generations (2).

Despite these advantages the widespread application of controlled release organotins in the environment has not been initiated due to the lack of knowledge of the exact biocidal mechanism of these compounds. The basic purpose of this report is to begin the elucidation of the TBTO biocidal mechanism in molluscs. It is believed that biochemical activity in the snail is different not only in degree but in kind from that observed in mammals and fish; and that the response to ultralow organotin concentrations is fairly specific to molluscs and a few species of insects. Data would be useful in further differentiation of the chronic intoxication syndrome observed in the snail and its apparent lack in other life forms. The ultimate goal of this effort is to determine the hazards of using controlled release organotin molluscicides and allow the computation of benefit/risk ratios for a given environmental situation. The experiments comprising the work performed were keyed to the determination of TBTO accumulation sites within the snail.

Examination of snails exposed to lethal concentrations of TBTO shows massive internal hemmorhage. Light microscopy from such snails indicates cell wall destruction in selective organs, e.g. connective tissues, whereas muscle and nerve tissue remain unchanged (4). Electron microscopy indicates little, if any, morphological alterations in mitochondia. Ongoing studies *in vitro* indicate that TBTO combines with selective amino acids and polypeptides, the rate of reaction showing considerable variance (to be published). Several investigators have postulated that TBTO complexes with histidine in living systems thus blocking critical transamination processes (5). Amine levels in tissue are suppressed by tributyltin compounds (6). Histidine, along with alanine, arginine, cystine, proline, and gultamic acid react readily with TBTO *in vitro* forming an ester linkage (7). Histidine is present in snail hemolymph (8). It is also noted that the primary degradation product of TBTO, dibutyltin oxide, reacts with sulfhydryl groups possibly interfering with $\alpha$ ketoacid oxidation (5). The apparent toxicological manifestations of TBTO in molluscs is thus a combination of organotin with proteins. The snail is unlikely to develop resistance to this form of attack because development of resistance would mean alteration of the protein (amino acid) structure itself. In an effort to further identify the biocidal mechanism, the accumulation of TBTO in snail tissue and subcellular components was determined. Studies were also performed on the effects of TBTO on snail hemolymph cells and total hemolymph protein.

## II. ACCUMULATION OF 1-$C^{14}$ TBTO IN SNAIL TISSUE

### A. *Experimental*

TBTO, labelled with $C^{14}$ at the α position, was formulated in the BioMet™ SRM recipe and processed as 8 mm diameter, 3 mm thick pellets (2). In order to quantify experimental work, a standard solution of 1-$C^{14}$ labelled TBTO was prepared. By measuring the CPM, at known concentrations in Multisol scintillation fluid, a calibration curve was determined. The detection limit of TBTO was found to be about 35 ppb.

Since the varying tissue compositions contribute non-uniformly to the degree of chemical and color quenching, standardization was performed. Samples of 100 mg snail tissue were solubilized in Unisol®, and the quench factors determined as follows:

| | |
|---|---|
| Head-foot tissue | 1.1 |
| Digestive system | 2.0 |
| Reproductive system | 1.1 |
| Renal tissue | 2.6 |
| Hemolymph | 1.3 |

Adult *Biomphalaria glabrata*, Puerto Rico albino strain, were exposed to water containing known amounts of 1-$C^{14}$ TBTO for varying periods. Snails were removed and immediately dissected under a 20x binocular microscope into four distinct sections; head-foot tissue, digestive system, reproductive system, and the renal system. The tissue was solubilized and the scintillation cocktail added. Twenty-minute counts were made on a Beckman 3100T Liquid Scintillation Counter. Exposure was in the BioMet SRM pellet immersion water (2).

### B. *Results and Discussion*

The following tables demonstrate a time profile of accumulation in the specific tissue. Measurements of the test water were taken at each time point. Tissue from five snails

*TABLE I. 1-$C^{14}$ TBTO Accumulation in Head-Foot Tissue*

| | TBTO concentration in tissue (ppm/ppm) | | | | |
|---|---|---|---|---|---|
| Exposure time (hr) | Pellet conc. 10 ppm | Pellet conc. 1 ppm | Pellet conc. 0.5 ppm | Pellet conc. 0.1 ppm | Pellet conc. 0.05 ppm |
| 0.5 | — | 5.6±0.9 | 9.1±1.7 | 13.0±1.1 | 4.8±1.0 |
| 1 | 1.4±0.2 | 9.8±1.9 | 9.7±2.2 | * | * |
| 4 | 3.2±0.8 | 20.9±5.8 | 23.4±2.2 | * | 5.2±0.7 |
| 8 | 3.3±2.1 | — | 18.2±2.9 | — | 31.7±2.2 |
| 24 | 3.9±1.1 | 2.4±0.3 | 10.0±2.9 | 13.3±3.8 | 16.5±3.4 |
| 72 | 14.5±2.9 | — | 4.6±1.2 | 47.0±2.0 | 3.0±0.6 |
| 96 | + | 28.6±5.3 | — | — | — |
| 120 | + | + | 4.2±1.2 | 23.7±2.2 | 47.5±7.2 |
| 168 | + | + | 12.9±3.0 | — | — |

* Readings at or below background
+ No measurement taken, due to snail mortality
— Experiment not performed

were pooled at each determination, and the results are expressed in terms of (ppm of TBTO in the tissue)/(ppm of TBTO in the test water).

At 10 ppm (pellet dosage, 600 ppb active) and 1 ppm (pellet dosage, 60 ppb active) snail response is almost immediate. There is cessation of browsing activity and mobility, with deep retraction. Death follows within a few days. The lack of $C^{14}$ label in the digestive system and high levels in the head-foot tissue indicate that entrance of TBTO is not through ingestion. The intoxication mechanism involved is either acute or subacute and may be quite dissimilar from the chronic syndrome observed at lower exposure levels.

Snails exposed to 0.5 ppm (30 ppb active) do not cease activity for several days and gradually become morbid, succumbing in the 8th to 12th day. Whether this is sub-acute or can be considered chronic intoxication is unknown. In this instance we see a rapid accumulation of label in the head-foot region, a maximum being reached at approximately 8 hours. Thereafter, there is an increase in the reproductive organs and the digestive system. A similar sequence is noted at

*TABLE II. 1-$C^{14}$ TBTO Accumulation in the Reproductive System*

| | TBTO concentration in tissue (ppm/ppm) | | | | |
|---|---|---|---|---|---|
| Exposure time (hr) | Pellet conc. 10 ppm | Pellet conc. 1 ppm | Pellet conc. 0.5 ppm | Pellet conc. 0.1 ppm | Pellet conc. 0.05 ppm |
| 0.5 | — | * | 3.8±0.1 | 15.3±3.0 | * |
| 1 | 0.08±.003 | * | * | * | * |
| 4 | 0.02±.003 | 4.7 0.57 | * | * | * |
| 8 | 0.61±0.11 | — | 1.7±0.27 | — | 4.2±0.1 |
| 24 | 0.50±0.06 | * | 5.72±3.6 | 7.5±0.7 | 4.2±0.3 |
| 72 | 0.75±0.02 | — | 3.0±0.47 | 14.6±1.5 | 9.2±1.2 |
| 96 | + | 3.96 0.4 | — | — | — |
| 120 | + | + | 3.1±1.5 | 10.6±1.1 | 24.5±7.6 |
| 168 | | + | 17.5±0.0 | — | — |

* Readings at or below background
\+ No measurement taken, due to snail mortality
— Experiment not performed

0.1 ppm (6 ppb active) except that the head-foot accumulation peaks at around 3 days and then clearance initiates. Such snails do not succumb for at least two weeks and usually do not die until day 10. Again as head-foot clearance commences, label build up is observed in the digestive system. The relatively high levels seen in the reproductive system may be related to the genetic involvement postulated as a rationale for the decrease in organotin tolerance noted in S1 and S2 progeny. At 0.5 ppm exposure (3 ppb active) the mortality threshold is approached. Continuous exposure usually results in death in the 3rd or 4th week. The sequence noted is similar to 0.1 ppm exposure, except that the head-foot build up is slower and movement into the other organ systems examined is delayed.

The mode of TBTO entrance into the snail may be by absorption along exposed surfaces of the head-foot and the tissue surrounding the mantle cavity and at the pseudo branch. However the mode of organotin entry under these experimental conditions where snails are generally not browsing may differ from that of feeding snails. A higher label concentration in

TABLE III. 1-$C^{14}$ TBTO Accumulation in the Digestive System

| | TBTO concentration in tissue (ppm/ppm) | | | | |
|---|---|---|---|---|---|
| Exposure time (hr) | Pellet conc. 10 ppm | Pellet conc. 1 ppm | Pellet conc. 0.5 ppm | Pellet conc. 0.1 ppm | Pellet conc. 0.05 ppm |
| 0.5 | — | * | * | * | * |
| 1 | * | * | * | * | * |
| 4 | * | * | * | * | * |
| 8 | * | — | * | — | * |
| 24 | * | * | * | * | * |
| 72 | * | — | 2.2±0.2 | 12.6±1.0 | 9.8±2.2 |
| 96 | + | * | — | — | — |
| 120 | + | + | 3.8±0.7 | 14.5±1.0 | 28.9±5.7 |
| 168 | + | + | 12.8±0.23 | — | — |

* Readings at or below background
\+ No measurement taken, due to snail mortality
\- Experiment not performed

the alimentary canal would be expected if the target snail were ingesting soil and/or vegetable matter containing TBTO.

As these results show, the mode of entry and resulting intoxication of TBTO appears to be different from that observed with copper salts (9,10).

## III. DETERMINATION OF TBTO IN SUBCELLULAR FRACTIONS OF *BIOMPHALARIA GLABRATA* HEAD-FOOT TISSUE

### A. *Experimental*

The distribution of TBTO was determined in the following experiments. Forty snails were exposed to a 10 ppm aqueous solution of $^{14}C$ labeled TBTO for a 24 hour period. The head-foot was removed and stored in a cold 0.25 M sucrose solution. The homogenate was prepared by mincing the tissue and homogenating for 10 minutes in a 0.25 M sucrose solution with $1.8 \times 10^{-4}$ M $CaCl_2$. The head-foot tissue proved very difficult to homogenate and had to be done in sections with several homogenates prepared. The homogenate was then fractionated using the procedure of Hogeboom (11). The homogenate was carefully layered over a 0.34 M sucrose solution with 1.8 x $10^{-5}$ M $CaCl_2$ and centrifuged to obtain the nuclear fraction. The supernatant was removed and resuspended in a 0.25 M sucrose solution and centrifuged to obtain the mitochondria. The resulting supernatant was then centrifuged to obtain the microsomal fraction. Each fraction was placed in a tared scintillation vial and weighed. One ml of Unisol® scintillation cocktail was added and the contents digested for 24 hours. After digestion, 0.5 ml of methanol was added to each vial to prevent precipitate formation. The digest was solubilized in 9 ml of Unisol® Compliment. The mitochondrial fraction was decolorized by the addition of 0.2 ml of 30% $H_2O_2$. Controls were prepared in order to determine the degree of quenching for each fraction. Standard $^{14}C$ TBTO solution, 4.0 µl, was added to each control sample and the activity determined. All vials were counted in the scintillation counter for 20 minutes. The counts per minute (CPM) for the exposed tissue were corrected for quenching and background with the results obtained from the standards. The CPM were then converted to millimoles of TBTO per 100 mg of tissue.

## B. *Results and Conclusions*

Although the homogeneity of the fractions obtained was undetermined it is relatively certain that they contained the specified cellular components. Assuming that the $^{14}C$ has remained attached to the TBTO, the CPMs are an accurate determination of the presence of this substance. These results are presented in Table IV.

The high concentration in the nuclear fraction would tend to support the evidence obtained from morphogenicity tests, which show a very low tolerance and low development rate of subsequent generations previously exposed to TBTO. This possibility indicates alterations in certain nuclear tissues. Although relatively high TBTO loads are noted in the mitochondria, electron micrography failed to detect any visible alterations.

*TABLE IV. Accumulation of TBTO in Subcellular Fractions*

| *Fraction* | *Millimoles TBTO/100 mg tissue* |
|---|---|
| *Nuclear* | $2.2 \times 10^{-6}$/100 mg |
| *Mitochondria* | $1.6 \times 10^{-6}$/100 mg |
| *Microsomal* | $4.4 \times 10^{-7}$/100 mg |

## IV. HEMOLYMPH PROTEIN INVOLVEMENT IN TBTO INTOXICATION

It has been demonstrated that organotins combine readily with specific amino acids and proteinaceous tissue (4,7,12). If organotin intoxication arises from blockage of transamination processes, then alteration in hemolymph protein ought to be seen as a function of exposure time.

### A. *Experimental*

Total hemolymph protein levels were determined for control snails and for snails exposed to solutions of 10, 1 and 0.5 ppm total available 1-$^{14}$C TBTO concentrations in controlled release BioMet SRM$^{TM}$ (2). The TBTO concentration in the hemolymph was determined by the previously described liquid scintillation counting methods using 100 mg samples of freshly drawn hemolymph. Tests were conducted using five snails per test in 100 ml of TBTO solution.

Hemolymph for total protein determinations were obtained from the Puerto Rican albino strain of *Biomphalaria glabrata*. Test snails measuring ten to fifteen millimeters in diameter were air-dried and then punctured in the heart region. Hemolymph was collected in capillary tubes and immediately centrifuged for five minutes in a hematocrit blood centrifuge and frozen until used.

Total protein concentrations were determined by Bailey's (13) modification of the method of Lowry *et al.* (14). A standard curve of concentration vs. absorption was plotted using bovine serum albumin standards. All determinations were done in duplicate. Two microliters (2 $\lambda$) of cell-free snail hemolymph were mixed with 0.5 ml of deionized water and five ml of Lowry's solution C. The mixture was permitted to stand for 10 minutes after which 0.5 ml of a 1 N solution of the Folin-Cioucalteau reagent was added and mixed thoroughly. The absorbance was measured at 750 nm after thirty minutes. Both a Bausch and Lomb Spec 20 spectrophotometer and a Beckman Model 25 recording spectrophotometer were used.

### *B. Results and Discussion*

Determinations of 1-$C^{14}$ TBTO in hemolymph indicated that TBTO accumulates in the hemolymph (Table 5) with its concentration being proportional to the water concentration of TBTO and also to the exposure times.

The average total hemolymph protein concentration of unexposed snails was 14.3 mg/ml which is quite close to Lee and Cheng's (15) value of 14.2 mg/ml. Individual concentrations for unexposed snails ranged from 12.0 to 16.0 mg/ml.

Hemolymph protein levels in snails exposed to 10 ppm TBTO[1] exhibited a steady decrease from zero to 12 hours exposure, followed by a sharp increase in protein levels eventually rising above the average level of the control snails. This increase can be attributed to release of cellular proteins from damaged tissues. A minimum protein level of 9.2 mg/ml was reached at 12 hours exposure. Exposure to 1 ppm TBTO[1] also resulted in a gradual decrease in hemolymph protein levels at zero to 12 hours exposure, with a minimum level of 7.6 mg/ml at 12 hours. After 12 hours, hemolymph protein levels slowly increased but never exceeded the average control level. When exposed to 0.5 ppm TBTO[1] hemolymph protein levels dropped to 9.6 mg/ml in 2 hours. The protein level is seen to fluctuate close to the average for up to 48 hours, then exhibit a large decrease at 72 hours post exposure. In all cases the protein levels decreased greatly after initial exposure to TBTO with minimum values in the range of 7.6-9.6 mg/ml. This indicates that TBTO combines quickly with specific hemolymph proteins of the snail, as exhibited by *in vitro* tests. Subsequent increases in protein levels may be due to an attempt by the snail to replace specific proteins which have combined with TBTO, though no clear evidence of this is presently available. The large increase in the protein concentration in the 10 ppm test is believed to be due to cellular disruption preceding death.

[1]*Total available concentration*

TABLE V.

| TBTO concentration[a] (ppm) | Exposure time (hr) | Average Total hemolymph protein (mg/ml) | Change in concentration (mg/ml) | TBTO level in hemolymph (ppm) |
|---|---|---|---|---|
| 10 | 0.5 | 14.4±4.9 | +0.01 | --- |
| 10 | 2 | 13.2±3.0 | -1.1 | 0.49 |
| 10 | 3 | 14.0±1.6 | -0.3 | 0.81 |
| 10 | 12 | 9.2±1.9 | -5.1 | 0.59 |
| 10 | 24 | 18.8±7.7 | +4.2 | 0.82 |
| 1 | 1 | 12.5±2.9 | -1.8 | --- |
| 1 | 3 | 11.7±3.9 | -2.6 | --- |
| 1 | 12 | 7.6±1.8 | -6.7 | 0.29 |
| 1 | 24 | 9.7±6.6 | -4.6 | 0.51 |
| 1 | 48 | 13.0±5.2 | -1.3 | 0.62 |
| 1 | 72 | 12.2±2.2 | -2.1 | 0.53 |
| 0.5 | 2 | 9.6±2.0 | -4.7 | --- |
| 0.5 | 4 | 13.1±0.6 | -1.2 | --- |
| 0.5 | 12 | 15.2±3.2 | +0.9 | 0.31 |
| 0.5 | 24 | 12.4±2.7 | -1.9 | 0.52 |
| 0.5 | 48 | 15.4±1.4 | +1.1 | 0.44 |
| 0.5 | 72 | 12.1±2.0 | -4.2 | 0.61 |
| 0 | Unexposed | 14.3±2.3 | 0 | 0 |

[a]Total available concentration

## V. ALTERATIONS IN HEMOLYMPH CHARACTERISTICS

### A. *Experimental*

Normal snail hemolymph was collected from Puerto Rican *Biomphalaria glabrata* with a 1.0 cc syringe and streaked across the slide in a thick smear. The slides were air dried and stained with Wright's stain (3 minutes) using standard blood techniques. The slides were studied extensively for characteristic cell morphology and hemolymph inclusions. Aqueous solutions of TBTO were prepared in varying concentrations from formulated BioMet SRM® pellets. Hemolymph was collected from snails immersed in the toxicant for varied time periods and stained and studied for morphological effects. In order to distinguish the effects resulting from the death of the animal from the effects of the toxicant, slides were made of hemolymph drawn from normal live snails, freshly killed snails, exposed live snails, and fatally exposed snails. The total available (ta) concentrations used were 10 ppm, 1 ppm, and 0.5 ppm. Times of exposure were varied at 2, 6, 12, 24, 48, 72, and 96 hours. Photographs were taken of normal and test hemolymph slides using a Ziess Universal Light Microscope with Nomarski Phase Contrast at 1000X.

To determine the total cell volume, hemolymph was drawn from normal snails using a Wintrobe hematocrit tube. The tubes were inserted through a small hole made in the shell near the mantle cavity area to insure a high percentage of cells using the technique described by Malek (16). The hemolymph was centrifuged for 10 minutes and then the total cell volume was recorded. Test snails were exposed to 10 ppm, 1 ppm and 0.5 ppm ta concentrations for varying time periods and tested in the same manner for total cell volume.

### B. *Results and Discussion*

Two different types of cells were observed in normal snail hemolymph: granulocytes and hyalinocytes. Using Wright's stain, the granulocytes appeared to have dark blue or purple cytoplasm with dark purple or magenta nuclei. Small granules located only within the central part of the cytoplasm stained slightly darker than the cytoplasm. The nuclei are ovoid to bean-shaped, and the pseudopodia are characteristically filamentous, sometimes forming threadlike connections between cells during aggluination. Clusters of agglutinated granulocytes were found most often, but occasionally single cells were seen. Hyalinocytes were more often found as isolated individual cells.

The hylinocytes are more regularly spherical in shape, but do have lobose pseudopodia. The cytoplasm is light purple stained; the roughly spherical nucleus (larger than granulocyte nuclei) stains dark purple. Hyalinocytes are much less common than granulocytes, and often contain ingested particles resembling bacteria. Hemolymph drawn from the mantle cavity sinus area had a larger number of cells than that drawn from any general body region.

Hemolymph inclusions in normal snail hemolymph were mostly single cells, cell fragments, and small clusters of cells of many types, epithelial being particularly numerous. Hemolymph obtained from snails by crushing the shell and collecting pools was much lower in blood cells and higher in inclusions (shell fragments, etc.) than that obtained by directing the syringe into sinus areas.

Short exposure to sub-lethal concentrations of the toxicant showed essentially no morphological changes in cells or inclusions. During early tests in which all snails were exposed to a fatal dosage of the toxicant, the hemolymph took on a brownish color and a thicker consistency than normal. Stained slides showed very few blood cells or agglomeration of blood cells. Hemolymph inclusions consisted of a vast littering of cellular material including some whole cells but most cell fragments, organelles, "ghosts" of ruptured cells, bacteria, and assorted unidentified cellular dibris. To determine whether this effect was caused by the organotin exposure or by the death of the animal, slides of hemolymph from normal live, freshly killed, exposed alive, and fatally exposed snails were compared. Snails which were killed by placing them in hot (85$^{\circ}$C) water exhibited many of the same effects as the organotin-killed snails. The hemolymph once again took on a brownish color, and showed an increase in the number and types of inclusions, especially numerous cocci bacteria. Fatally exposed snails' hemolymph exhibited the same accumulation of cell fragments and degenerating cells. The blood cells often showed dark stained large granules or particles, and in some cases these granules closely resembled some newly ingested cocci similar to numerous extracellular cocci in the surrounding hemolymph. Nonfatally exposed snails also showed a degeneration of cells in hemolymph inclusions, but not to the extent of the dead snails, and the hemolymph retained its red color. Inclusions from these snails seemed to consist mostly of cellular fragments and "ghosts" of ruptured cells. Also found within the hemolymph of exposed snails were small clumps of tissue (not blood cells) containing spherical, yellow-colored inclusions very similar to those reportedly found in connective tissue of organotin-killed snails (4). The hemolymph cells, especially granulocytes, showed an increase in the presence of phagocytized particles, even though the extracellular material was free from surrounding cocci. These particles ranged from small and granular to large irregularly shaped particles, all very dark stained. It has long been known that molluscan hemolymph cells exhibit ameboid qualities and are phagocytic (17). Foreign substances such as India ink particles, bacteria, pollen, and polystyrene spheres (18, 19) have been phagocytized by hemolymph cells. However, this is generally accompanied by an increase in the number of hemolymph cells in the plasma, as the snails' immunological system recognizes the substance as "non-self".

Hematocrit readings from normal snails ranged from 0.5 to 1.0% but never exceeded 1%. Cell volumes were observed for exposed snails in varying concentrations, and in all tests for determination of total protein. In most cases the cell volume was within this range and there was no correlation between exposure time and percentage of cells present. Since increasing cell volume is a characteristic response to foreign materials in the body, it seems that the animal does not elicit an immunological response to toxicant.

## VI. CONCLUSIONS

Several sites of TBTO accumulation in the snail are indicated by these studies. These include the cell membrane, the nucleus, the mitochondria, and hemolymph proteins.

High concentrations of TBTO in the nuclear fraction of snail tissue may be due to either the combination of TBTO with proteins or phospholipids in the cell membrane or combination with nuclear material. Disruption of the cell membrane is supported by morphological studies showing massive hemorrhage and cell breakdown in snails exposed to TBTO, particularly in connective tissue (4). Nuclear involvement is implicated by TBTO accumulation in reproductive tissue and morphogenicity studies showing drastic decreases in tolerance of successive generations to TBTO (2).

Organotins have been shown in inhibit oxidation in *Biomphalaria alexandria* and *Biomphalaria truncatus* (20), and also to inhibit oxidation in intact mitochondria (21). Accumulation of TBTO in the mitochondrial fraction of exposed snails provides further evidence of inhibition of mitochondrial oxidation, although the exact mechanism is unknown.

Accumulation of TBTO in hemolymph with simultaneous alterations in the hemolymph protein levels suggests a combination of TBTO with specific hemolymph proteins. Separation and analysis of hemolymph proteins will show what specific proteins are decreasing and what, if any, new proteins are found in the hemolymph.

Phagocytic action of granulocyte and hyalinocyte blood cells, a common response to foreign material in the hemolymph, is found in snails exposed to TBTO. However, the lack of increase in cell volume with exposure indicates that the animal as a whole does not elicit an immunological response to the toxicant.

The toxicity of TBTO in *Biomphalaria glabrata* appears to result from several processes. In all cases the effects of TBTO may be due to its combination with proteins. It remains to be determined what the sites of protein attachment are and what the biochemical effects of such combinations are.

## ACKNOWLEDGMENTS

The authors would like to thank the World Health Organization for sponsoring this investigation and also Mr. Nate Cardarelli, Mr. William Evans, and Mr. Thomas Quick for their invaluable assistance.

## REFERENCES

1. N.F. Cardarelli, Controlled Release Molluscicide, Monogr. Publ., University of Akron, Akron, Ohio, May 1977, pp. 136.
2. N.F. Cardarelli, Laboratory and Field Evaluations of Controlled Release Molluscicides and Schistolarvicides, Annual Report (unpublished), Edna McConnell Clark Foundation, 276-0091, New York, July 1, 1977.
3. N.F. Cardarelli, Controlled Release Pesticide Formulations, CRC Press, Cleveland, Ohio (1976).
4. N.F. Cardarelli, *et al.*, "A Method of Destroying Snail Vectors of Schistosomiasis," Final Report (Univ. of Akron), World Health Organization, November 10, 1969.
5. P.J. Smith and L. Smith, Chem. in Brit., 11, 208-212, 226, (1975).

6. I. Robinson, Food and Cosm. Tox., 7, 47-52, (1969).
7. S.V. Kanakkanatt, University of Akron, personal communication, (1979).
8. E.D. Gilbertson, *et al.*, J. Parasit, 53, 565-568, (1967).
9. T.C. Cheng and J.T. Sullivan, "Mode of Entry, Action and Toxicity of Copper Molluscicides," in Molluscicides in Schistosomiasis Control, T.C. Cheng (Ed.), pp. 89-153.
10. C.M. Yager and H.W. Harry, Malacologia, 1, 339-353, (1964).
11. C.M. Hageboom, "Fractionation of Cell Components of Animal Tissue," in Methods of Enzymology, Colowich and Kaplan (Eds.), pp. 16-19, Academic Press, (1955).
12. J.J. Zuckerman (Ed.), Organotin Compounds, Adv. in Chem., Ser. No. 157, Amer. Chem. Soc., Washington, D.C., pp. 229, (1976).
13. J.L. Bailey, Techniques in Protein Chemistry, American Elsevier, New York, (1967).
14. O. Lowry, N. Rosenbough, A. Farr, and R. Randall, "Protein Measurement with Folin-Wu Phenol Reagent," J. of Biol. Chem. 193, pp 265-275, (1951).
15. T.C. Cheng and O.L. Lee, "*Schistosoma mansoni:* Alterations in Total Protein and Hemoglobin in the Hemolymph of Infected *Biomphalaria glabrata,*" Expt. Parasitology, 31, pp. 203-216 (1972).
16. E.A. Malek, Laboratory Guide and Notes for Medical Malacology, Burgess Publ., Minneapolis, (1962).
17. M. Kollman, Ann. Sci. Nat. Zool. (Paris), 8, 1-240, (1908).
18. L.A. Stauber, "The Fate of India Ink Injected Intracardially into Oyster, *Ostea virginica,*" Gmelin. Brol. Bull, 98, pp. 227-241, (1950).
19. M.R. Tripp, J. Parasit., 47, 745-751, (1961).
20. A.A. Sharaf, *et al.*, Egyptian J. of Bilharzia, 1, No. 2, 227-237, (1974).
21. M.J. Selwyn, *et al.*, Europ. J. of Biochem., 15, 242, (1970).

# MECHANISM OF RELEASE OF ORGANOTIN FROM THERMOSET POLYMERS

*K. N. Somasekharan*
*R. V. Subramanian*

Department of Materials Science and Engineering
Washington State University
Pullman, Washington

## I. INTRODUCTION

We have reported extensively on antifouling formulations in which the tributyltin carboxylate group is chemically anchored to a polymer chain (1-3). In these controlled release formulations, the prepolymers were prepared by partial esterification of linear polymers carrying carboxylic acid or anhydride groups by bis(tri-n-butyltin) oxide (TBTO). The residual free carboxylic acid or anhydride groups were then reacted with diepoxides to form thermoset organotin-epoxy polymers. Many variations of this scheme were investigated, including one which provided for simultaneous vinyl polymerization and carboxyl-epoxide reactions. Curing was accomplished at high temperatures (about 150$^{o}$C) in all these cases.

There exists, however, a need for antifouling coatings that can be cured at room temperature. Preliminary results on the incorporation of organotin groups into room-temperature cured polyesters, urethanes and aziridines have also been reported (4). New organotin-epoxy systems curable at room temperature have been developed and characterized (5).

In this paper, the mechanism of release of organotin species from a polymer matrix is examined in detail. Various steps involved in the transport mechanism are taken into account in constructing theoretical models and in arriving at predictions based on them.




ISBN 0-12-074450-3

*A. Controlled Release Systems*

Laboratory study of biotoxicity (6) and performance tests in marine environments (3,4) reveal that resistance to fouling is influenced by matrix characteristics. Antifouling performance of organotin carboxylate polymers indicates that their mode of action corresponds to the bulk abiotic bond cleavage model (7). We have carefully considered all the possible controlling mechanisms, viz.,

(a) diffusion of water (and possibly chloride) into the polymer matrix from sea water;

(b) hydrolysis of tributylin carboxylates to produce TBTO (or TBTCl);

(c) diffusion, from the matrix to the surface, of the mobile species produced;

(d) phase transfer of the organotin species;

(e) its migration across the boundary layer; and

(f) possible mechanical loss of the tributyltin species from the surface.

The tributyltin group undergoes fast chemical exchange, and hence a hydrolytic equilibrium is rapidly established between TBT carboxylates and TBTO (8). Laboratory determination of the release rate, under laminar flow conditions, shows that the phase transfer and migration across the boundary layer are also relatively fast (9). Thus we have come to the conclusion that diffusion, from the matrix to the surface, of the mobile TBT species produced, is the rate controlling factor in release from epoxy systems (9).

As the mobile species produced (TBTO) diffuses out, the hydrolysis is expected to proceed at a concentration-dependent rate. The model developed by Godbee and Joy for predicting the leachability of radionuclides from cementitious grouts (10) closely represents this situation. Based on their equations, the rate of release of tin (dq/dt in $g/cm^2/sec$) from the surface at time, t, should be

$$dq/dt = C_m D^{\frac{1}{2}} K^{\frac{1}{2}} \left[\mathrm{erf}(K^{\frac{1}{2}} t^{\frac{1}{2}}) + \frac{\exp(-Kt)}{(\pi Kt)^{\frac{1}{2}}}\right] \quad (1)$$

where $C_m$ is the concentration of the mobile species in $g/cm^3$, D is the effective diffusivity in the matrix in $cm^2/sec$, and K is the concentration-dependent hydrolysis rate constant in $sec^{-1}$.

The most important feature of this model is that when Kt becomes large, $\mathrm{erf}(Kt)^{\frac{1}{2}}$ approaches unity, and the rate becomes independent of time. This zero order rate is the coveted characteristic of controlled release systems. However, we have not been able to realize this ideal behavior in the epoxy systems (9). This may partly be due to the tight matrix in these epoxy systems. Past systems have considered only TBT carboxylate closely bound to the backbone of the polymer. By varying the length of the chain holding the TBT moiety, greater mobility of the TBT group is to be expected. A lower $T_g$ is actually observed as this chain is extended (5), which means a lower activation energy for this motion. The decrease in matrix constraints allowing this motion may also result in decreased resistance to the diffusion of TBTO. The synthesis and mechanical characterization of these systems have been the first steps in pursuing the idealized controlled release from epoxy systems.

## II. CONVENTIONAL ANTIFOULING SYSTEMS

In the conventional antifouling compositions, the organotin compound (TBTO, TBTF, TBTCl, TBTOAc, TBTS) is mechanically mixed into the paint vehicle. When the TBT species is completely soluble in the polymer matrix, factors (a) and (b) become unimportant in most cases. The mobile species is already present; its diffusion in the matrix, phase transfer and migration across the boundary layer into ocean environment may be represented by Figure 1. When the organotin compound forms a dispersed second phase, the rate of its dissolution in the polymer matrix becomes another factor to consider.

### A. *Diffusion in Matrix*

Figure 1 represents the concentration profile of the tin species during the service life of the coating. During service, the organotin species is lost from the surface. The diffusion in the polymer matrix is represented by Fick's second law for nonsteady state flow:

$$\partial C_1/\partial t = D_1 \; (\partial^2 C_1/\partial x^2) \qquad -\infty < x < 0 \qquad (2)$$

This refers to the depletion of the tin species at a given point in the matrix as a function of time. ($D_1$, the effective diffusivity, is expressed in $cm^2/sec$; $C_1$, the concentration, in $g/cm^3$; x, the distance, in cm; and t, time, in sec.) Based on this, the transport equation for a semi-infinite medium of uniform initial concentration of mobile

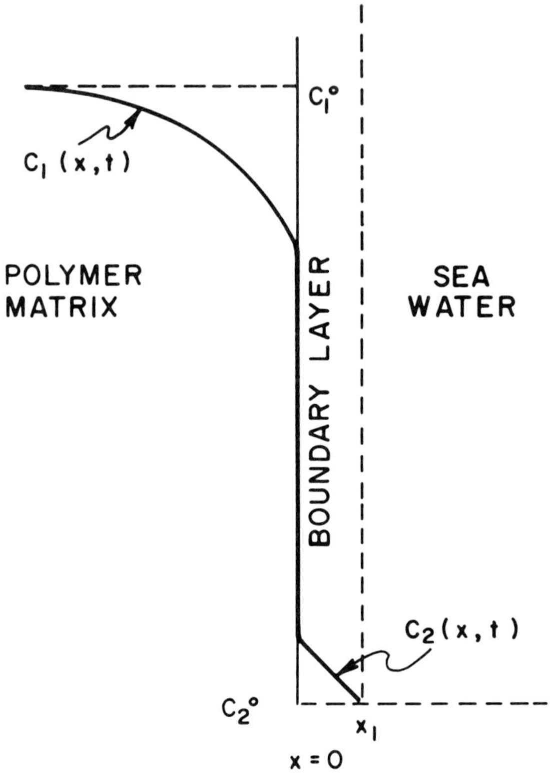


*FIGURE 1. Release of organotin from polymer matrix under service conditions.*

species $C_1^o$, with the surface concentration equal to zero for time greater than zero, is given by Crank (11). The concentration profile for this model is:

$$C_1 = C_1^o \text{ erf } x/2(D_1t)^{\frac{1}{2}} \tag{3}$$

The rate of mass transfer at the surface (in g/cm$^2$/sec) is:

$$dq/dt = C_1^o \ (D_1/\pi t)^{\frac{1}{2}} \tag{4}$$

and the quantity of tin lost from the matrix (Q in g/cm$^2$):

$$Q = 2C_1^o \ (D_1 t/\pi)^{\frac{1}{2}} \tag{5}$$

### *B. Phase Transfer*

A resistance to mass transfer can be expected at the interface. Cardarelli observed that antifouling vulcanized rubber in the partially toxicant-depleted state showed no measurable toxicant gradient (12), indicating that diffusion is very fast in the rubber matrix. Also, tests performed on an underwater rotating device showed no significant increase with velocity in the rate of loss of TBTO (12), suggesting that the boundary layer effects are also not very significant. Thus the controlling factor for the rate of release of toxicant becomes the phase transfer stage:

$$dq/dt = -K_1C_1 \tag{6}$$

where the rate is expressed in g/cm$^2$/sec and $C_1$, the concentration in the matrix, in g/cm$^3$.

The conventional method of treating this situation, however, is by assuming an interfacial equilibrium:

$$\text{at } x = 0, \qquad C_2 = mC_1 \qquad \text{for all } t > 0 \tag{7}$$

where $C_2$ is the concentration in the aqueous phase and m is the distribution coefficient of TBTO between the polymer phase and the secondary aqueous phase. Based on the reported solubility of TBTCl in sea water (13), "m" may be assigned a value of 5 x $10^{-5}$. An assumption is being made here that the equilibrium is fast. Since Cardarelli has pointed out the possibility of a rate-controlling interfacial transfer, we have considered the phase transfer rate rather than interfacial equilibrium in the model reported earlier (9).

### *C. Boundary Layer*

There exists a quiescent layer through which the organotin species must diffuse before being carried by the seawater flow past the surface of the coating. The thickness of the boundary layer (L) is defined (14) under laminar flow conditions by:

$$L = \zeta \frac{\eta^{0.5} \; x^{0.5}}{U^{0.5} \; d^{0.5}} \tag{8}$$

and under turbulent flow conditions by:

$$L = \xi \frac{\eta^{0.9} \; x^{0.1}}{U^{0.9} \; d^{0.9}} \tag{9}$$

where $\eta$ = the viscosity,
$x$ = the distance from the leading edge,
$U$ = the fluid velocity,
$d$ = the density,
and $\zeta$ & $\xi$ are constants.

The diffusion across the boundary layer is given by Fick's first law for stationary state flow:

$$P = -D_2 \ (\partial C_2/\partial x) \qquad 0 < x < x_1 \tag{10}$$

This refers to the permeation of the tin species through the boundary layer as a function of the concentration gradient. ($D_2$, the effective diffusivity in the boundary layer fluid, is expressed in $cm^2/sec$; $C_2$, the concentration, in $g/cm^3$; $x$, the distance from the polymer surface, in cm; and $t$, time, in sec.).

Ketchum *et al.*, who evaluated the leaching of copper from paint matrices, have reported substantial changes in leaching rate caused by agitation (15). However, Marson predicts less significant differences in the leaching rate of copper under laminar and turbulent flow conditions (14). We have analyzed the data of Ketchum *et al.* in detail and found that the boundary layer effects are significant only in the early stages of copper leaching and become negligible as matrix effects become dominant. Tests with antifouling rubber performed on an underwater rotating device showed no significant increase with velocity in the rate of loss of TBTO (12).

### *D. Nonsteady-Rate-Steady (N-R-S) Model*

Since the matrix, interface and boundary layer are visualized as offering resistance in succession to mass transport, the resultant resistance is obtained by combining the individual resistances in series. On the basis of this, we have derived an expression for the overall rate of release (9):

$$\frac{dq}{dt} = \frac{C_1^0 \ D_1^{\frac{1}{2}} \ K_1 D_2}{K_1 D_2 \ \pi^{\frac{1}{2}} \ t^{\frac{1}{2}} + D_1^{\frac{1}{2}} \ D_2 + D_1^{\frac{1}{2}} \ K_1 L} \tag{11}$$

### *E. Nonsteady-Equilibrium-Steady (N-E-S) Model*

As pointed out earlier, the conventional method of treating the situation existing at the interface in Figure 1 is by assuming an equilibrium between $C_1$ and $C_2$. We then wish to solve the diffusion Equations (2) and (10), under the following initial and boundary conditions:

$$\text{at } t = 0, \quad C_1 = C_1^0 \qquad -\infty < x < 0 \qquad (12)$$

$$\text{at } t = 0, \quad C_2 = C_2^0 \qquad +\infty > x > 0 \qquad (13)$$

$$\text{at } x = 0, \quad C_2 = m\, C_1 \qquad \text{for all } t > 0 \qquad (7)$$

$$\text{at } x = 0, \quad -D_1 \frac{\partial C_1}{\partial x} = -D_2 \frac{\partial C_2}{\partial x} \qquad \text{for all } t > 0 \qquad (14)$$

$$\text{at } x = -\infty, \quad C_1 = C_1^0 \qquad \text{for all } t \qquad (15)$$

$$\text{at } x = x_1, \quad C_2 = C_2^0 \qquad \text{for all } t \qquad (16)$$

Laplace transformation now yields the concentration profile in the matrix:

$$C_1 = C_1^0 + \left(\frac{C_2^0}{m} - C_1^0\right)\left[\text{erfc}\, \frac{x}{(4D_1t)^{1/2}} - \exp\{K_2x + D_1tK_2^2\}\, \text{erfc}\left\{\frac{x}{(4\, D_1\, t)^{1/2}} + K_2(D_1\, t)^{1/2}\right\}\right] \qquad (17)$$

where $K_2 = \dfrac{D_2\, m}{D_1\, L}$

Substitution of the concentration gradient obtained for boundary layer in Fick's first law yields the rate of release of tin into the environment (in $g/cm^2/sec$):

$$\frac{dq}{dt} = (mC_1^0 - C_2^0)\, \frac{D_2}{L}\, \{\exp\, (K_2^2D_1t)\ \text{erfc}\ K_2(D_1t)^{1/2}\} \qquad (18)$$

Integration of the above equation over time gives the amount of tin released (in $g/cm^2$):

$$Q = \frac{mC_1^0 - C_2^0}{K_2m}\, \{2K_2(D_1t/\pi)^{1/2} - 1 + \exp(K_2^2D_1t)\text{erfc}\ K_2(D_1t)^{1/2}\} \qquad (19)$$

### *F. Nonsteady-Equilibrium-Nonsteady (N-E-N) Model*

The situation existing during service life in marine environments is illustrated in Figure 1. To simulate the conditions in the laboratory for the direct determination of the release rate, it is essential to maintain a flow and to

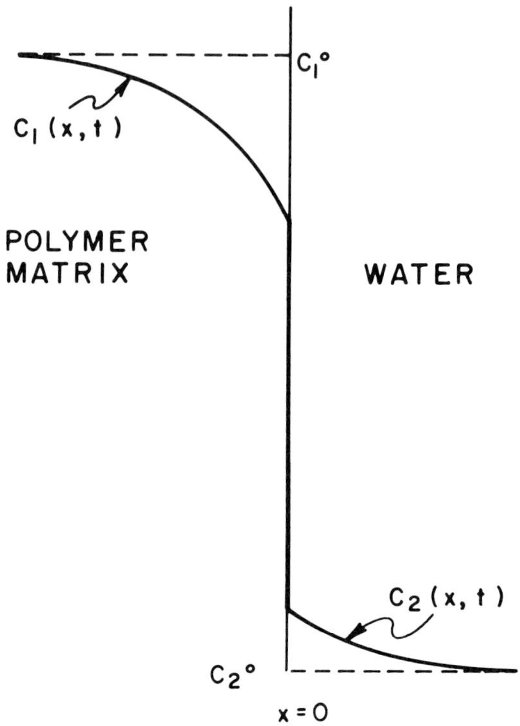


*FIGURE 2. Release under static experimental conditions.*

keep the concentration of TBTCl in the bulk near zero. Static experiments approach the situation described in Figure 2. The diffusion in the polymer matrix is again represented by Fick's second law:

$$\frac{\partial C_1}{\partial t} = D_1 \frac{\partial^2 C_1}{\partial x^2} \qquad -\infty < X < 0 \tag{2}$$

The diffusion in water also will have to be represented by Fick's second law:

$$\frac{\partial C_2}{\partial t} = D_2 \frac{\partial^2 C_2}{\partial x^2} \qquad 0 < x < +\infty \tag{20}$$

The boundary conditions may be set as follows:

$$\text{at } t = 0 \quad C_1 = C_1^0 \qquad -\infty < x < 0 \tag{12}$$

$$\text{at } t = 0 \quad C_2 = C_2^o \qquad 0 < x < +\infty \qquad (13)$$

$$\text{at } x = 0 \quad C_2 = mC_1 \qquad \text{for all } t > 0 \qquad (7)$$

$$\text{at } x = 0 \quad -D_1 \frac{\partial C_1}{\partial x} = -D_2 \frac{\partial C_2}{\partial x} \qquad \text{for all } t > 0 \qquad (14)$$

$$\text{at } x = -\infty \quad C_1 = C_1^o \qquad \text{for all } t \qquad (15)$$

$$\text{at } x = +\infty \quad C_2 = C_2^o \qquad \text{for all } t \qquad (21)$$

This problem has been solved by Bird *et al.* by Laplace transform techniques (16). The concentration profiles are:

$$\frac{C_1 - C_1^o}{C_2^o - mC_1^o} = \frac{1 + \operatorname{erf} x/(4D_1t)^{\frac{1}{2}}}{m + (D_1/D_2)^{\frac{1}{2}}} \qquad (22)$$

$$\frac{C_2 - C_2^o}{C_1^o - (C_2^o/m)} = \frac{1 - \operatorname{erf} x/(4D_2t)^{\frac{1}{2}}}{1/m + (D_2/D_1)^{\frac{1}{2}}} \qquad (23)$$

Differentiation of these equations, evaluation at $x = 0$, and substitution in Fick's first law provides the mass transfer rate at the interface:

$$\frac{dq}{dt} = \frac{D_1^{\frac{1}{2}}}{(\pi t)^{\frac{1}{2}}} \; \frac{C_2^o - m\,C_1^o}{m + (D_1/D_2)^{\frac{1}{2}}} \qquad (24)$$

The amount of tin released from the polymer matrix at the interface is obtained by integrating the rate over time:

$$Q = 2\,(D_1/\pi)^{\frac{1}{2}} \; \frac{C_2^o - m\,C_1^o}{m + (D_1/D_2)^{\frac{1}{2}}} \; t^{\frac{1}{2}} \qquad (25)$$

## III. MODEL PREDICTIONS

The salient features of the three models are summarized in Table I, and compared with the models of Crank (11), and Cardarelli (12).

TABLE I. Comparison of the Features of Various Models.

| | Crank[a] | N-E-N | N-R-S | N-E-S | Cardarelli[b] |
|---|---|---|---|---|---|
| Matrix | Diffusion, Fick's 2nd law, nonsteady | Diffusion, Fick's 2nd law, nonsteady | Diffusion, Fick's 2nd law, nonsteady | Diffusion, Fick's 2nd law, nonsteady | Diffusion, Fick's 2nd law, fast |
| Interface | No account | Equilibrium at interface | Rate of interface transfer | Equilibrium at interface | Rate of interface transfer |
| Boundary Layer | No account | Diffusion, Fick's 2nd law, nonsteady | Diffusion, Fick's 1st law, steady | Diffusion, Fick's 1st law, steady | Not significant |

[a] Crank, Reference 11.
[b] Cardarelli, Reference 12.

*A. Effect of Time on Rate of Release (Table II)*

It can be readily seen from Equations (4) and (24) that the rate of release is inversely proportional to the square root of time in both Crank's and N-E-N models. The dependence of the rate on time is more complex in the N-E-S model:

$$\text{Rate} \propto \exp(K_2^2\ D_1\ t)\ \text{erfc}\ (K_2 D_1^{\frac{1}{2}}\ t^{\frac{1}{2}}) \qquad (26)$$

which approximates to Rate $\propto t^{-\frac{1}{2}}$

at $\dfrac{D_2\ m}{\sqrt{D_1} L} \simeq 1.6 \times 10^{-2}$.

High rate of flow, high solubility, high diffusivity in water will help approach this condition.

*B. Effect of $D_1$ on the Rate of Release (Table III)*

The rate is proportional to the square root of $D_1$ in Crank's model. However, the relationship is highly insensitive to the matrix diffusivity in the N-E-N model, if the solubility is very low. Consequently, this crucial matrix parameter cannot be determined by static experiments.

*TABLE II. Effect of Time on the Rate of Release*

| | Rate ($10^{-8}$ g/cm²/day) | | |
|---|---|---|---|
| *Time (days)* | *Crank* | *N-E-S (L = 1 cm)* | *N-E-N* |
| 1 | 524 | 4.30 | 25.0 |
| 3 | 303 | 4.28 | 14.4 |
| 5 | 235 | 4.27 | 11.2 |
| 7 | 198 | 4.26 | 9.4 |
| 9 | 175 | 4.25 | 8.3 |

$C_1^o = 0.1$ g/cm³; $C_2^o = 0$; $m = 5 \times 10^{-5}$
$D_1 = 10^{-13}$ cm²/sec; $D_2 = 10^{-7}$ cm²/sec

*TABLE III. Effect of $D_1$ on the Rate of Release*

| $D_1$ $cm^2/sec$ | Rate ($10^{-8}$ $g/cm^2/day$) Crank | N-E-S (L = 1 cm) | N-E-N |
|---|---|---|---|
| $10^{-13}$ | 198 | 4.26 | 9.44 |
| $10^{-12}$ | 627 | 4.30 | 9.76 |
| $10^{-11}$ | 1982 | 4.31 | 9.86 |
| $10^{-10}$ | 6268 | 4.32 | 9.90 |

$C_1^o = 0.1\ g/cm^3;\ C_2^o = 0;\ m = 5 \times 10^{-5};$

$D_1 = 10^{-7} cm^2/sec;\ t = 7$ days

The N-E-S model will also fail to show a strong dependence on matrix diffusion, if the thickness of the boundary layer is high (for a species of low solubility). This problem, however, can be easily overcome by maintaining a good flow of the fluid.

### C. *Effect of $D_2$ on the Rate of Release (Table IV)*

Diffusivity in water does not figure in Crank's expressions. Increase in $D_2$ helps to overcome the stagnation in the boundary layer in the N-E-S model. The N-E-N model begins to approach Crank's model only at unrealistically high values of $D_2$.

### D. *Effect of Solubility on the Rate of Release (Table V)*

If the solubility of the diffusing species is very high (over 5%), both N-E-N and N-E-S models approach Crank's equations. However, when the solubility is low as in the case of TBTCl (50 ppm), boundary layer effects are very significant.

TABLE IV. Effect of $D_2$ on the Rate of Release

| $D_2$ $cm^2/sec$ | Rate ($10^{-8}$ $g/cm^2/day$) Crank | N-E-S (L = 1 cm) | N-E-N |
|---|---|---|---|
| $10^{-7}$ | 198 | 4.3 | 9.4 |
| $10^{-6}$ | 198 | 37.8 | 27.1 |
| $10^{-5}$ | 198 | 160.8 | 66.1 |
| $10^{-4}$ | 198 | 197.6 | 121.4 |

$C_1^o = 0.1\ g/cm^3$; $C_2^o = 0$; $m = 5 \times 10^{-5}$

$D_1 = 10^{-13}\ cm^2/sec$; $t = 7$ days

TABLE V. Effect of Solubility on the Rate of Release

| m (equilibrium constant) | Rate ($10^{-8}$ $g/cm^2/day$) Crank | N-E-S (L = 1 cm) | N-E-N |
|---|---|---|---|
| $5 \times 10^{-5}$ | 198 | 4 | 9 |
| $5 \times 10^{-4}$ | 198 | 38 | 66 |
| $5 \times 10^{-3}$ | 198 | 161 | 165 |
| $5 \times 10^{-2}$ | 198.2 | 197.6 | 195.3 |

$C_1^o = 0.1\ g/cm^3$; $C_2^o = 0$; $D_2 = 10^{-7}\ cm^2/sec$;

$D_1 = 10^{-13}\ cm^2/sec$; $t = 7$ days

### E. *Effect of Boundary Layer Thickness on the Rate of Release (Tables VI and VII)*

As the fluid velocity increases, the boundary layer thickness decreases; the rate of release increases in the N-E-S model. The N-E-S model approaches the predictions of Crank's model, at a boundary layer thickness of $10^{-3}$ cm.

## IV. CONCLUSIONS

The coveted characteristic of controlled release systems is the zero order delivery of the active agent. It is acknowledged that reservoir devices, especially when membrane encapsulated, are capable of steady-state release. Erodible devices of proper geometry can also approach a constant rate of delivery. Matrix devices, however, are generally expected to show a $\sqrt{t}$ relation in the release profile. It is shown in this paper that a time-independent rate of release of organotin is possible from polymer monoliths in which trialkyltin carboxylate groups are chemically attached to the polymer network.

A mathematical model (N-E-S) is developed to describe the release of organotin from conventional antifouling systems under service conditions. Nonsteady-state diffusion in the polymer matrix, equilibrium at the interface, and steady-state diffusion in the boundary layer are the features of this (N-E-S) model.

A different model (N-E-N) is required to represent the release, when experimental determinations are carried out under static conditions. Nonsteady-state diffusion in the matrix, equilibrium at interface and nonsteady-state diffusion in the liquid are the characteristics of this (N-E-N) model.

Analysis of the two models (N-E-S and N-E-N) reveals that, experimental data collected under static conditions do not yield information on the performance under flow conditions. Again, static testing does not yield information on matrix characteristics in cases where the solubility of the released species is very low in the fluid. On the other hand, release rate obtained at one flow rate can be easily converted to release rates at other flow rates.

The models are general and can be applied to drug release from polymer matrices. Thus Chien *et al.*, have shown that a matrix device may show a $\sqrt{t}$ release profile in *in vitro* testing, but might exhibit a zero-order release *in vivo* (17).

TABLE VI. Effect of Boundary Layer Thickness on the Rate of Release

| Time (days) | Rate ($10^{-8}$ g/cm$^2$/day) | | |
|---|---|---|---|
| | L = 1 cm, N-E-S | L = 0.1 cm, N-E-S | L = 0.01 cm, N-E-S |
| 14 | 4.24 | 35.9 | 124. |
| 28 | 4.20 | 33.4 | 92. |
| 42 | 4.18 | 31.7 | 77. |
| 56 | 4.16 | 30.4 | 67. |

$C_1^o$ = 0.1 g/cm$^3$; $C_2^o$ = 0; m = 5 x $10^{-5}$
$D_1$ = $10^{-13}$ cm$^2$/sec; $D_2$ = $10^{-7}$ cm$^2$/sec.

TABLE VII. Effect of Boundary Layer Thicknesses on the Rate of Release

| Time (days) | Rate ($10^{-8}$ g/cm$^2$/day) | |
|---|---|---|
| | L = 0.001, N-E-S | Crank |
| 1 | 513. | 524. |
| 3 | 300 | 303. |
| 5 | 233. | 235. |
| 7 | 197.6 | 198.2 |

$C_1^o$ = 0.1 g/cm$^3$; $C_2^o$ = 0; m = 5 x $10^{-5}$
$D_1$ = $10^{-13}$ cm$^2$/sec; $D_2$ = $10^{-7}$ cm$^2$/sec.

The N-E-S model actually predicts this behavior: $\sqrt{t}$ release profile at high rates of flow and a steady-state release at low flow (Tables VI and VII).

## ACKNOWLEDGMENTS

We thank Ashok Damle, Department of Chemical Engineering, WSU, for the Laplace Transform.

## REFERENCES

1. R. V. Subramanian and M. Anand in Chemistry and Properties of Crosslinked Polymers, S. S. Labana, (Ed.), Academic Press, New York, 1977.
2. R. V. Subramanian, B. K. Garg, J. J. Jakubowski, J. Corredor, J. A. Montemarano, and E. C. Fischer, Am. Chem. Soc., Div. Org. Coat. Plast. Chem., Pap., 36(2), 660 (1976).
3. R. V. Subramanian and B. K. Garg in Proc. Controlled Release Pesticide Symposium, Oregon State Univ., Corvallis, Oregon, 1977.
4. R. V. Subramanian, B. K. Garg, and K. N. Somasekharan, Am. Chem. Soc., Div. Org. Coat. Plast. Chem., Pap., 39, 572 (1978).
5. R. S. Williams, Ph.D. Thesis, Washington State Univ., Pullman, Washington (1979); R. V. Subramanian, R. S. Williams, and K. N. Somasekharan, Am. Chem. Soc., Div. Org. Coat. Plast. Chem., Pap., 41, 38 (1979).
6. R. V. Subramanian, B. K. Garg, and J. Corredor in Organometallic Polymers, C. E. Carraher Jr., J. E. Sheats, and C. U. Pittman Jr., (Eds.), Academic Press, New York, 1978.
7. V. J. Castelli and W. L. Yeager in Controlled Release Polymeric Formulations, D. R. Paul and F. W. Harris, (Eds.), American Chemical Society, Washington, D.C. 1976.
8. K. N. Somasekharan and R. V. Subramanian, Am. Chem. Soc., Div. Org. Coat. Plast. Chem., Pap., 40, 167 (1979).
9. K. N. Somasekharan and R. V. Subramanian in Modification of Polymers, C. E. Carraher Jr., (Ed.), Academic Press, New York, 1979 (in press).

10. H. W. Godbee and D. S. Joy, Assessment of the Loss of Radioactive Isotopes from Waste Solids to the Environment. Part 1: Background and Theory, Oak Ridge National Laboratory, Oak Ridge, Tennessee, 1974.
11. J. Crank, The Mathematics of Diffusion, 2nd Edition, Clarendon Press, Oxford, 1975, p. 32.
12. N. Cardarelli, Controlled Release Pesticide Formulations, CRC Press, Cleveland, Ohio, 1976, p. 35-36.
13. R. F. Bennett and R. J. Zedler, J. Oil Colour Chem. Ass., 49, 928 (1966).
14. F. Marson, J. Appl. Chem., 19, 93 (1969).
15. B. H. Ketchum, J. D. Ferry, A. C. Redfield, and A. E. Burns Jr., Ind. Eng. Chem., 37, 457 (1945).
16. R. W. Bird, W. E. Stewart, and E. N. Lightfoot, Notes on Transport Phenomena, John Wiley, New York, 1958, Chap. 19.
17. Y. W. Chien and H. J. Lambert, J. Pharm. Sci., 63, 515 (1974).

# BEHAVIOURAL RESPONSES TO AMINO AND CARBOXYLIC ACIDS BY *BIOMPHALARIA GLABRATA*, ONE OF THE SNAIL HOSTS OF *SCHISTOSOMA MANSONI*

*J. D. Thomas*
*C. Cowley*

School of Biological Sciences
University of Sussex
Falmer, Brighton

*J. Ofusu-Barko*

Zoology Department
University of Ghana
Legon, Accra, Ghana

## I. INTRODUCTION

The serious problems associated with the application of broad spectrum molluscicides to aquatic habitats for the purpose of controlling diseases such as schistosomiasis and fasciolasis prompted Cardarelli (1) and his associates to investigate the possibility of using controlled release technology to increase the efficiency of mollusciciding. Of the three possible methods of applying this technology, illustrated in Figure 1, the first two have already been developed (1). However, the third is clearly the method of choice as it would allow target animals to be removed selectively without harming other organisms. Before such a formulation can be developed it is necessary to achieve the following objectives. Firstly, it will be necessary to identify potent species-specific factors that will serve consistently as attractants, arrestants, phagostimulants, and feeding incitants to the target animals over the range of chemical conditions that are likely to be encountered in natural water bodies. Some




ISBN 0-12-074450-3

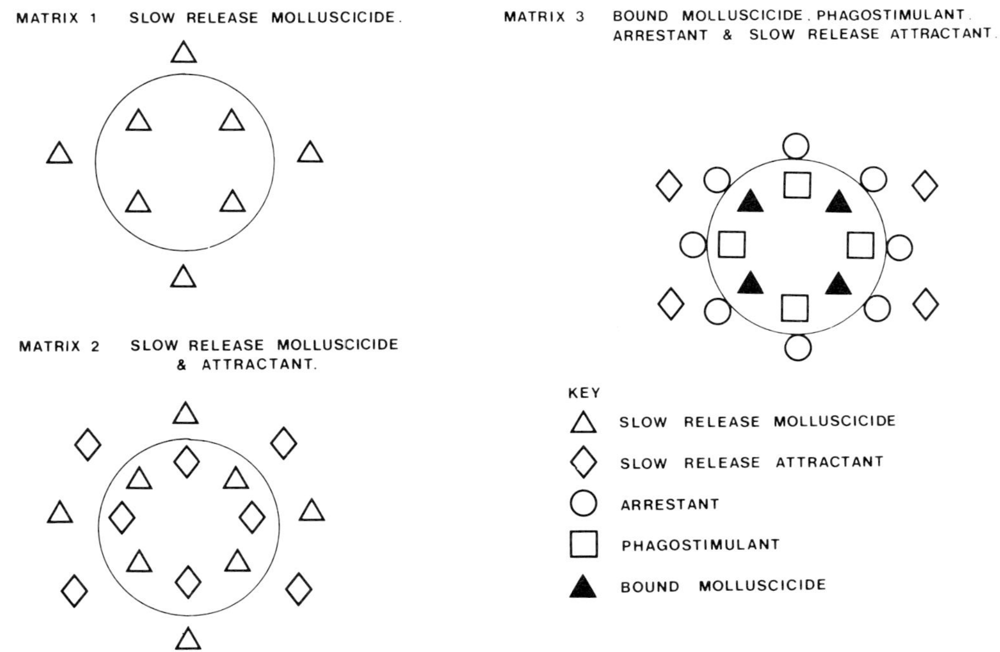


*FIGURE 1. Diagrammatic representation of the three basic controlled release systems.*

optimism regarding the outcome of this approach seems justified because there is an accumulating body of information which shows that chemoreception plays a key role in influencing movement and feeding behavior of aquatic molluscs (2). Secondly, it will be necessary to develop methods for incorporating these active factors into an ingestible matrix containing a bound toxicant.

The work described in the present contribution was undertaken with a view to achieving the following short term objectives. The first was to identify chemical species which might serve as potent attractants and arrestants to adult *B. glabrata* from among amino and carboxylic acids. Both of these classes of compounds are known to be released by snails (3,4) and also by plants which might serve as food organisms (5-10). The second was to ascertain whether the responses to substances identified as potent attractants and arrestants, under the bioassay conditions, were sufficiently robust to make it possible to use them under natural conditions. This is an important consideration as consistency of response is a pre-requisite to the successful application of the proposed method.

## II. MATERIALS AND METHODS

Large, sexually mature, *Biomphalaria glabrata*, weighing approximately 250-350 mg were used in the bioassays. They were acclimated for two days prior to experimentation in standard snail water (SSW2) prepared as described by Thomas *et al.* (11). In order to minimize variation in behavior each of the 120 snails prepared for a series of experiments was kept separately in 100 ml SSW2 in 250 ml covered beakers. These were kept in an environmental unit maintained at a temperature of 26±1°C and a photoperiod of 12 hours light and 12 hours dark. When not used in experiments, each snail was provided with a daily ration of one to four, 3 cm diameter lettuce discs and the SSW2 changed every alternate day. However, on days preceding experimentation, they were placed in fresh SSW2 without food at approximately 17-18 hours. Under normal conditions, on week days, the snails were used for experiments on alternate days. They were provided with standard rations of food at 17-18 hours at the end of each series of experiments.

The type of olfactometer used in the experiments involving amino acids is illustrated in Figure 2 and has already been described (12). A slightly longer, narrower olfactometer with a central chamber measuring 8 x 2 x 1.8 cm and two

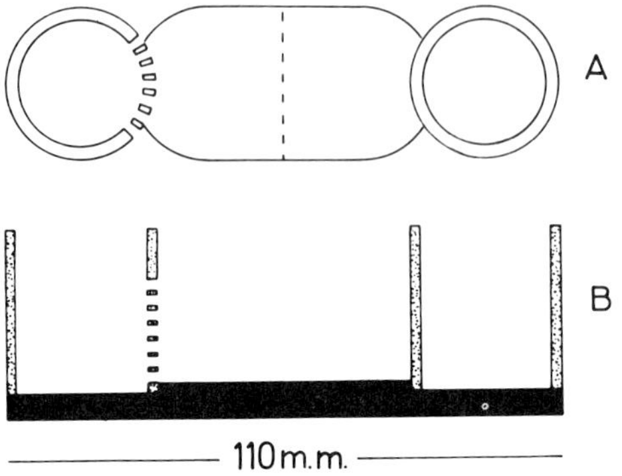


*FIGURE 2. Scale drawing of the large diffusion olfactometer used for studying chemoreception of snails to amino acids.*

end chambers measuring 2.4 cm diameter and a height of 0.2 cm was used to study the chemoreception of snails to carboxylic acids. This was specially designed to give a chemical gradient similar in conformation to that in the small diffusion olfactometers used to study chemoreception in juvenile *B. glabrata* (13). The same volume, 20 ml of 0.22 μ filtered SSW2 was introduced into both types of olfactometer prior to use. Discs of Whatman No. 1 filter paper, which had been cut to fit exactly into the end chambers, were used to introduce the test and control chemicals. Sufficient test chemical was placed on the discs to give an effective concentration of 4-5 x $10^{-4}$M in the test or positive end. Discs soaked in SSW2 were generally used as controls. In the case of the amino acids, it was shown that steep chemical gradients could be maintained for the duration of the experiment, even in the presence of snails, provided they did not exceed 350 mg in weight (13).

As two perspex blocks of 20 olfactometers were used in each assay, every treatment was replicated forty times. Treatments and controls were alternated in successive olfactometers to neutralize any directional bias that might have been caused by gradients in light intensity or temperature. The position of each assay snail was recorded at intervals of 2.0 - 2.5 mins over 16 successive time intervals. Snails were recorded as being + or ⊕ if they were on the test side or test disc respectively and - or ⊖ if they were on the control side or control disc respectively. This made it possible to calculate the mean number of time units spent by the snails on each of these. Paired 't' tests (14) and Wilcoxon paired tests (15) were used to ascertain whether the assay snails spent significantly more or less time on the test sides or the test discs compared with controls. Chemical species which caused the snails to spend significantly more time on the test disc or test side compared with controls were deemed to be attractants or arrestants, respectively. Conversely, if the snails spent significantly more time on the control side or control disc, the test chemicals were classed as repellents. Control experiments in which SSW2 soaked discs were placed on both ends of the olfactometer were carried out periodically to check on any bias not attributable to treatments.

## III. RESULTS

Figures 3 and 4 enable comparisons to be made between the effectiveness of amino acids, carboxylic acid, and related compounds as attractants, arrestants, and repellents to adult *B. glabrata*. Of the amino-acids tested only four, namely aspartic acid, glutamic acid, proline and hydroxyproline were found to be statistically significant attractants. However, as can be seen from Table I, none was consistently effective. This was also the case with lysine and diamino-butyric acid as they were found to be weak arrestants in only some of the trials. Ten amino acids; alanine, phenylalanine, methionine, serine, homoserine, cysteine, penicillamine, tyrosine, ornithine, and tryptophan were found to be statistically significant repellents. This was also the case with indole. The remaining amino acids were neutral as they failed to elicit any statistically significant effects.

In contrast to the amino acids, the simple straight chain aliphatic acids were found to be consistently effective both as attractants and arrestants in all the trials (Table I). The propionate ion proved to be very potent as it was found to be effective, both as an attractant and as an arrestant, at concentrations as low as 5 x $10^{-7}$M over a pH range of 6-8 (Table II; Figure 5). It also continued to be effective even after repeated testing, although some attractants such as β hydroxy-butyric acid showed a classical habituation response when it was subjected to the same treatments (Figure 6). The hydroxy acids; glycolic acid, L(+) lactic acid, and α hydroxybutyric acids were also statistically significant attractants or arrestants. However, the response indices were consistently lower than those of the corresponding acid in all three

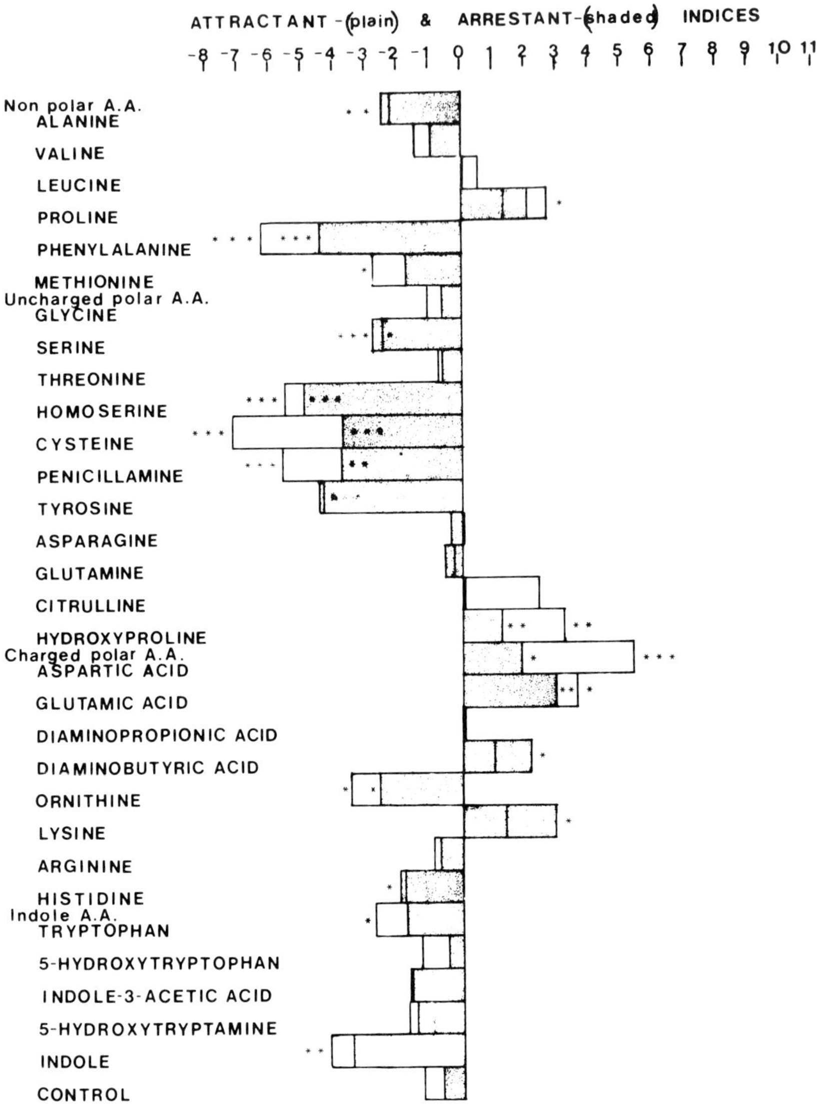


*FIGURE 3. The attractant and arrestant indices obtained when various amino acids and related compounds were tested after being placed on paper discs on the positive (+) or test side of the olfactometer to give effective concentrations of 2-5 x $10^{-4}$M. The asterisks *, **, *** indicate significant levels of P<0.05, P<0.01 and P<0.001 respectively. In the control, the paper discs on both positive and negative sides were soaked in standard snail water.*

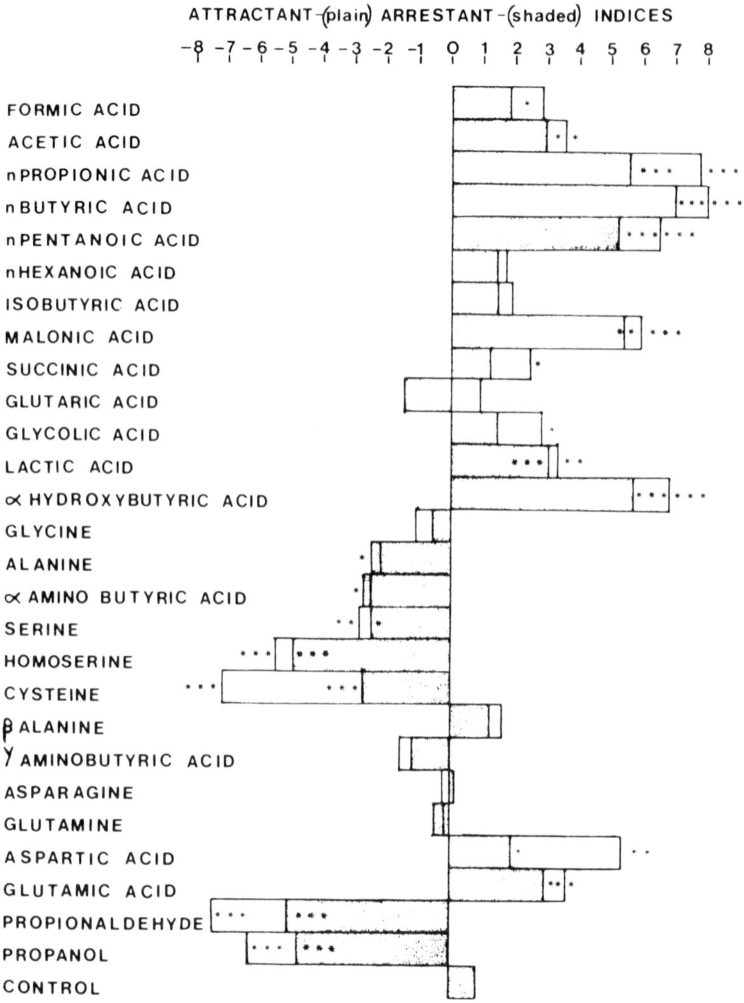


*FIGURE 4. The attractant and arrestant indices obtained when carboxylic acids and related compounds were tested after being placed on paper discs on the positive or test side (+) of the olfactometer to give effective concentrations of 5 x $10^{-4}$M. The P and control values are as indicated in Figure 3.*

cases. Of the dicarboxylic acids tested, malonic acid proved to be a strong attractant but increasing the chain length to give succinic and glutaric acids was followed by a progres-

TABLE I. The Percentage of Trials in Which Various Compounds Were Found to be Statistically Significant Attractants or Arrestants

| Chemical species | Glutamic acid | Aspartic acid | Proline | Hydroxyproline | Propionic acid | Butyric acid |
|---|---|---|---|---|---|---|
| Number of tests | 5 | 5 | 6 | 7 | 5 | 5 |
| Percent of trials in which they were effective as attractants | 20 | 40 | 17 | 29 | 100 | 100 |
| Percent of trials in which they were effective as arrestants | 60 | 60 | 0 | 29 | 100 | 100 |

sive weakening of the response. The aldehyde or alcohol derivatives of the mono-carboxylic acids proved to be strong repellents.

The structure-activity relationship studies suggest that the following subsites are important in determining whether the chemical species involved would act as repellents, attractants, or arrestants: COOH, or $COO^-$, $CH_2$, $NH_2$, $ONH_2$, OH, SH, S, the phenyl and indole group. The propositions which arise from this hypothesis and the supporting evidence are given below:

1. The two key subsites needed to generate consistent attractant or arrestant responses are the hydrophilic, dissociated, carboxylic group on the α carbon and a single, hydrophobic terminal methyl group at the other end of a straight chain molecule. However, the distance between these two subsites is critical. The evidence in support of these two statements is as follows. Firstly, the maximum activity levels were obtained with *n* propionic and *n* butyric acids with the methyl groups on the β and γ carbons respectively. Either increasing or decreasing the chain length to give other acids in the homologous series was followed by a decrease in response. Hexanoic acid was, in fact, inactive. Secondly, the shape of the molecule is clearly important as the snails responded strongly to *n* butyric acid but not to *iso* butyric acid.

2. Substitution of other subsites in the simple aliphatic acid configuration usually results in either a weakening, loss, or reversal of the attractant effect. The evidence in support of this statement is as follows:

a) Substitutions of a hydroxy group for one of the hydrogens on the α-carbon of acetic, propionic and butyric acids to give glycolic, L(+) lactic and α hydroxy-butyric acid leads to a weakening of the response indices in all three cases, although they remain statistically significant.

b) Unlike 'a' above, substitution of an amino group for one of the hydrogen atoms on the α carbon of acetic, propionic and butyric acid to give glycine, alanine and α aminobutyric acid respectively is followed by either a loss or reversal of the attractant effects.

c) Substitution of an imino group for one of the hydrogen atoms on the α carbon has a less drastic effect, as both proline and hydroxy-proline were statistically significant attractants, in some of the tests.

d) Substitution of OH, SH, phenyl or indole groups for one of the H atoms in the distal carbon of an amino acid give chemical species which are strongly repellent, cysteine, penicillamine, serine, homoserine, tyrosine, phenylalanine, and tryptophan.

TABLE II. *The Effectiveness of Sodium Propionate as an Attractant and as an Arrestant to Adult B. glabrata*

| Effective concentration (mols) | $10^{-8}M$ | $10^{-7}M$ | $10^{-6}M$ | $10^{-5}M$ | $10^{-4}M$ |
|---|---|---|---|---|---|
| | $\bar{x}$ SE | $\bar{x}$ SE | $\bar{x}$ SE | $\bar{x}$ SE | $\bar{x}$ SE |
| Attractant index | $1.90 \pm 1.0$ | $3.00^{a} \pm 1.3$ | $5.55^{c} \pm 1.2$ | $5.35^{c} \pm 0.8$ | $7.35^{c} \pm 1.2$ |
| Arrestant index | $1.90 \pm 0.8$ | $2.60^{a} \pm 1.2$ | $5.08^{c} \pm 1.1$ | $4.55^{c} \pm 0.7$ | $6.85^{c} \pm 1.1$ |

$\bar{x}$ = mean
SE = standard error

$^{a}p < 0.05$
$^{b}p < 0.01$
$^{c}p < 0.001$

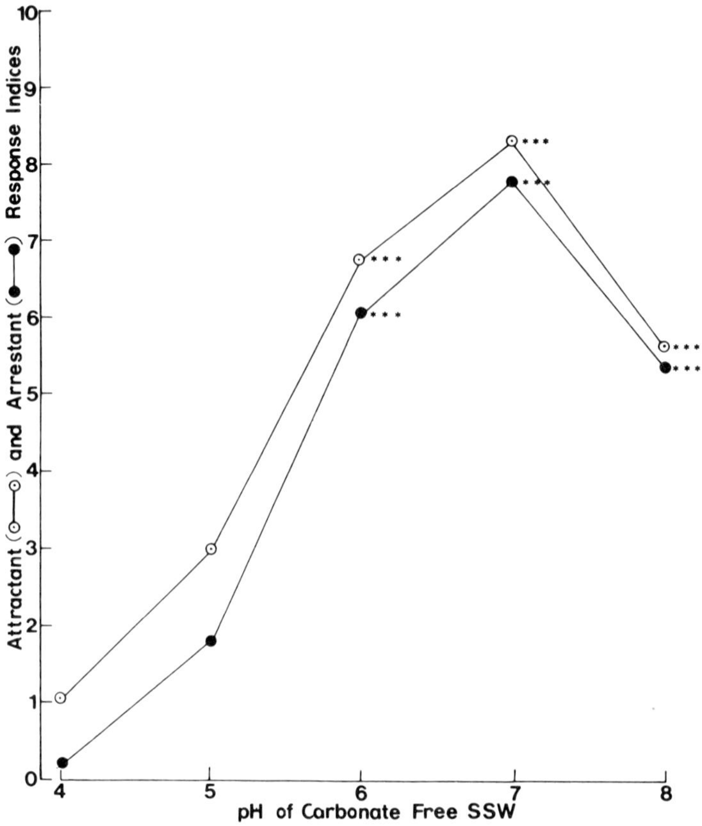


*FIGURE 5. The effects of varying the pH of carbonate free SSW2 on the attractant and arrestant indices obtained when testing sodium propionate at effective concentrations of 5 x $10^{-4}$M. P values are as indicated in Figure 3.*

e) Substitution of amide or amino groups for one of the hydrogen atoms on the distal carbon of amino acids have less drastic effects than in 'b' above. Thus asparagine, glutamine, and citrulline are neutral whereas lysine and diaminobutyric acids are weak arrestants in some of the tests.

f) The presence of the sulphur atom attached to the γ-carbon as in methionine seems to be associated with its repellent action.

g) The substitition of a hydroxy substituent for one of the hydrogen atoms on the β carbon of an amino acid results in a compound such as threonine which is neutral, whereas

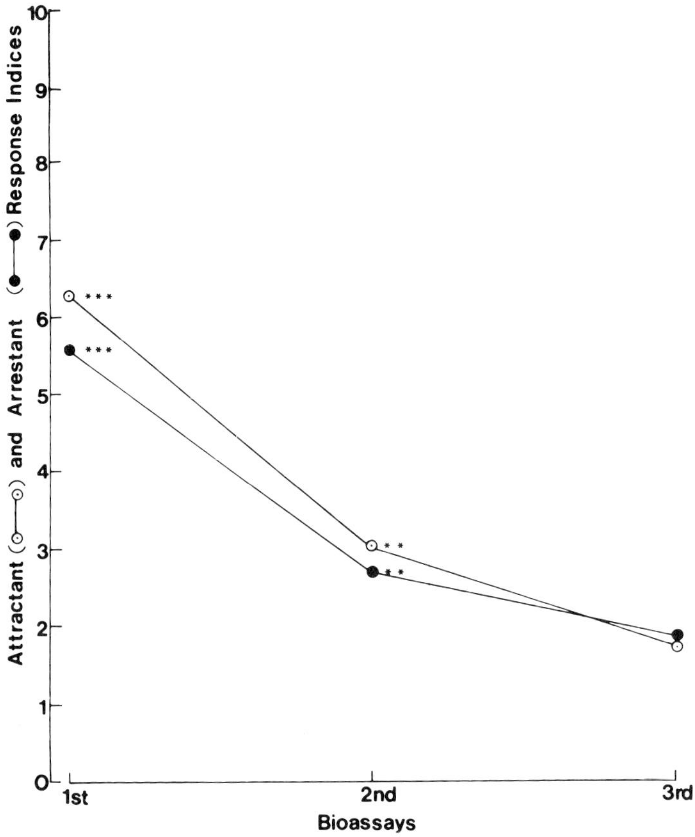


*FIGURE 6. The effects of repeated testing on the responses of adult b. glabrata to β hydroxybutyric acid. P values are as indicated in Figure 3.*

the presence of a hydroxy group on the distal carbon is associated with repellent effects.

h) Substitution of a carboxyl group for one of the hydrogen atoms on the distal carbon of an amino acid can restore the attractant and arrestant properties of the molecule. Thus aspartic and glutamic acids were statistically significant attractants or arrestants in some of the tests. Dicarboxylic acids such as malonic acid, which lack the amino group are also attractants although the effect weakens as the chain length increases in this series.

i) Substitution of hydroxy or aldehyde groups for the carboxyl groups of simple aliphatic acids gives compounds which have very strong repellent effects.

## IV. DISCUSSION

So far as chemoreception to amino and aliphatic acids is concerned there is very little overlap between the niches of adult *B. glabrata* and their neonate conspecifics. They were found to resemble each other only in finding propionic and butyric acids to be attractants and indole and penicillamine to be repellents. The following are some of the major differences. Firstly, none of the ten amino acids identified as being strong attractants to neonates (13) were found to be so for adults. Secondly, adult snails are attracted to aspartic acid, glutamic acid, proline and hydroxy-proline, whereas neonates are not. Thirdly, the response indices of juvenile snails to amino acids, tend to be higher and more consistent than was the case with adults. Fourthly, adults respond to a wider range of aliphatic acids than the juveniles.

Two hypotheses can be advanced to explain why snails find some amino acids attractive and also why neonates and adults have different preferences. Firstly, it is possible that snails are attracted to exogenous amino acids because they are utilized as growth factors or nutrients. Observations which show that pulmonate snails can take up amino acids actively from the medium even when they are present at low concentrations (16,17) support this hypothesis. If it is correct, it is to be expected that neonates would respond more strongly than adults as their relatively larger surface area/volume ratio would favor the uptake of amino acids. Secondly, they may respond to amino and aliphatic acids because they serve as indicators of the presence of conspecifics (pheromones) or of food organisms (kairomones). As adult snails differ from neonates in their feeding niches (13) and in being reproductively active, it is to be expected that they would respond to different sets of chemical messengers.

Glutamic acid, proline, aspartic acid and at least 17 other amino acids are known to be present in media conditioned by *B. glabrata* (4) and it has been suggested by Uhazy *et al.* (18) that the first two may act as pheromones. However, there are several observations that militate against this hypothesis. Firstly, these amino acids are also common constituents of plants, including lettuce (18) used to feed the snails. They are, therefore, likely to be released into the medium particularly when the snails are feeding and they may, therefore, serve as kairomones. Secondly, aspartic and glutamic acids are amongst the commonest constituents of surface water and serve as attractants or feeding incitants to a variety of other aquatic organisms including *Aplysia*, several species of crabs, lobsters and barnacles (19). Thirdly, conspecific neonates do not respond to these amino acids and the adult snails are not attracted to them consistently. The variability in the responses of the adults has yet to be explained. There appears to be no correlation between the extent to which adult snails respond to these amino acids and the level of sexual activity (Thomas and Cowley, unpublished). However, the possibility that their responses to amino acids may be related to the amino acid composition of their diet or the conditioned medium in which they live requires investigation.

The strong repellent effects of many of the amino acids, indole, aldehydes and alcohols also require explanation. One possibility is that the snails avoid them because they are toxic. In this connection, it is well known that many non-protein amino acids may serve as feeding inhibitors or toxins to herbivores (20). Janzen *et al.* (21) have recently shown that even protein amino acids may be toxic if they are not normally encountered in the diet because they can disrupt the supply-demand balance in protein synthesis.

Stronger arguments can be advanced to support the hypothesis that propionic and butyric acids are acting as pheromones than was the case with the amino acids. Firstly, it has been shown that these acids are present in mucus secreted by the foot glands of pulmonate snails (3), including *B. glabrata*. (C. Simpson and R. Chamberlain; personal communication). Secondly, both neonate and adult *B. glabrata* respond to them consistently. Thirdly, they are effective over a pH range of 6-8 which is that normally encountered in freshwater ecosystems. Fourthly, they remain active at concentrations as low as $5 \times 10^{-7}$M. Fifthly, they are able to discriminate between these acids and related forms on the basis of size and shape.

Aliphatic acids are also released by aquatic plants (9), the commonest ones being glycolic, formic, lactic and keto-acids although dicarboxylic acids such as oxalic and succinic acids have been recorded from root exudates (22). As it has been shown that many of these are attractants or arrestants to the snails they may, therefore, act as kairomones.

It has been shown that various species of carboxylic acids can convey information to other organisms as well as snails. For example, they serve as attractants to fungi (23), planaria (24), and miracidia of trematodes (25) including *Schistosoma mansoni* (26), which uses *B. glabrata* as its snail host. As it has been shown that *n* butyric acid serves as an attractant to this miracidium (26), it would appear that it uses the pheromones of the host to identify its target. Formic acid is used by ants as an alarm pheromone or as a repellent (allomone) (27). Other aliphatic acids have been shown to serve as sex pheromones for species of insects and higher organisms including primates (28).

Both propionic and butyric acids satisfy several of the criteria that would make it possible to use them as attractants and arrestants in controlled release formulations. Thus they are effective at low concentrations, over a pH range that is normally encountered in nature and the snails do not habituate very readily to them. Further work is, however, needed to ascertain whether they are sufficiently specific in their action. If they also satisfy this requirement, one possible approach would be to incorporate them in a cellulose matrix containing bound phagostimulants, feeding incitants and a toxicant such as the naturally occurring dialdehyde warburgenal.

## ACKNOWLEDGMENTS

This investigation received financial support from the UNDP/World Bank/WHO Special Programme for Research Training in Tropical Diseases. We are also grateful to Professor R.J. Andrew and Dr. B. Goodwin for providing facilities, to Dr. F. McCapra for advice and encouragement and Miss F. McAllister for technical assistance.

## REFERENCES

1. N. Cardarelli, Controlled Release Molluscicides, Univ. of Akron, Akron, Ohio (1977).
2. J. D. Thomas and J. D. Bousfield, Proc. Controlled Release Pesticide Symposium, National Bureau of Standards, Gaithersburg, Maryland, August 1978.
3. R. A. Wilson, Comp. Biochem. Physiol., 24, 629 (1968).
4. A. J. MacInnis, M. W. Bethel, and E. M. Cornford, Nature, 248, 361 (1974).
5. R. C. Mertz, R. G. Zehnpfennig, and R. J. Klima, J. Water Poll. Control Fed., 34, 103 (1962).
6. R. G. Wetzel, Bioscience, 19, 539 (1969).
7. G. E. Fogg, Arch. Hydrobiol., 15, 1 (1971).
8. R. G. Wetzel and B. A. Manny, Arch. Hydrobiol., 18, 162 (1972).
9. S. A. Hellebust, in Algal Physiology and Biochemistry in Botanical Monographs, 10, 836, D. P. Stewart (Ed.), Blackwell Sci., Oxford University Press, 1974.

10. J. E. Hutchinson, A Treatize in Limnology 3. Limnological Botany, Wiley Interscience, New York and London, 1975.
11. J. D. Thomas, A. S. Lough, and R. W. Lodge, J. Appl. Ecol., 12, 421 (1975).
12. J. D. Thomas and B. Assefa, Proc. Controlled Release Symposium, pp. 4.1, Natural Bureau of Standards, Gaithersburg, Maryland, August 1978.
13. J. D. Thomas and B. Assefa, Comp. Biochem. Physiol., 63C, 99-108 (1979).
14. N. I. J. Bailey, Statistical Methods in Biology, E.U.P. London (1959).
15. S. Seigel, Non-Parametric Statistics for the Behavioural Sciences, McGraw Hill, London 1956.
16. T. W. Potts, Biol. Rev., 42, 1 (1967).
17. D. E. Gilbertson and K. G. Jones, Comp. Biochem. Physiol. 42A, 621 (1972).
18. L. S. Uhazy, R. D. Tanaka, and A. J. MacInnis, Science, 201, 924 (1978).
19. J. E. Bardach, in Olfaction and Taste, D. A. Denton and J. P. Coughlan (Eds.), International Symposium Series V, 121, Academic Press, London (1975).
20. A. Navon and E. A. Bernays, Comp. Biochem. Physiol., 59A, 161 (1978).
21. D. H. Janzen, B. H. Juster, and A. E. Bell, Phytochem., 16, 233 (1977).
22. V. Vacuna and A. Hovadik, Plant and Soil, 22, 21 (1965).
23. J. N. Cameron and M. J. Carlile, Nature, 271, 448 (1978).
24. D. R. Mason, Anim. Behav., 23, 460 (1975).
25. R. A. Wilson and J. Denison, Comp. Biochem. Physiol., 32, 511 (1970).
26. A. J. MacInnis, J. Parasit., 51, 731 (1965).
27. E. O. Wilson and F. E. Regnier, Amer. Nat., 105, 279 (1971).
28. J. B. Harborne, Introduction to Ecological Biochemistry, Academic Press, London 1977.

# FOCAL CONTROL OF *SCHISTOSOMA HAEMATOBIUM* - TRANSMITTING SNAILS, *BULINUS (Ph.) ABYSSINICUS*, USING CONTROLLED RELEASE TRI-N-BUTYLTIN FLUORIDE AND COPPER SULPHATE

*E. S. Upatham*

Department of Biology
Faculty of Science
Mahidol University
Bangkok, Thailand

*M. Koura*
*M. A. Dagal*
*A. H. Awad*
*M. D. Ahmed*

Schistosomiasis Control Project
Ministry of Health
Mogadishu, Somalia

## I. INTRODUCTION

In Somalia, *Schistosoma haematobium*, the form of the disease which affects mainly the urinary bladder, is endemic in the settled areas of the Shebelli and Juba Rivers, and *Bulinus (Ph.) abyssinicus* snails serve as its intermediate host.

In 1976, a Schistosomiasis Control Project, organized by the Ministry of Health and assisted by the World Health Organization, was begun in Somalia. Koryole and Merca Districts were chosen, respectively, as the Project's Pilot Control Area and the Project's Additional Working Area.

Prevalence rates of *S. haematobium* were 48.2% in Koryole District and 60.7% in Merca District (1,2), and reported snail infection rates of 1.7% and 1.6%, respectively, for




ISBN 0-12-074450-3

Koryole and Merca Districts, According to the investigators, snails were found mainly in standing waterbodies such as inundated depressions, flooded fields, and water reservoirs or ponds. Some of the large water depressions and water reservoirs, where water remained most of the year, supported stable snail populations and served as snail seeding sources for smaller waterbodies during the rains and flooding. These smaller waterbodies were dried up within a short period of time, and would not have a significant impact on the schistosome transmission. However, some of them were kept in existence with water overflowing from canals nearby, and this continuous mismanagement of water sources helped create suitable snail habitats. Therefore, the transmission of schistosomiasis was maintained owing to frequent human contact with infectious waters.

This paper reports field trials of controlled release molluscicides against *Bul. abyssinicus* in Koryole and Merca Districts.

## II. STUDY AREAS

Koryole District lies 125 km south of Mogadishu. The District covers an area of approximately 5,255 $km^2$; the Shebelli River divides it into two. There are approximately 85,000 inhabitants, grouped into 20 large villages, of whom 43,000 are nomads. The settlers reside in villages along the river banks and till the land using irrigation canals at the period of high flow. The irrigation canals are not operated systematically: they are opened and closed according to the farmers' needs. The crops consist mainly of maize, sesame, sorghum, bananna, and vegetables.

Merca District, 100 km from Mogadishu, is the Project's Additional Area. It lies east of and adjoining to Koryole District and covers approximately 1,428 $km^2$ in area. The population is approximately 125,000, residing in 192 villages, of whom 62,000 are nomads. Agricultural practices and main crops are similar to those in Koryole District.

## III. MOLLUSCICIDING OPERATION AND RESULTS

The mollusciciding operation was designed to control, and to eradicate, snails in breeding sites of particular hazard to the community.

Prior to the mollusciciding operation, snail populations and *S. haematobium* transmission patterns were routinely monitored for eighteen months. At each site, snails were sampled with a standard wire mesh scoop for 40 scoops per month. After the mollusciciding operation, a monthly snail surveillance program was set up to monitor snail populations and the parasite transmission in both treated and untreated sectors of waterbodies in the study areas.

Snails were found mainly near the edge of a habitat, and their abundance was closely related to plants and weeds growing there. Infected snails were found mainly at water contact sites. Snail habitats are of mud bottom, and their depths vary from 1 m to over 10 m, depending on the types of waterbodies. The snail-infested waterbodies contain grasses and sedges and aquatic plants such as *Nymphaea lotus*, *N. coerulea*, *Elodea canadensis*, *Ceratophyllum demersum*, *Pistia stratiotes*, etc.

### A. *Preliminary Field Trials*

Three types of controlled release molluscicides were used to conduct preliminary field trials: CBL-9B rubber pellets (tributyltin fluoride, 20% A.I.), EC-1000 ethylene propylene pellets (copper sulphate monohydrate, 43% A.I.), and Incracide 10 A stylene-butadiene polymer pellets (copper sulphate monohydrate, 18% A.I.).

In Shalambood Village, Merca District, CBL-9B pellets were used to treat one whole pond (area treated = 756 $m^2$) and one whole depression (area treated = 924 $m^2$), situated next to each other and connected to each other with a water channel, at a pellet dosage of 10 $g/m^2$ each, on 24 September, 1977. In Genale Village, Merca District, EC-1000 pellets were used to treat edges of 2 depressions (areas treated; 200 $m^2$ and 60 $m^2$) and a whole ditch (area treated; 192 $m^2$) at a pellet dosage of 20 $g/m^2$, on 5 October, 1977. Incracide 10 A pellets were used to treat 2 whole depressions, one in Bufow Village (area treated; 1,500 $m^2$) and another in Genale Village (area treated; 2,000 $m^2$) both in Merca District, at a pellet dosage of 30 $g/m^2$, on 1 December, 1977. Attempts were made to broadcast molluscicidal pellets uniformly into treated sectors of waterbodies. The formulations sank a few hours later.

The results are shown in Table I. With CBL-9B pellets, snail populations started to decline 2 weeks after treatment, and a complete control was achieved one month after treatment. In November 1978, both treated habitats were almost dried up. As water receded, many dead shells of *Bul. abyssinicus* were

TABLE I. *Snail Populations and Infection Rates Before and After Treatment of Habitats with Controlled Release Molluscicides.*

| Snail sampling time | Total number of *B. abyssinicus* collected 40 scoops per habitat per month | | | | | |
|---|---|---|---|---|---|---|
| | | Shalambood CBL-9B | | Bufow Incracide 10 A | | Genale EC-1000 |
| 1977 | | | | | | |
| April | | 300 (0.7)[a] | | | | |
| May | | 225 (1.8) | | 158 (9.5)[a] | | |
| June | | 300 (2.0) | | 210 (4.8) | | 119 (7.6)[a] |
| July | | 210 (3.8) | | 125 (6.4) | | 143 (6.3 |
| August | | 157 | | 107 (0.9) | | 155 |
| September | | 180 (2.2) | | 31 | | 125 (8.0) |
| October | T[b] | 50 | WR[c] | 50 (2.0 | T | 75 (1.3) |
| November | | 0 | | 42 | | 17 |
| December | | 0 | T | 17 | WR | 27 |
| 1978 | | | | | | |
| January | | 0 | | Dry | | Dry |
| February | | 0 | | Dry | | Dry |
| March | WR | 0 | | Dry | | Dry |
| April | | 0 | | Dry | NW | 5 |
| May | NW[d] | 0 | | Dry | | 24 |
| June | | 0 | | Dry | | Dry |
| July | | 0 | NW | 0 | | Dry |
| August | | 0 | | 43 | NW | 15 |
| September | | 0 | | 33 (15.2) | | 13 |
| October | WR | 0 | | Dry | | Dry |
| November | | 0 | | Dry | | Dry |
| December | | 0 | | Dry | | Dry |

[a]=Percentage snail infection rate.
[b]=Treatment started.
[c]=Water receded.
[d]=Water refilled.

seen in the treated sectors. Owing to heavy rains and flooding in May and June 1979, the habitats were refilled with water and snails reappeared. Hence, the CBL-9B pellets kept snails under a complete control for approximately 19 months. Aquatic weeds and plants as well as immature stages of insects and fish did not appear to be affected during the period where both habitats still contained water.

With EC-1000 pellets, snail population also declined after treatment, but a complete control was not achieved. The water started to recede one month after treatment, and the habitats dried up 2 months later. Two weeks after the habitats were refilled with water, snails reappeared.

Two depressions were treated with Incracide 10 A. One depression was treated just before it was drying up. All snails were killed during the 6 month dry-up period of the habitat. About one month after the habitat was refilled with water, snails reappeared and infections occurred.

### B. *Pilot Field Trials*

The mollusciciding operation was started in Koryole and Merca Districts at the beginning of December 1978 and was finished in the last week of December. One adviser/malacologist, one biologist, and field assistants were involved in the operation. Controlled release molluscicides CBL-9B rubber granules, EC-1000 pellets and E-51 rubber rods (copper sulphate monohydrate, 50% A.I.) were used to treat mainly standing waterbodies. These molluscicides were contained in drums of 100 lbs. each, were easy to handle, and were convenient to be transported to the field. Application techniques and methods were simple and unsophisticated. Field assistants applied molluscicides by broadcasting them to sectors of habitats intended to be treated at a pellet dosage of 20 $g/m^2$. As a precautionary measure, all field assistants wore gloves and face masks during the applications. In large habitats where focal treatments were intended,the sectors up to approximately 10 m from the edge were treated. The total amount of controlled release molluscicide used and the total area treated in Koryole and Merca Districts are summarized in Table II.

The results are shown in Table III. All three types of controlled release molluscicides used achieved complete control of snail populations one month after treatment. Up to the time of writing this paper, CBL-9B has kept snails under control for 6 months, whereas snails in habitats treated with EC-1000 reappeared after three months, and in habitats treated with E-51 snails reappeared, after six

*TABLE II. Total Amount of Controlled Release Molluscicide used and Total Area Treated in Koryole and Merca Districts.*

| *Molluscicide* | *Amount used (kg)* | *Area treated ($m^2$)* |
|---|---|---|
| *CBL-9B* | *889* | *44,450* |
| *EC-1000* | *67* | *3,350* |
| *EC-51* | *30* | *1,500* |

months. Water levels in some habitats were low before the initiation of mollusciciding, which accounted for low snail populations. After the mollusciciding operation, water levels in all snail habitats started to recede and some habitats dried up completely, especially those treated with EC-1000 and E-51. In June 1979, the areas were flooded and all habitats were refilled with water, owing to heavy rains (180 mm rainfall in June). Consequently, many snail-freed habitats may have been reseeded with snails from untreated habitats nearby. Thus, it will be interesting to know whether the 6-month old controlled release molluscicide CBL-9B can prevent the invading snails from colonizing the habitats. The experiment is still continuing.

## IV. DISCUSSION

Controlled release molluscicides have opened up the way to new concepts in snail control by interrupting the cycles of transmission of some snail-borne trematode parasites. The toxicants are released continuously into particular parts of a snail habitat over a considerable period of time, which results in snail control with minimum effects on the biotope.

Controlled release materials can be produced in a variety of shapes such as granules, pellets, rods, sheets, strips, etc., which allows snail control authorities to select the appropriate materials to suit the water systems to be treated. The materials can be made to sink or to float. The advantages of these materials lie in the fact that they can be conveniently transported to the field, require simple and

TABLE III. Populations of *Bulinus abyssinicus* and Transmission Patterns of *Schistosoma haematobium* in Pilot Control Area Before and After Treatment with Controlled Release Molluscicides, from January 1978 to June 1979.

| Snail Sampling time | Total number of Bulinus abyssinious collected 40 scoops per habitit per month | | |
|---|---|---|---|
| | CBL-9B | EC-1000 | E-51 |
| 1978 | | | |
| January | 308 (1.6)[a] | 48 | 41 |
| February | 322 (0.9 | 44 | 19 |
| March | 372 | 106 | 32 |
| April | 391 (1.3) | 57 (5.3) | 18 (5.6) |
| May | 166 (3.6) | 44 (2.3) | 24 |
| June | 122 | 30 | 62 |
| July | 131 | 44 | 103 |
| August | 146 | 79 | 70 (1.4) |
| September | 68 (5.9) | 69 (2.9) | 61 |
| October | 41 (2.4) | 6 | 23 (4.3) |
| November | 69 | 4 | 37 |
| December | Treatment - no snail sampling | | |
| 1979 | | | |
| January | 0 | 0 | 0 |
| February | 0 | 0 | 0 |
| March | 0 | 0 | 0 |
| April | - | - | - |
| May | 0 | 8 | 0 |
| June | 0 | 1 | 9 |

[a]Percentage snail infection rate

unsophisticated application techniques, can be applied by unskilled personnel without handling risk, can be used to control snails focally, and the need for repeated applications is avoided.

There have been a few small-scale field trials of controlled release organotin molluscicides against the schistosome-bearing snails. In Iran, Mansouri (3) applied BioMeT SRM rubber pellets (6% bis (tri-n-butyltin) oxide) to snail-infested ponds at various dosages, and obtained 100% mortality in *Bul. truncatus* at 5.4 mg/l active agent 37 days after treatment. In Palo, Leyte, the Philippines, Santos *et al.* (4) applied 6 kg of BioMeT SRM rubber pellets to an irrigation canal, with a flow of 0.4 cu. m/sec, and reported 100%, 93%, and 93% mortality of *Oncomelania quadrasi*, respectively, at 50 m, 100 m, and 150 m downstream from the point of release 2 months after treatment.

In Brazil, Castleton (5) treated ponds, pits, a well, and drainage systems infested with *Biomphalaria tenagophila* with BioMeT SRM rubber pellets at dosages ranging from 3 to 12 mg/l active agent. He obtained a complete elimination of the snail populations within one week in all static waterbodies, and the molluscicide remained active up to 9 months. In the drainage systems, only partial control was achieved. In a subsequent experiment Gilbert *et al.* (6) incorporated 11% TBTO in the extruded rubber-clay composition, and applied this formulation to *Biomp. tenagophila*-infested static and flowing sites in Rio de Janeiro at dosages of 5, 10 and 15 $g/m^2$. At dosages of 10 and 15 $g/m^2$, snails were under complete control in ditches and marshes for 12 to 18 months after treatment. In São Desidério, Bahia, the investigators also applied the same formulation to a 3.5 km canal and primitave irrigation systems and 1,400 $m^2$ of drainage canals in a banana plantation, all infested with *Biomp. glabrata* and *Biomp. straminea*. The TBTO formulation, corresponded to 15 $g/m^2$, was deposited into perforated polypropylene-fiber sacks and placed at the extreme upstream points. Snails were eliminated from the main canal, with a flow rate of 900 l/min, and from irrigation systems approximately 800 m downstream for nine months after treatment. In the banana plantation's drainage canals, snails were eliminated four weeks after treatment, approximately 500 m downstream from points of release.

Extensive field trials were performed in Rhodesia (7-9). In the first experiment, they applied 9 $m^2$ of 6% TBTO antifouling rubber strips to 300 $m^2$ of a 950 $m^2$ farm reservoir (Coldstream) perimeter. In addition, they applied 6% TBTO rubber pellets to one area (480 $m^2$) of Rothbury Dam at a pellet dosage of 20 $g/m^2$ and to two areas of Balgownie School

Dam: one area of 2,092 $m^2$ at 20 g/$m^2$ and the other area of 850 $m^2$ at 10 g/$m^2$. At a pellet dosage of 20 g/$m^2$,the snails *Bul. globosus* and *Biomp. pfeifferi* were brought under control within three months after treatment, and the pellets remained effective for over a year. In the subsequent experiments, they used 6% TBTO rubber pellets to treat 430 $m^2$ area of a 1,320 $m^2$ dam on a farm in Hippo Valley at 30 g/$m^2$ and 2 areas of a dam at the Experimental Farm in Chiredzi. One area, 810 $m^2$, was treated at a pellet dosage of 25 g/$m^2$ and the other area, 600 $m^2$, at 15 g/$m^2$. Snail control was also achieved within three months after treatment. Therefore, the pellet dosage level required to control snails is between 15 and 25 g/$m^2$.

In preliminary field trials in Somalia, the authors treated one marsh (924 $m^2$) and one pond (756 $m^2$) with CBL-9B rubber pellets at a dosage of 10 g/$m^2$. The snails *Bul. abyssinicus* were kept under control for 19 months after treatment. In standing waterbodies treated with EC-1000 pellets (452 $m^2$) at a dosage of 20 g/$m^2$ and Incracide 10 A pellets (3,500 $m^2$) at 30 g/$m^2$, only partial snail control was achieved. In pilot field trials, the authors treated 44,450 $m^2$ areas with CBL-9B rubber granules, 3,350 $m^2$ areas with EC-1000 pellets, and 1,500 $m^2$ areas with E-51 rubber rods, all at a dosage of 20 g/$m^2$. The *Bul. abyssinicus*-infested habitats consisted of lakes (huge inundated depressions), marshes, flooded fields, small inundated depressions, and water reservoirs and ponds. In habitats treated with CBL-9B granules, the snails have been kept under control for six months (up to the time of writing); whereas the habitats treated with EC-1000 pellets and E-51 rods, the snails reappeared three and six months, respectively, after treatment.

The concept of treating an area of habitat's bottom rather than the entire waterbody is based on the browsing habit of snails. Snails feed on detritus and, because of this, they will spend some time browsing on the bottom. Eventually, snails will come into contact with the treated part of the habitat (9).

According to our experience in Somalia, treatment of a snail-infested sector in any large waterbody should first be performed approximately 10 m from the bank towards the open water. This allows the controlled release material to remain in water for at least one month, especially during summer time, before the field operator revisited the same area on his routine surveillance-treatment duty. In some large waterbodies, the water may recede beyond 10 m-treated sectors within a few months during summer time; whereas in small waterbodies, they may be dried up completely within 2 to 3 months.

For controlling snails, controlled release TBTO and TBTF are more effective and stay effective longer in the field than controlled release copper sulphate. Analysis of residual tin in controlled release TBTO and TBTF pellets immersed in various field sites indicate a long release life, in years. BioMeT SRM pellets immersed in Brazil (6) and Rhodesia (7) show a release life of four to six years, whereas CBL-9B pellets immersed in Rhodesia (7) and in Somalia show a release life of 8 to 21 years. So far, no controlled release organotin molluscicides have been observed to release for over two years under field conditions. This may be due to the fact that the formulations are covered under the mud.

According to Shiff and Yiannakis (personal communication), mud samples taken just below the BioMeT SRM pellets contained 0.09 to 5.07 mg/l tin, whereas both clear bottom water samples contained no detectable tin. Regarding the effects of controlled release TBTO on the biotope, studies in Rhodesia (7,9), show no discernable effects on mud dwelling organisms, on plants, on fish, and on other aquatic fauna. For the controlled release TBTF, it appears that the molluscicide causes no ill effects on plant and aquatic fauna. Under natural conditions, if fish are irritated by the toxicant, they can swim towards the open water, away from the treated sectors of a waterbody. This aspect of not harming fish of the controlled release molluscicides is very important in Somalia because Malaria Control Project utilizes various kinds of fish such as *Tilapia zillii, Aphanius dispar, Notobranchius palmquisti, N. guantheri,* etc., to control mosquito larvae, especially those of *Anopheles gambiae,* in standing waterbodies. These field studies on the environmental impact of the controlled release molluscicides were performed for approximately one year. Therefore, there is a need to investigate further the chronic effects of the chemicals on fish and the biotope.

ACKNOWLEDGMENTS

The authors express their thanks to Mr. Muse G. Mohamed, Director-General, Ministry of Health, Somalia for permission to publish this paper, and to staff of the Schistosomiasis Control Project for their field assistance. They also wish to thank Professor Nate Cardarelli at the Environmental Management Laboratory of the University of Akron in Ohio and International Copper Research Association in New York for providing control release molluscicides to conduct preliminary field trials. Much appreciation is also expressed to UNICEF for providing molluscicides to the Project for pilot field trials.

## REFERENCES

1. M. Koura, E. S. Upatham, A. H. Awad, and M. D. Ahmed, Prevalence of *Schistosoma haematobium* in Koryole and Merca Districts, Somali Democratic Republic. Submitted to Annals of Tropical Medicine and Parasitology.
2. E. S. Upatham, M. Koura, M. D. Ahmed, and A. H. Awad, Studies on the Transmission of *Schistosoma haematobium* and on the Bionomics of *Bulinis (Ph.) abyssinicus* in Somali Democratic Republic. Submitted to Annals of Tropical Medicine and Parasitology.
3. A. Mansouri, Proc. Controlled Release Pesticide Symposium, Wright State Univ., Dayton, Ohio, September 8-10, 1975, pp. 196-199.
4. A. T. Santox, M. J. Santos, B. L. Blas, and E. A. Banez, Proc. Controlled Release Pesticide Symposium, Oregon State Univ., Corvallis, Oregon, August 22-24, 1977, pp. 114-123.
5. C. W. Castleton, Proc. Controlled Release Pesticide Symposium, Univ. Akron, Akron, Ohio, September 16-18, 1974, pp. 22.1-22.9.
6. B. Gilbert, J. V. Toledo, M. S. Bulhões, C. S. M. da Silva, J. A. da Silva Netto, L. A. Paes Lems, C. W. Castleton, and R. P. Martins, Proc. Controlled Release Pesticide Symposium, Univ. of Akron, Akron, Ohio, September 13-15, 1976, pp. 5.9-5.15.
7. C. J. Shiff, in Molluscicides in Schistosomiasis Control, T. C. Cheng, (Ed.), Academic Press, New York, pp. 241-248.
8. C. J. Shiff, C. Yiannakis, and A. C. Evans, Proc. Controlled Release Pesticide Symposium, Wright State Univ. Dayton, Ohio, September 8-10, 1975, pp. 177-188.
9. C. J. Shiff and A. C. Evans, Proc. Controlled Release Pesticide Symposium, Univ. of Akron, Akron, Ohio, September 13-15, 1976, pp. 5.18-5.26.

# LABORATORY TRIALS OF CONTROLLED RELEASE MOLLUSCICIDES ON *BULINUS (Ph.) ABYSSINICUS*, THE INTERMEDIATE HOST OF *SCHISTOSOMA HAEMATOBIUM* IN SOMALIA

*E. S. Upatham*

Department of Biology
Faculty of Science
Mahidol University
Bangkok, Thailand

*M. Koura*
*M. D. Ahmed*
*A. H. Awad*

Schistosomiasis Control Project
Ministry of Health
Mogadishu, Somalia

## I. INTRODUCTION

The controlled release concept allows a low continuous level of molluscicides to be maintained in an aqueous environment. Frequently molluscicides are incorporated into rubber from which they slowly migrate into the water phase (1). Attempts have also been made to incorporate molluscicides into other elastomeric matrices, and the organotin compounds appear to particularly lend themselves to this kind of formulation.

Organotin compounds have been studied by several workers. Thus Hopf and Muller (2) found organotin compounds to have molluscicidal properties on *Australorbis glabratus*. Ritchie *et al.* (3) studied the molluscicidal time-concentration relationships for several organotin compounds, using *A. glabratus*, and found that butyltin oxides were superior to phenyltin and Webbe and Sturrock (4) found that tri n-butyl-




ISBN 0-12-074450-3

tin acetate was comparable in activity to triphenyl tin acetate when they tested the chemicals on *Biomphalaria sudanica tanganyicensis* and *Bulinus (Physopsis) nasutus productus*. Of all the organotin compounds tested (5-8) bis (tri n-butyltin) oxide seems to be the best organotin molluscicide.

Work on controlled release organotin compounds has been concentrated on bis(tri n-butyltin) oxide (TBTO), tributyltin fluoride (TBTF), bis(tri n-butyltin) sulphide (TBTS), tributyltin acetate (TBTA), tributyltin adipate (TBTAd), and tributyltin resinate (TBTR). The preferred base elastomers for molluscicides are natural rubber and ethylene-propylene rubber (9,10). Thus, Berrios-Duran and Ritchie (11) tested TBTO rubber formulation on *A. glabratus,* and found the chemical to be molluscicidal after daily washings over an 18-week period. Hopf and Goll (12) exposed *Biomp. glabrata* to controlled release TBTO, TBTR, TBTS, and TBTO/niclosamide, and reported that several materials remained biologically active for over 26 months. Mansouri (13) obtained efficacious results when exposing *Bulinus truncatus* to controlled release TBTO (BioMet SRM). Santos *et al.* (14) reported that controlled release TBTO and TBTF (CBL-9B) were effective against *Oncomelania hupensis quadrasi*.

Under laboratory conditions, controlled release copper sulphate (E-51), was reported to be effective on *Biomp. glabrata* (15,16), and on *Biomp. pfeifferi* and *Bul. truncatus* (amin, 1976); whereas under preliminary field trials, the chemical was effective on *Bul. globosus* and *Biomp. pfeifferi* for 2 months (17), and on *Biomp. pfeifferi* and *Bul. truncatus* for a month (18). This paper reports laboratory trials with controlled release BioMet SRM (TBTO), CBL-9B (TBTF), EC-13 (TBTF), and E-51 (copper sulphate) on *Bul. abyssinicus*.

## II. MATERIALS AND METHODS

*Bul. abyssinicus* used in these experiments were the offspring of field snails. Snails were tested in three liter enamel containers, and were fed with local lettuce and water lily leaves (*Nymphaea lotus*). The following three sets of experiments were performed.

### A. *Experiment 1—Effects of Controlled Release Molluscicides on Bul. abyssinicus*

BioMet SRM rubber pellets (6% bis(tri n-butyltin) oxide), CBL-9B rubber pellets (20% tributyltin fluoride), EC-13 floating ethylene propylene copolymer pellets (30% tributyltin fluoride), and E-51 rubber rods (50% copper sulphate monohydrate) were tested against *Bul. abyssinicus*. Four concentrations (0.1, 1.0, 10.0, and 100 mg/l) were used. Prior to testing against snails appropriate doses of the molluscicide pellets were prewashed in 1,500 liters of pond water for 7 days to remove excess surface chemicals. Thirty snails were used in each test, 10 small: <5mm, 10 medium: 5-10mm, and 10 large: >10mm. In one set of experiments, the water was maintained unchanged at a constant level; in a second set of experiments, the water was changed daily. The containers were examined daily for the presence of dead snails which were removed and counted. All experiments were done in duplicate.

### B. *Experiment 2—Effects of Low Concentrations of Controlled Release Molluscicides on Bul. abyssinicus*

In this experiment, the effect of BioMet SRM, CBL-9B, EC-13, and E-51 molluscicides at concentrations of 0.03 and 0.3 mg/l on snails was determined. The experimental design was similar to Experiment 1 except that the appropriate doses of molluscicides were prewashed in 4,500 liters of water, and that the test water was not changed throughout the experiment. As before, thirty snails were tested at each concentration for each molluscicide, and all tests were done in duplicate.

### C. *Experiment 3—Effects of Controlled Release Molluscicides in the Presence of Mud and Plants*

Snail-infested habitats have mud bottoms, and are also infested with plants and vegetation. An experiment was therefore performed to determine the effects of mud and plants on molluscicide effectiveness. The procedure used was the same as in Experiment 1 except that the test water was not changed throughout the test. In one set of these tests, a layer of mud was spread on the bottom of the container to a depth of approximately one cm. In another set of tests, 30 g (wet weight) of *Elodea canadensis* and 30 g (wet weight) of *Ceratophyllum demersum*, were added to each container. Two molluscicide concentrations were used, 1.0 and 10.0 mg/l. As before thirty snails were used for each test at each concentration for each molluscicide, and the experiments were done in duplicate.

## III. RESULTS AND DISCUSSION

### A. *Experiment 1—Effects of Controlled Release Molluscicides on Bul. abyssinicus*

The results of Experiment 1 are shown in Table I. The results indicate that the organotin compounds killed all of the snails within 1 to 2 days at 100 mg/l and within 5-7 days at 1.0 mg/l in the set of experiments were water remained unchanged throughout the test. Furthermore, there were no significant differences among the 3 organotin compounds (BioMet SRM, CBL-9B, and EC-13). E-51 (copper sulphate) was also effective, but took a much longer time to kill the snails (3 days at 100 mg/l, and 28 days at 1.0 mg/l). In the experiments in which the water was changed daily, the organotin compounds killed all snails within 3 to 5 days at 100 mg/l and within 6 to 8 days at 1.0 mg/l, and there were no significant differences among the 3 organotin compounds. In these experiments, E-51 took 6 and 36 days to kill all snails at 100 and 1.0 mg/l, respectively. The effect of changing the test water daily was not significant, therefore, with the organotin compounds but significantly decreased the effectiveness of the E-51, copper sulphate containing molluscicide.

### B. *Experiment 2—Effects of Low Concentrations of Controlled Release Molluscicides on Bul. abyssinicus*

The results are illustrated in Table II. At the concentration of 0.03 mg/l, the organotin compounds required 36 to 40 days to kill all the snails, whereas E-51 had killed only 33% of the snails at the termination of the experiment on day 46. At 0.3 mg/l, BioMet SRM and CBL-9B required 9 to 10 days to kill all snails, but EC-13 required a much longer time of 22 days. Again, E-51 was less effective and at this same concentration, and killed only 60% of the snails by day 46.

### C. *Experiment 3—Effects of Controlled Release Molluscicides in the Presence of Mud and Plants*

The results are shown in Table III. It required only a few days for each organotin compound to kill all snails, exposed in various variables, at the concentrations tested, 1.0 and 10.0 mg/l. For CBL-9B and EC-13, there were no

TABLE I. *Showing Effects of Controlled Release Molluscicides on B. abyssinicus Snails*

| Concentration active ingredient (mg/l) | Days required for 100% snail mortality | | | | | | | |
|---|---|---|---|---|---|---|---|---|
| | Water unchanged | | | | Water changed daily | | | |
| | BioMet SRM | CBL-9B | EC-13 | E-51 | BioMet SRM | CBL-9B | EC-13 | E-54 |
| 100 | 1 | 2 | 1 | 3 | 3 | 5 | 3 | 6 |
| 10 | 4 | 4 | 4 | 19 | 5 | 5 | 5 | 25 |
| 1 | 7 | 5 | 5 | 28 | 8 | 8 | 6 | 36 |
| 0.1 | 12 | - | - | - | 14 | - | - | - |
| Control | 20.0% mortality at day 28 | | | | 26.7% mortality at day 36 | | | |

TABLE II. Showing Effects of Controlled Release Molluscicides on B. abyssinicus Snail

| Concentration active ingredient (mg/l) | Days required for 100% snail mortality | | | |
|---|---|---|---|---|
| | BioMet SRM | CBL-9B | EC-13 | E-51 |
| 0.3 | 10 | 9 | 22 | 60[a] |
| 0.03 | 38 | 36 | 40 | 33[a] |
| Control | 27% mortality at day 46 | | | |

[a]Percent mortality of snails at day 46

TABLE III. Effects of Controlled Release Molluscicides on B. abyssinicus Snails

| Molluscicide type and concentration of active ingredient (mg/l) | | Days required for 100% snail mortality | | |
|---|---|---|---|---|
| | | Water alone | Water and mud | Water and plants |
| BioMet SRM | 10 | 4 | 3 | 3 |
| | 1 | 7 | 10 | 10 |
| CBL-9B | 10 | 4 | 4 | 4 |
| | 1 | 5 | 5 | 5 |
| EC-13 | 10 | 4 | 3 | 3 |
| | 1 | 5 | 3 | 3 |
| E-51 | 10 | 19 | 5 | 5 |
| | 1 | 28 | --[a] | 21 |
| Control | | 27% mortality at day 30 | | |

[a]47% mortality at day 30

significant differences regarding the time required to kill all snails between the two concentrations tested. Similarly, the time required to kill the snails in the three environments did not differ significantly. For BioMet SRM, snails exposed in 10.0 mg/l died sooner than those exposed in 1.0 mg/l. At 1.0 mg/l, snails exposed in water containing molluscicide alone required a slightly shorter time to kill all snails than those exposed in water containing molluscicide and mud and in water containing molluscicide and plants.

For E-51, the time required to kill all snails, exposed in different environments, differed significantly at the two concentrations tested. At 10.0 mg/l, snails exposed in water containing molluscicide and mud and in water containing molluscicide and plants were all killed much sooner than those exposed in water containing molluscicide alone. At 1.0 mg/l, it required a shorter time to kill all the snails exposed in water containing molluscicide and plants than those exposed in water containing molluscicide alone. Forty-seven percent of the snails exposed in water containing molluscicide and mud were killed at day 30 when the experiment was terminated.

At the concentration of 1.0 mg/l, the mud reduced copper concentration to low levels, which resulted in low snail mortality. This phenomenon has been shown by several workers (16, 19-21). Although the reasons for this reduction in copper concentration is unclear, Paulini (22) has postulated that copper may be bound by clay through an ion exchange reaction. At higher concentration, 10.0 mg/l, all snails exposed in the molluscicide and mud and in the molluscicide and plants were killed within 5 days, much sooner than those exposed in the molluscicide alone (19 days). This may be because the snails obtained a high dose of the molluscicide when browsing on mud and plants which had adsorbed the molluscicide from the water.

## ACKNOWLEDGMENTS

The authors thank Mr. Muse G. Mohamed, Director-General, Ministry of Health, Somalia, for permission to publish this paper, and the staff of the Schistosomiasis Control Project for their field assistance. Much appreciation is expressed to Professor Nate Cardarelli at the Environmental Management Laboratory of the University of Akron in Ohio for providing controlled release molluscicides to perform laboratory trials.

REFERENCES


1. S. V. Kanakkanatt, Use of Polymeric Matrices in the Control of Schistosomiasis, OAU Symposium on Schistosomiasis, Addis Ababa, 1970, 205-210.
2. H. S. Hopf and R. L. Muller, Bull. of World Health Organiz., 27, 783-789 (1962).
3. L. S. Ritchie, L. A. Berrios-Duran, L. P. Frick, and J. Fox, Bull. of World Health Organiz., 31, 147-149 (1964).
4. G. Webbe and R. F. Sturrock, Annals of Trop. Med. and Parasitology, 58, 234-239 (1964).
5. J. P. DeVillers and J. G. MacKenzie, Bull. of World Health Organiz., 29, 424-427 (1963).
6. L. P. Frick and W. de. Jimenez, Bull. of World Health Organiz., 31, 429-432 (1964).
7. H. S. Hopf, J. Duncan, S. Beasley, D. J. Webley, and R. F. Sturrock, Bull. of World Health Organiz., 36, 955-961 (1967).
8. T. P. Pesigan, A. T. Santos, R. G. Garcia, E. Banez, and M. Santos, Screening of Chemical Coumpounds for Molluscicidal Activity Against *Oncomelania quadrasi*, WHO Informal Report MOL/INF/67.23, 1967.
9. N. F. Cardarelli, U.S. Patent 3,417,181 (1968).
10. N. F. Cardarelli and H. F. Neff, U.S. Patent 3,639,583 (1972).
11. L. A. Berrios-Duran and L. S. Ritchie, Bull. of World Health Organiz., 39, 310-312 (1968).
12. H. S. Hopf and P. H. Goll, Biocidal Rubber: a Two Year Assessment of the Organotin Containing Formulations, WHO document, PD/MOL/70.14, 1970.
13. A. Mansouri, Proc. Controlled Release Pesticide Symposium, Wright State Univ., Dayton, Ohio, September 1975.
14. A. T. Santos, B. L. Blas, E. A. Banez, and C. delos Reyes, Proc. Controlled Release Pesticide Symposium, Oregon State University, Corvallis, Oregon, August 1977.
15. K. E. Walker and N. F. Cardarelli, Development of Slow Release Copper Sulfate as a Molluscicide, Annual Report, Creative Biology Laboratory, Inc., Barberton, Ohio, 1973.
16. J. D. Christie, M. A. Prentice, E. S. Upatham, and G. Barnish, Amer. J. of Trop. Med. and Hygiene, 27, 616-622 (1978).
17. C. C. Appleton and G. Stiles, Proc. Controlled Release Pesticide Symposium, University of Akron, Akron, Ohio, September 1976.

18. M. Amin, Proc. Controlled Release Pesticide Symposium, Univ. of Akron, Akron, Ohio, September 1976.
19. W. H. Greany, Annals of Trop. Med. and Parasitology, 46, 250-267 (1952).
20. W. Rudolfs and A. L. Zuber, Sewage Industrial Wastes, 25, 142-154 (1953).
21. K. Y. Chu, Bull. of World Health Organiz., 54, 417-420 (1976).
22. E. Paulini in Molluscicides in Schistosomiasis Control, T. C. Cheng, (Ed.), Academic Press, New York, 1974.

# INDEX

## G



## H



## I



## J



## K



## L



## M



## N



## O



## P

## R



## S



## T



## W